Lecture Notes in Computer Science   9371

Commenced Publication in 1973
Founding and Former Series Editors:
Gerhard Goos, Juris Hartmanis, and Jan van Leeuwen

More information about this series at http://www.springer.com/series/7409

Giuseppe Amato · Richard Connor
Fabrizio Falchi · Claudio Gennaro (Eds.)

Similarity Search
and Applications

8th International Conference, SISAP 2015
Glasgow, UK, October 12–14, 2015
Proceedings

 Springer

Editors
Giuseppe Amato
ISTI-CNR
Pisa
Italy

Richard Connor
University of Strathclyde
Glasgow
UK

Fabrizio Falchi
ISTI-CNR
Pisa
Italy

Claudio Gennaro
ISTI-CNR
Pisa
Italy

ISSN 0302-9743   ISSN 1611-3349   (electronic)
Lecture Notes in Computer Science
ISBN 978-3-319-25086-1   ISBN 978-3-319-25087-8   (eBook)
DOI 10.1007/978-3-319-25087-8

Library of Congress Control Number: 2015950012

LNCS Sublibrary: SL3 – Information Systems and Applications, incl. Internet/Web, and HCI

Printed on acid-free paper

Springer International Publishing AG Switzerland is part of Springer Science+Business Media
(www.springer.com)

Preface

This volume contains the papers presented at the 8th International Conference on Similarity Search and Applications (SISAP 2015) held in Glasgow, Scotland, UK, during October 12–14, 2015. The International Conference on Similarity Search and Applications (SISAP) is an annual forum for researchers and application developers in the area of similarity data management. It focuses on technological problems shared by many application domains, such as data mining, information retrieval, computer vision, pattern recognition, computational biology, geography, biometrics, machine learning, and many others that need similarity searching as a necessary supporting service.

Traditionally, SISAP conferences have put emphasis on distance-based searching, but in general the conference concerns both the effectiveness and efficiency aspects of any similarity search approach, welcoming contributions that range from theoretical aspects to innovative developments for which similarity search plays the central role.

The call for papers welcomed research papers (full or short papers) presenting previously unpublished research contributions, poster papers presenting innovative work in progress, and demonstration papers presenting applications of similarity search techniques.

We received 65 submissions. The Program Committee (PC) comprised 64 researchers from 21 different countries. Each submission was assigned to at least three PC members. Reviews were discussed by the chairs and PC members when the reviews diverged and no consensus had been reached. The final selection of papers was made by the PC chairs based on the reviews received for each submission. Finally, the conference program included 19 full papers, 11 short papers, three posters, and two demonstrations, resulting in an acceptance rate of 29 % for full papers and 48 % cumulative for full and short papers.

The conference program and the proceedings are organized into six parts.

The first part comprises papers proposing improvements to different methods and techniques for similarity search. The second part is devoted to papers dealing with issues related to metrics and evaluation. The third part focuses on applications of similarity search to specific domains, such as image search, face retrieval, banknote recognition, object recognition, text retrieval. The fourth part comprises papers devoted to implementation and engineering solutions for similarity search. The fifth part consists of posters and demonstration papers.

The conference program also included three invited talks from outstanding scientists from industry and academia. The first, "Large-Scale Similarity Joins with Guarantees" by Prof. Rasmus Pagh from the University of Copenhagen, gives an overview of randomized techniques for high-dimensional similarity search. The second, "Directions for Similarity Search in Television Recommender Systems" by Dr. Billy Wallace from Think Analytics, discusses how similarity search can be exploited to build recommender systems for television programs. The third, "Deep Learning and Similarity

Search" by Dr. Bobby Jaros from Yahoo Labs, discusses how deep learning can help in identifying more subtle notions of similarity.

As in previous editions, the proceedings are published by Springer in the *Lecture Notes in Computer Science* series. A selection of the best papers presented at the conference were recommended for publication in the *Information Systems Journal*. The selection of best papers was made by the PC, based on the reviews received by each paper, and on the discussion during the conference.

SISAP conferences are organized by the SISAP initiative (www.sisap.org), which aims to become a forum for the exchange of real-world, challenging, and innovative examples of applications, new indexing techniques, common test-beds and benchmarks, source code, and up-to-date literature through its Web page, serving the similarity search community.

We would like to thank all the authors who submitted papers to SISAP 2015. We would also like to thank all members of the PC and the external reviewers, for the enormous amount of work they have done. We want to express our gratitude to the PC members for their effort and contribution to the conference. All the submission, reviewing, and proceedings generation processes were carried out through the Easy-Chair platform.

October 2015

Giuseppe Amato
Richard Connor
Fabrizio Falchi
Claudio Gennaro

Organization

Program Committee Chairs

Richard Connor   University of Strathclyde, Glasgow, UK
Giuseppe Amato   ISTI-CNR, Pisa, Italy

Program Commitee Members

Agma Traina	University of Sao Paulo, Brazil
Ahmet Sacan	Drexel University, USA
Alfredo Ferro	University of Catania, Italy
Andrea Esuli	ISTI-CNR, Italy
Andreas Zuefle	Ludwig-Maximilians-Universität München, Germany
Andreas Rauber	Vienna University of Technology, Austria
Apostolos N. Papadopoulos	Aristotle University of Thessaloniki, Greece
Benjamin Bustos	University of Chile, Chile
Bjorn Thor Jonsson	Reykjavik University, Iceland
Claudio Gennaro	ISTI-CNR, Italy
Costantino Grana	Università degli Studi di Modena e Reggio Emilia, Italy
Daniel Keim	University of Konstanz, Germany
David Mount	University of Maryland, USA
Deepak P.	IBM Research, India
Dimitrios Tzovaras	Informatics and Telematics Institute/Centre for Research and Technology Hellas, Greece
Dong Deng	Tsinghua University, China
Edgar Chavez	CICESE, Mexico
Eduardo Valle	University of Campinas, Brazil
Elaine Sousa	University of Sao Paulo - ICMC/USP, Brazil
Fabrizio Falchi	ISTI-CNR, Italy
Fabrizio Silvestri	Yahoo Labs, Spain
Franco Maria Nardini	ISTI-CNR, Italy
Giuseppe Amato	ISTI-CNR, Italy
Gonzalo Navarro	University of Chile, Chile
Hanghang Tong	City College, CUNY, USA
Henning Müller	HES-SO, Switzerland
Jakub Lokoc	Charles University in Prague, Czech Republic
Joao Eduardo Ferreira	University of Sao Paulo, Brazil
Joe Tekli	Lebanese American University, USA
Johannes Niedermayer	LMU Munich, Germany
Kaoru Yoshida	Sony Computer Science Laboratories, Inc., Japan

K. Selcuk Candan	Arizona State University, USA
Laurent Amsaleg	CNRS-IRISA, France
Leonid Boytsov	N/A, USA
Luisa Mico	University of Alicante, Spain
Magnus Lie Hetland	Norwegian University of Science and Technology, Norway
Marcela Ribeiro	Federal University of São Carlos - UFSCar, Brazil
Marco Patella	DEIS - University of Bologna, Italy
Maria Luisa Sapino	Università di Torino, Italy
Michael E. Houle	National Institute of Informatics, Japan
Michel Crucianu	CNAM, France
Nieves R. Brisaboa	Universidade da Coruna, Spain
Oscar Pedreira	Universidade da Coruna, Spain
Panagiotis Bouros	Humboldt-Universität zu Berlin, Germany
Paolo Ciaccia	University of Bologna, Italy
Peter Stanchev	Kettering University, USA
Petros Daras	Information Technologies Institute, Greece
Raffaele Perego	ISTI-CNR, Italy
Renata Galante	UFRGS, Brazil
Renato Fileto	UFSC, Brazil
Richard Connor	University of Strathclyde, UK
Richard Chbeir	LIUPPA Laboratory, France
Robert Moss	University of Strathclyde, UK
Robson Cordeiro	ICMC-USP, Brazil
Rodrigo Paredes	Universidad de Talca, Chile
Rosalba Giugno	University of Catania, Italy
Rui Mao	Shenzhen University, China
Stephane Marchand-Maillet	Viper Group - University of Geneva, Switzerland
Thomas Seidl	RWTH Aachen University, Germany
Virendra Bhavsar	University of New Brunswick, Canada
Walid Aref	Purdue University, USA
Walter Allasia	Eurix, Italy
Yoshiharu Ishikawa	Nagoya University, Japan
Yasin Silva	Arizona State University, USA
Vincent Oria	New Jersey Institute of Technology, USA

Keynotes

Large-Scale Similarity Joins with Guarantees

Rasmus Pagh

IT University of Copenhagen, Copenhagen, Denmark

The ability to handle noisy or imprecise data is becoming increasingly important in computing. In the information retrieval community the notion of similarity join has been studied extensively, yet existing solutions have offered weak performance guarantees. Either they are based on deterministic filtering techniques that often, but not always, succeed in reducing computational costs, or they are based on randomized techniques that have improved guarantees on computational cost but come with a probability of not returning the correct result.

The aim of this talk is to give an overview of randomized techniques for high-dimensional similarity search, and then proceed to discuss two recent advances. First, we consider ways of improving the locality of data access by using a recursive approach. This provably lowers the I/O cost of large-scale similarity joins. Second, we consider new methods for eliminating the probability of error inherent in classical locality-sensitive hashing methods for similarity join in Hamming space, while almost matching their theoretical performance.

The research leading to these results has received funding from the European Research Council under the European Union's Seventh Framework Programme (FP7/2007-2013)/ERC grant agreement no. 614331.

Directions for Similarity Search in Television Recommender Systems

Billy Wallace

Founding Developer, Think Analytics, Glasgow, UK

Recommender systems require similarity search in order to find a movie or tv show that is similar to another. There are interesting constraints however, that differentiate this application from a pure similarity search. Just finding similar content does not give good recommendations, as we are trying to fulfil a business use-case such as up-selling paid-for content or exposing users to content on channels they don't normally watch. Instead, we use similarity almost as a bloom filter, where we populate a "candidate set" using similarity search and then use a second pass to select good recommendations based on the requirements of the use-case. It is common that we can't find enough recommendations to fulfil a request from the candidate set unless we supply some hints to the indexes being used to execute the similarity search, for example, prefer new content, prefer popular content or candidates must be in the user's "package". Measuring the success of such recommender systems is difficult. There are no standard test sets available, and it is difficult to convince broadcasters that they should share data that they may not own outright, or which may present privacy issues if shared. We will discuss an approach that we are starting to look at. Although the scale of the catalogues indexed is modest, with numbers of items in the hundred thousands rather than millions, there are scalability concerns due to the number of requests - millions of customers requiring thousands of requests per second with sub-second response times - and also the fact that the catalogue changes frequently - usually in it's entirety several times per day. It is hoped that by sharing insights from current commercial work in this area, that new research directions, or applications of existing research are suggested.

Deep Learning and Similarity Search

Bobby Jaros

Yahoo Labs, San Francisco, California, USA

Deep Learning has received tremendous attention recently thanks to its impressive results in computer vision, speech, medicine, robotics, and beyond. Although many of the highly visible results have been in a classification setting, a prime motivation for deep learning has been to learn rich feature vectors that are useful across a wide array tasks. One goal of such features — for example, in perception-oriented tasks — might be that items deemed similar by humans would have mathematically similar feature vectors. As deep learning continues to advance, we can expect continued improvement in our ability to identify more and more interesting and subtle notions of similarity. In the other direction, similarity search can also empower deep learning, as recent work invokes similarity search as a core module of deep learning systems.

Contents

Applications and Specific Domains

Implementation and Engineering Solutions

Posters

Demo Papers

Improving Similarity Search Methods and Techniques

Approximate Furthest Neighbor
in High Dimensions

Rasmus Pagh, Francesco Silvestri, Johan Sivertsen, and Matthew Skala[⊠]

IT University of Copenhagen, Copenhagen, Denmark
{pagh,fras,jovt,mska}@itu.dk

Abstract. Much recent work has been devoted to approximate nearest neighbor queries. Motivated by applications in recommender systems, we consider *approximate furthest neighbor* (AFN) queries. We present a simple, fast, and highly practical data structure for answering AFN queries in high-dimensional Euclidean space. We build on the technique of Indyk (SODA 2003), storing random projections to provide sublinear query time for AFN. However, we introduce a different query algorithm, improving on Indyk's approximation factor and reducing the running time by a logarithmic factor. We also present a variation based on a query-independent ordering of the database points; while this does not have the provable approximation factor of the query-dependent data structure, it offers significant improvement in time and space complexity. We give a theoretical analysis, and experimental results.

1   Introduction

Similarity search is concerned with locating elements from a set S that are close to a given query q. The query q can be thought of as criteria we would like returned items to satisfy approximately. For example, if a customer has expressed interest in a product q, we may want to recommend other, similar products. However, we do not want to recommend products that are *too* similar, since that would not significantly increase the probability of a sale. Among the points that satisfy a near neighbor condition ("similar"), we would like to return those that also satisfy a furthest-point condition ("not too similar"), without explicitly computing the set of all near neighbours and then searching it.

In this paper we focus on the problem of returning a furthest point, referred to as "furthest neighbor" by analogy with nearest neighbor. In particular, we consider point sets in d-dimensional Euclidean space (ℓ_2^d). We argue that the exact version of this problem would also solve exact similarity search in d-dimensional Hamming space, and thus is as difficult as that problem. The reduction follows from the fact that the complement of every sphere in Hamming space is also a sphere. That limits the hope we may have for an efficient solution to the exact version, so we consider the *c-approximate furthest neighbor* (c-AFN) problem where the task is to return a point x' with $d(q, x') \geq \max_{x \in S} d(q, x)/c$, with $d(x, u)$ denoting the distance between two points. We will pursue randomized solutions having a small probability of not returning a c-AFN. The success probability can be made arbitrarily close to 1 by repetition.

© Springer International Publishing Switzerland 2015
G. Amato et al. (Eds.): SISAP 2015, LNCS 9371, pp. 3–14, 2015.
DOI: 10.1007/978-3-319-25087-8_1

Another use for AFN comes from data structures that solve the approximate near neighbor $((c, r)$-ANN) problem via locality-sensitive hashing. A known issue with the locality-sensitive hash approach to ANN is that if a point happens to be very close to the query, then nearly every hash function will return it. These duplicates cost significant time to filter out. AFN could offer a way to reduce such duplication. If each hash function preferentially returns the points furthest away from the query among the points it would otherwise return, then perhaps we could reduce the cost of filtering out duplicate results while still hitting all desired points with good probability.

We describe the data structure and query algorithm in section 2.2. The data structure itself is closely similar to one proposed by Indyk [10], but our query algorithm differs. It returns the c-approximate furthest neighbor, for any $c > 1$, with probability at least 0.72. When the number of dimensions is $O(\log n)$, our result requires $\tilde{O}(n^{1/c^2})$ time per query and $\tilde{O}(n^{2/c^2})$ total space, where n denotes the input size. The $\tilde{O}()$ notation omits polylog terms. Theorem 2 gives bounds in the general case.

Our data structure requires more than linear space when $c < \sqrt{2}$, and there are theoretical reasons why $\sqrt{2}$ may be an important boundary for all data structures that solve this problem. In section 2.2 we give a preliminary result showing that a data structure for c-AFN must store at least $\min\{n, 2^{\Omega(d)}\} - 1$ data points when $c < \sqrt{2}$.

In section 3 we provide experimental support of our data structure by testing it, and some modified versions, on real and randomly-generated data sets. In practice, we can achieve approximation factors significantly below the $\sqrt{2}$ theoretical result, even with a simplified version of the algorithm that saves time and space by examining candidate points in a query-independent order. We can also achieve good approximation in practice with significantly fewer projections and points examined than the worst-case bounds suggested by the theory. Our techniques are much simpler to implement than existing methods for $\sqrt{2}$-AFN, which generally require convex programming [6, 15]. Our techniques can also be extended to general metric spaces.

1.1   Related Work

Exact Furthest Neighbor. In two dimensions the furthest neighbor problem can be solved in linear space and logarithmic query time using point location in a furthest point Voronoi diagram (see e.g. de Berg et al. [3]). However, the space usage of Voronoi diagrams grows exponentially with the number of dimensions, making this approach impractical in high dimensions. Indeed, an efficient data structure for the *exact* furthest neighbor problem in high dimension would lead to surprising algorithms for satisfiability [19], so barring a breakthrough in satisfiability algorithms we must assume that such data structures are not feasible.

Further evidence of the difficulty of exact furthest neighbor is the following reduction from Goel, Indyk, and Varadarajan [9]: Given a set $S \subseteq \{0, 1\}^d$ and a query vector $q \in \{0, 1\}^d$, a furthest neighbor (in Euclidean space) from $-q$ is a vector in S of minimum Hamming distance to q. That is, exact furthest neighbor is at least as hard as exact nearest neighbor in d-dimensional Hamming space, generally believed to be hard for large d and worst-case data.

Approximate Furthest Neighbor. Bespamyatnikh gives a dynamic data structure for the *c-approximate* furthest neighbor problem; however, its query time is exponential in the dimension [4]. Indyk [10] avoids this exponential dependence. More precisely, Indyk showed how to solve a *fixed radius* version of the problem where given a parameter r the task is to return a point at distance at least r/c given that there exist one or more points at distance at least r. He then gives a solution to the furthest neighbor problem with approximation factor $c+\delta$, where $\delta > 0$, by reducing it to queries on many copies of that data structure. The overall result is space $O(dn^{1+1/c^2} \log^{(1-1/c)/2}(n) \log_{1+\delta}(d) \log\log_{1+\delta}(d))$ and query time $O(dn^{1/c^2} \log^{(1-1/c)/2}(n) \log_{1+\delta}(d) \log\log_{1+\delta}(d))$. While our new data structure uses the same basic method as Indyk, multiple random projections to one dimension, we are able to avoid the fixed radius version entirely and get a single and simpler data structure that works for all radii. Moreover, being interested in static queries, we are able to reduce the space to $\tilde{O}(dn^{2/c^2})$.

Methods Based on an Enclosing Ball. Goel et al. [9] show that a $\sqrt{2}$-approximate furthest neighbor can always be found on the surface of the minimum enclosing ball of S. More specifically, there is a set S^* of at most $d+1$ points from S whose minimum enclosing ball contains all of S, and returning the furthest point in S^* always gives a $\sqrt{2}$-approximation to the furthest neighbor in S. This method is *query independent* in the sense that it examines the same set of points for every query. Conversely, Goel et al. [9] show that for a random data set consisting of n (almost) orthonormal vectors, finding a c-approximate furthest neighbor for a constant $c < \sqrt{2}$ gives the ability to find an $O(1)$-approximate near neighbor. Since it is not known how to do that in time $n^{o(1)}$ it is reasonable to aim for query times of the form $n^{f(c)}$ for approximation $c < \sqrt{2}$.

Applications in Recommender Systems. Several papers on recommender systems have investigated the use of furthest neighbor search [16,17]. However, these works are not primarily concerned with (provable) efficiency of the search. Other related works in recommender systems include those of Abbar et al. [1] and Indyk et al. [11], which use core-set techniques to return a small set of recommendations no two of which are too close. In turn, core-set techniques also underpin works on approximating the minimum enclosing ball [2,13].

2   Algorithms and Analysis

2.1   Provably Good Furthest Neighbor Data Structure

Our data structure works by choosing a random line and storing the order of the data points along it. Two points far apart on the line are at least as far apart in the original space. So given a query we can find the points furthest from the query on the projection line, and take those as candidates to be the furthest point in the original space. We build several such data structures and query them in parallel, merging the results.

Given a set $S \subseteq \mathbb{R}^d$ of size n (the input data), let $\ell = 2n^{1/c^2}$ (the number of random lines) and $m = 1 + e^2 \ell \log^{c^2/2 - 1/3} n$ (the number of candidates to be examined at query time), where $c > 1$ is the desired approximation factor. We pick ℓ random vectors $a_1, \ldots, a_\ell \in \mathbb{R}^d$ with each entry of a_i coming from the standard normal distribution $N(0, 1)$. We use $\arg \max_{x \in S}^m f(x)$ for the set of m elements from S that have the largest values of $f(x)$, breaking ties arbitrarily.

For any $1 \le i \le \ell$, we let $S_i = \arg \max_{x \in S}^m a_i \cdot x$ and store the elements of S_i in sorted order according to the value $a_i \cdot x$. Our data structure for c-AFN consists of ℓ subsets $S_1, \ldots, S_\ell \subseteq S$, each of size m. Since these subsets come from independent random projections, they will not necessarily be disjoint in general; but in high dimensions, they are unlikely to overlap very much. This data structure is essentially that of Indyk [10]; our technique differs in the query procedure, given by Algorithm 1.

Algorithm 1. Query-dependent approximate furthest neighbor

1: initialize a priority queue of (point, integer) pairs, indexed by real keys
2: **for** $i = 1$ to ℓ **do**
3:     compute and store $a_i \cdot q$
4:     create an iterator into S_i, moving in decreasing order of $a_i \cdot x$
5:     get the first element x from S_i and advance the iterator
6:     insert (x, i) in the priority queue with key $a_i \cdot x - a_i \cdot q$
7: **end for**
8: $rval \leftarrow \bot$
9: **for** $j = 1$ to m **do**
10:     extract highest-key element (x, i) from the priority queue
11:     **if** $rval = \bot$ or x is further than $rval$ from q **then**
12:         $rval \leftarrow x$
13:     **end if**
14:     get the next element x' from S_i and advance the iterator
15:     insert (x', i) in the priority queue with key $a_i \cdot x' - a_i \cdot q$
16: **end for**
17: return $rval$

Our algorithm succeeds if and only if S_q contains a c-approximate furthest neighbor. We now prove that this happens with constant probability.

We make use of the following standard lemmas that can be found, for example, in the work of Datar et al. [7] and Karger, Motwani, and Suden [12], respectively.

Lemma 1 (See Section 3.2 of Datar et al. [7]). *For every choice of vectors* $x, y \in \mathbb{R}^d$:

$$\frac{a_i \cdot (x - y)}{||x - y||_2} \sim N(0, 1). \tag{1}$$

Lemma 2 (see Lemma 7.1.3 in Karger, Motwani, and Suden [12]). *For every $t > 0$, if $X \sim N(0,1)$ then*

$$\frac{1}{\sqrt{2\pi}} \cdot \left(\frac{1}{t} - \frac{1}{t^3} \right) \cdot e^{-t^2/2} \leq \Pr[X \geq t] \leq \frac{1}{\sqrt{2\pi}} \cdot \frac{1}{t} \cdot e^{-t^2/2} \tag{2}$$

The next lemma follows, as suggested by Indyk [10, Claims 2-3].

Lemma 3. *Let p be a furthest neighbor from the query q with $r = ||p - q||_2$, and let p' be a point such that $||p' - q||_2 < r/c$. Let $\Delta = rt/c$ with t satisfying the equation $e^{t^2/2} t^{c^2} = n/(2\pi)^{c^2/2}$ (i.e., $t = O\left(\sqrt{\log n}\right)$). Then, for a sufficiently large n, we get*

$$\Pr_a \left[a \cdot (p' - q) \geq \Delta \right] \leq \frac{\log^{c^2/2 - 1/3} n}{n} \tag{3}$$

$$\Pr_a \left[a \cdot (p - q) \geq \Delta \right] \geq (1 - o(1)) \frac{1}{n^{1/c^2}} . \tag{4}$$

Proof. Let $X \sim N(0,1)$. By Lemma 1 and the right part of Lemma 2, we get for a point p' that

$$\Pr_a \left[a \cdot (p' - q) \geq \Delta \right] = \Pr_a \left[X \geq \Delta/||p' - q||_2 \right] \leq \Pr_a \left[X \geq \Delta c/r \right]$$

$$\leq \frac{1}{\sqrt{2\pi}} \frac{e^{-t^2/2}}{t} \leq \left(t\sqrt{2\pi} \right)^{c^2 - 1} \frac{1}{n} \leq \frac{\log^{c^2/2 - 1/3} n}{n}.$$

The last step follows because $e^{t^2/2} t^{c^2} = n/(2\pi)^{c^2/2}$ implies that $t = O\left(\sqrt{\log n}\right)$, and holds for a sufficiently large n. Similarly, by Lemma 1 and the left part of Lemma 2, we have for a furthest neighbor p that

$$\Pr_a \left[a \cdot (p - q) \geq \Delta \right] = \Pr_a \left[X \geq \Delta/||p - q||_2 \right] = \Pr_a \left[X \geq \Delta/r \right]$$

$$\geq \frac{1}{\sqrt{2\pi}} \left(\frac{c}{t} - \left(\frac{c}{t} \right)^3 \right) e^{-t^2/(2c^2)} \geq (1 - o(1)) \frac{1}{n^{1/c^2}}.$$

$\square$

Theorem 1. *The data structure when queried by Algorithm 1 returns a c-AFN of a given query with probability $1 - 2/e^2 > 0.72$ in $O(n^{1/c^2}(d + \log^{c^2/2 + 2/3} n))$ time per query. The data structure requires $O(n^{1+1/c^2}(d + \log n))$ preprocessing time and total space*

$$O\left(\min\left\{ dn^{2/c^2} \log^{c^2 - 1/3} n, dn + n^{2/c^2} \log^{c^2 - 1/3} n \right\} \right). \tag{5}$$

Proof. The space required by the data structure is the space required for storing the ℓ sets S_i. If for each set S_i we store the $m \leq n$ points and the projection values, then $O(\ell m d)$ memory words are required. On the other hand, if pointers to the input points are stored, then the total required space is $O(\ell m + nd)$.

Both representations are equivalent, and the best one depends on the value of n and d. The claim on the space requirements follows. The preproceesing time is dominated by the computation of the $n\ell$ projection values and by the sorting for computing the sets S_i. Finally, the query time is dominated by the at most $2m$ insertion or deletion operations on the priority queue and by the search in S_q for the furthest neighbor.

We now upper bound the success probability. As in the statement of Lemma 3, we let p denote a furthest neighbor from q, $r = ||p-q||_2$, p' be a point such that $||p' - q||_2 < r/c$, and $\Delta = rt/c$ with t such that $e^{t^2/2}t^{c^2} = n/(2\pi)^{c^2/2}$. The query succeeds if: (i) $a_i(p - q) \geq \Delta$ for at least one projection vector a_i, and (ii) the (multi)set $S_n = \{p'|\exists i : a_i(p' - q) \geq \Delta, ||p' - q||_2 < r/c\}$ contains at most $m - 1$ points (i.e., there are at most $m - 1$ near points whose distances from the query is at least Δ in some projections). If (i) and (ii) hold, then the set S_Q must contain the furthest neighbor p since there are at most $m - 1$ points near to q with projection values larger than the maximum projection value of p. Note that we do not consider points at distance larger than r/c but smaller than r: they are c-approximate furthest neighbors of q and can only increase the success probability of our data structure.

By Lemma 3, event (i) happens with probability $1/n^{1/c^2}$. Since there are $\ell = 2n^{1/c^2}$ independent projections, this event does not happen with probability at most $(1 - 1/n^{1/c^2})^{2n^{1/c^2}} \leq 1/e^2$. For a point p' at distance at most r/c from q, the probability that $a_i(p' - q) \geq \Delta$ is less than $(\log^{c^2/2 - 1/3} n)/n$ for Lemma 3. Since there are ℓ projections of n points, the expected number of such points is $\ell \log^{c^2/2 - 1/3} n$. Then, we have that S has size larger than $m - 1$ with probability at most $1/e^2$ by the Markov inequality. Note that a Chernoff bound cannot be used since there exists a dependency among the projections under the same random vector a_i. By a union bound, we can therefore conclude that the algorithm succeeds with probability at least $1 - 2/e^2 \geq 0.72$. $\square$

2.2   A Lower Bound on the Approximation Factor

In this section, we show that a data structure aiming at an approximation factor less than $\sqrt{2}$ must use space $\min\{n, 2^{\Omega(d)}\} - 1$ on worst-case data. The lower bound holds for those data structures that compute the approximate furthest neighbor by storing a suitable subset of the input points.

Theorem 2. *Consider any data structure $\mathcal{D}$ that computes the c-AFN of an n-point input set $S \subseteq \mathbb{R}^d$ by storing a subest of the data set. If $c = \sqrt{2}(1 - \epsilon)$ with $\epsilon \in (0,1)$, then the algorithm must store at least $\min\{n, 2^{\Omega(\epsilon^2 d)}\} - 1$ points.*

Proof. We prove by contradiction that any data structure requiring less than $\min\{n, 2^{\Omega(\epsilon^2 d)}\} - 1$ input points cannot return a $\sqrt{2}(1 - \epsilon)$-approximation. Suppose there exists a set S' of size $r = 2^{\Omega(\epsilon'^2 d)}$ such that for any $x \in S'$ we have $(1 - \epsilon') \leq ||x||_2^2 \leq (1 + \epsilon')$ and $x \cdot y \leq 2\epsilon'$, with $\epsilon' \in (0,1)$. We will later prove that such a set exists.

Assume $n \leq r$. Consider the input set S consisting of n arbitrary points of S' and set the query q to $-x$, where x is an input point not in the data structure. The furthest neighbor is x and it is at distance at least $||x - (-x)||_2 \geq 2\sqrt{1 - \epsilon'}$. On the other hand, for $y \in S \setminus \{x\}$, we get

$$||y - (-x)||_2^2 = ||x||_2^2 + ||y||_2^2 + 2x \cdot y \leq 2(1 + \epsilon') + 4\epsilon'.$$

Therefore, the point returned is at least a c' approximation with

$$c' \leq \sqrt{2}\sqrt{\frac{1 - \epsilon'}{1 + 3\epsilon'}}. \tag{6}$$

The claim follows by setting $\epsilon' = \sqrt{(2\epsilon - \epsilon^2)/(1 + 3(1 - \epsilon)^2)}$.

Assume now that $n > r$. Without loss of generality, let n be a multiple of r. Consider as input set the set S containing n/r copies of each vector in S', each copy expanded by a factor i for any $i \in [1, n/r]$; specifically, let $S = \{ix | \forall x \in S', \forall i \in [1, n/r]\}$. Let the query q be $-hx$, where x is a point not in the data structure and h is the largest integer such that hy, with $y \in S'$, is in the data structure. The furthest neighbor in S is at distance at least $2h\sqrt{1 - \epsilon'}$. On the other hand, every point in the data structure is at distance at most $h\sqrt{2(1 + \epsilon') + 4\epsilon'^2}$. We then get the same approximation factor c' given in equation 6, and the claim follows by suitably setting ϵ'.

The existence of the set S' of size r follows from the Johnson-Lindenstrauss lemma [14]. Specifically, consider an orthornormal base $x_1, \ldots x_r$ of $\mathbb{R}^r$. Since $n = \Omega \left(\log r / \epsilon^2\right)$, by the Johnson-Lindenstrauss lemma there exists a linear map $f(\cdot)$ such that $(1 - \epsilon')||x_i - x_j||_2^2 \leq ||f(x_i) - f(x_j)||_2^2 \leq (1 + \epsilon')||x_i - x_j||_2^2$ and $(1 - \epsilon') \leq ||f(x_i)||_2^2 \leq (1 + \epsilon')$ for any i, j. We also have that $f(x_i) \cdot f(x_j) = (||f(x_i)||_2^2 + ||f(x_j)||_2^2 - ||f(x_i) - f(x_j)||_2^2)/2$, and hence $-2\epsilon \leq f(x_i) \cdot f(x_j) \leq 2\epsilon$. It then suffices to set S' to $\{f(x_1), \ldots, f(x_r)\}$. □

The upper bound on space translates into a lower bound for the query time in data structures for AFN which are query independent. Indeed, the lower bound translates into the number of points that must be read by each query. However, this does not apply for query dependent data structures.

3   Experiments

To test the algorithm and confirm both its correctness and practicality we implemented several variations in both the C and F# programming languages. This code is available on request. Our C implementation is structured as an alternate index type for the SISAP C library [8], returning the furthest neighbor instead of the nearest.

We selected four databases for experimentation: the "nasa" and "colors" vector databases from the SISAP library, and two randomly generated databases of 10^5 10-dimensional vectors each, one using a multidimensional normal distribution and one uniform on the unit cube. The 10-dimensional random distributions were intended to represent realistic data, but their intrinsic dimensionality is significantly higher than what we would expect to see in real-life applications.

For each database and each choice of ℓ from 1 to 30 and m from 1 to 4ℓ, we made 1000 approximate furthest neighbor queries. To provide a representative sample over the randomization of both the projection vectors and the queries, we used 100 different seeds for generation of the projection vectors, and did 10 queries (each uniformly selected from the database points) with each seed. We computed the approximation achieved, compared to the true furthest neighbor found by brute force, for every query. The resulting distributions for the uniform, normal, and nasa databases are summarized in Figures 1–3.

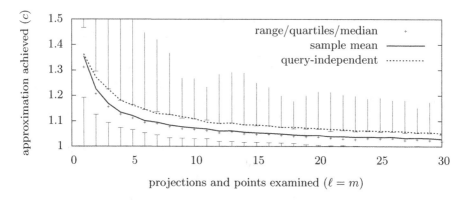

Fig. 1. Experimental results for 10-dimensional uniform distribution

We omit a similar figure from our experiment on the colors database because the result was of little interest: it apparently contains a few very extreme outliers, making the furthest neighbor problem too easy to meaningfully test the algorithm. We also ran some informal experiments on higher-dimensional random vector databases (with 30 and 100 dimensions, in particular) and saw approximation factors very close to those achieved for 10 dimensions.

ℓ *vs.* m *Tradeoff.* The two parameters ℓ and m both improve the approximation as they increase, and they each have a cost in the time and space bounds. The best tradeoff is not clear from the analysis. We chose $\ell = m$ as a typical value, but we also collected data on many other parameter choices.

Figure 4 offers some insight into the tradeoff: since the cost of doing a query is roughly proportional to both ℓ and m, we chose a fixed value for their product, $\ell \cdot m = 48$, and plotted the approximation results in relation to m given that. As the figure shows, the approximation factor does not change much with the tradeoff between ℓ and m.

Query-independent Ordering. The furthest-neighbor algorithm described in our theoretical analysis examines candidates for the furthest neighbor in a *query dependent* order. It seems intuitively reasonable that the search will usually

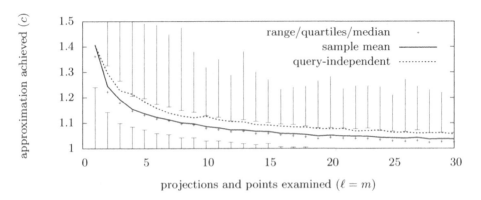

Fig. 2. Experimental results for 10-dimensional normal distribution

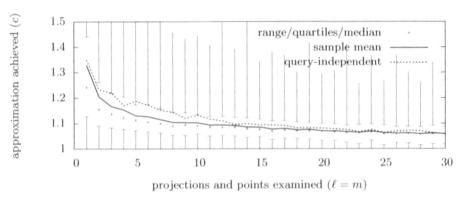

Fig. 3. Experimental results for SISAP nasa database

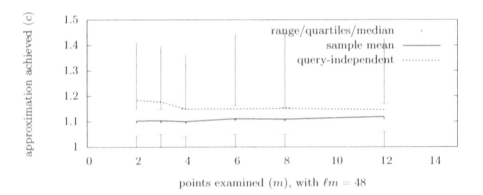

Fig. 4. The tradeoff between ℓ and m

examine points in a very similar order regardless of the query: first those that are outliers, on or near the convex hull of the database, and then working its way inward. Maybe there could be a single generic ordering of the points that would serve reasonably well for all queries?

We implemented a modified version of the algorithm in which the index stores a single ordering of the points. Given a set $S \subseteq \mathbb{R}^d$ of size n, for each point $x \in S$ let $key(x) = \max_{i \in 1...\ell} a_i \cdot x$. The key for each point is its greatest projection value on any of the ℓ randomly-selected projections. The data structure stores points (all of them, or enough to accomodate the largest m we plan to use) in order of decreasing key value: $x_1, x_2, \ldots$ where $key(x_1) \geq key(x_2) \geq \cdots$.

The query simply examines the first m points in this *query independent* ordering and returns the one furthest from the query point. Sample mean approximation factor for this algorithm in our experiments is shown by the dotted lines in Figures 1–4.

Variations on the Algorithm. We have experimented informally with a number of practical improvements to the algorithm. The most significant is to use the rank-based *depth* of projections rather than the projection value. In this variation we sort the points by their projection value for each a_i. The first and last point then have depth 0, the second and second-to-last have depth 1, and so on up to the middle at depth $n/2$. We find the minimum depth of each point over all projections and store the points in a query independent order using the minimum depth as the key. This approach generally yields slightly better approximations, but is more complicated to analyze. A further improvement is to break ties in the minimum depth by count of how many times that depth is achieved, giving more priority to investigating points that repeatedly project to extreme values.

The number of points examined m can be chosen per query and even during a query, allowing for interactive search. After returning the best result for some m, the algorithm can continue to a larger m for a possibly better approximation factor on the same query. The smooth tradeoff we observed between ℓ and m suggests that choosing an ℓ during preprocessing will not much constrain the eventual choice of m.

Discussion. The main experimental result is that the algorithm works very well for the tested datasets in terms of returning good approximations of the furthest neighbor. Even for small ℓ and m the algorithm returns good approximations. Another result is that the query independent algorithm returns points only slightly worse than the query dependent. The query independent algorithm is simpler to implement, it can be queried in time $O(m)$ as opposed to $O(m \log \ell + m)$ and uses only $O(m)$ storage. In many cases these advances more than make up for the slightly worse approximation observed in these experiments. However, by Theorem 2, to guarantee $\sqrt{2} - \epsilon$ approximation the query-independent ordering version would need to store and read $m = n - 1$ points.

In data sets of high intrinsic dimensionality the furthest point from a query may not be much further than any randomly selected point, and we can ask whether our results are any better than a trivial random selection from the database. The intrinsic dimensionality statistic ρ of Chávez and Navarro [5] provides some insight into this question. Skala gives a formula for its value on a multidimensional normal distribution [18, Theorem 2.10], which yields $\rho = 9.768\ldots$ for the 10-dimensional distribution used in Figure 2. With the definition $\rho = \mu^2/2\sigma^2$, this means the standard deviation of a randomly selected distance will be about 32% of the mean distance. Our experimental results come much closer than that to the true furthest distance, and so are non-trivial.

The concentration of distances in data sets of high intrinsic dimensionality reduces the usefulness of approximate furthest neighbor. Thus, although we observed similar values of c in higher dimensions to our 10-dimensional random vector results, random vectors of higher dimension may represent a case where c-approximate furthest neighbor is not a particularly interesting problem. Fortunately, vectors in a space with many dimensions but low intrinsic dimensionality, such as the colors database, are more representative of real application data, and our algorithms performed well on such data sets.

4 Conclusions and Future Work

We have proposed a data structure for AFN with theoretical and experimental guarantees. Although we have proved that it is not possible to use less than $\min\{n, 2^{\Omega(d)}\} - 1$ total space when the c approximation factor is less than $\sqrt{2}$, it is an open problem to close the gap between this lower bound and the space requirements of our result. Another interesting problem is to apply our data structure to improve the output sensitivity of near neighbor search based on locality-sensitive hashing. We conjecture that, by replacing each hash bucket with an AFN data structure with suitable approximation factors, it is possible to control the number of times each point in S is reported.

Our data structure extends naturally to general metric spaces. Instead of computing projections with dot product, which requires a vector space, we could choose some random pivots and order the points by distance to each pivot. The query operation would be essentially unchanged. Analysis and testing of this extension is a subject for future work.

Acknowledgement. The research leading to these results has received funding from the European Research Council under the European Union's Seventh Framework Programme (FP7/2007-2013) / ERC grant agreement no. 614331.

References

1. Abbar, S., Amer-Yahia, S., Indyk, P., Mahabadi, S.: Real-time recommendation of diverse related articles. In: Proc. 22nd International Conference on World Wide Web (WWW), pp. 1–12 (2013)
2. Bădoiu, M., Clarkson, K.L.: Optimal core-sets for balls. Computational Geometry **40**(1), 14–22 (2008)

3. de Berg, M., Cheong, O., van Kreveld, M., Overmars, M.: Computational Geometry: Algorithms and Applications, 3rd edn. Springer-Verlag TELOS (2008)
4. Bespamyatnikh, S.N.: Dynamic algorithms for approximate neighbor searching. In: Proceedings of the 8th Canadian Conference on Computational Geometry (CCCG 1996), pp. 252–257. Carleton University, August 12–15, 1996
5. Chávez, E., Navarro, G.: Measuring the dimensionality of general metric spaces. Tech. Rep. TR/DCC-00-1, Department of Computer Science, University of Chile (2000)
6. Clarkson, K.L.: Las Vegas algorithms for linear and integer programming when the dimension is small. Journal of the ACM (JACM) **42**(2), 488–499 (1995)
7. Datar, M., Immorlica, N., Indyk, P., Mirrokni, V.S.: Locality-sensitive hashing scheme based on p-stable distributions. In: Proc. 20 Annual Symposium on Computational Geometry (SoCG), pp. 253–262 (2004)
8. Figueroa, K., Navarro, G., Chávez, E.: Metric spaces library (2007) (online). http://www.sisap.org/Metric_Space_Library.html
9. Goel, A., Indyk, P., Varadarajan, K.: Reductions among high dimensional proximity problems. In: Proc. 12th ACM-SIAM Symposium on Discrete Algorithms (SODA), pp. 769–778 (2001)
10. Indyk, P.: Better algorithms for high-dimensional proximity problems via asymmetric embeddings. In: Proc. 14th ACM-SIAM Symposium on Discrete Algorithms (SODA), pp. 539–545 (2003)
11. Indyk, P., Mahabadi, S., Mahdian, M., Mirrokni, V.S.: Composable core-sets for diversity and coverage maximization. In: Proc. 33rd ACM SIGMOD-SIGACT-SIGART Symposium on Principles of Database Systems (PODS), pp. 100–108. ACM (2014)
12. Karger, D., Motwani, R., Sudan, M.: Approximate graph coloring by semidefinite programming. Journal of the ACM (JACM) **45**(2), 246–265 (1998)
13. Kumar, P., Mitchell, J.S., Yildirim, E.A.: Approximate minimum enclosing balls in high dimensions using core-sets. Journal of Experimental Algorithmics **8**, 1–1 (2003)
14. Matoušek, J.: On variants of the Johnson-Lindenstrauss lemma. Random Structures and Algorithms **33**(2), 142–156 (2008)
15. Matoušek, J., Sharir, M., Welzl, E.: A subexponential bound for linear programming. Algorithmica **16**(4–5), 498–516 (1996)
16. Said, A., Fields, B., Jain, B.J., Albayrak, S.: User-centric evaluation of a k-furthest neighbor collaborative filtering recommender algorithm. In: Proc. Conference on Computer Supported Cooperative Work (CSCW), pp. 1399–1408 (2013)
17. Said, A., Kille, B., Jain, B.J., Albayrak, S.: Increasing diversity through furthest neighbor-based recommendation. In: Proceedings of the WSDM Workshop on Diversity in Document Retrieval (DDR 2012) (2012)
18. Skala, M.A.: Aspects of Metric Spaces in Computation. Ph.D. thesis, University of Waterloo (2008)
19. Williams, R.: A new algorithm for optimal constraint satisfaction and its implications. In: Díaz, J., Karhumäki, J., Lepistö, A., Sannella, D. (eds.) ICALP 2004. LNCS, vol. 3142, pp. 1227–1237. Springer, Heidelberg (2004)

Flexible Aggregate Similarity Search in High-Dimensional Data Sets

Michael E. Houle[1]($\boxtimes$), Xiguo Ma[2], and Vincent Oria[3]

[1] National Institute of Informatics, Tokyo 101-8430, Japan
`meh@nii.ac.jp`
[2] Google, Mountain View, CA 94043, USA
`maxiguo@google.com`
[3] New Jersey Institute of Technology, Newark, NJ 07102, USA
`oria@njit.edu`

Abstract. Numerous applications in different fields, such as spatial databases, multimedia databases, data mining and recommender systems, may benefit from efficient and effective aggregate similarity search, also known as aggregate nearest neighbor (AggNN) search. Given a group of query objects Q, the goal of AggNN is to retrieve the k most similar objects from the database, where the underlying similarity measure is defined as an aggregation (usually *sum*, *avg* or *max*) of the distances between the retrieved objects and every query object in Q. Recently, the problem was generalized so as to retrieve the k objects which are most similar to a fixed proportion of the elements of Q. This variant of aggregate similarity search is referred to as 'flexible AggNN', or FANN. In this work, we propose two approximation algorithms, one for the *sum* and *avg* variants of FANN, and the other for the *max* variant. Extensive experiments are provided showing that, relative to state-of-the-art approaches (both exact and approximate), our algorithms produce query results with good accuracy, while at the same time being very efficient — even for real datasets of very high dimension.

1 Introduction

The aim of classical similarity search is to retrieve from the database a set of objects most similar to a specified query object, based on a single ranking criterion that is usually expressed in terms of a similarity function. In recent years, the use of *multiple* ranking criteria has been investigated, in which the final rankings of objects are obtained by combining the individual rankings according to some aggregation function (for example, *min*, *max*, *sum* or *avg*). The ranking criteria used in applications of this form of similarity search have differed greatly from area to area. In multimedia applications [1], ranking criteria have been defined in terms of several distance functions computed over different sets of discriminative features, such as color features and texture features. In keyword-based search [2], criteria have been defined with respect to each individual keyword used in the search; while in subspace similarity search [3], ranking criteria have been defined on each individual dimension of the targeted subspace dimensions.

© Springer International Publishing Switzerland 2015
G. Amato et al. (Eds.): SISAP 2015, LNCS 9371, pp. 15–28, 2015.
DOI: 10.1007/978-3-319-25087-8_2

Given a group of query objects Q, *aggregate similarity* or *aggregate nearest neighbor* (AggNN) search aims to retrieve the k objects from the database S that are most highly ranked with respect to Q, where the ranking criterion (similarity measure) is defined as an aggregation (usually *sum*, *avg* or *max*) of distances between the retrieved objects and every object in Q.

Due to its importance and generality, AggNN has received a considerable amount of attention in the literature. Specifically, it has been addressed in the contexts of content-based image retrieval [4], recommender systems [5], road networks [6], and for indexing in both Euclidean vector spaces [7–9], and metric spaces [10]. AggNN methods tend to favor only those objects that are similar to all query objects in Q, which may drastically limit its performance when the characteristics of the objects of Q vary greatly [11].

In many scenarios, it may be difficult (or impossible) to determine a set of objects that are similar to all the query objects of Q. For this reason, Li et al. [11] proposed a generalized problem, called flexible aggregate similarity search (FANN), in which the restrictiveness of AggNN is eased by calculating aggregate distances only over subsets of Q. More precisely, FANN aims to retrieve the k objects which are most similar to a subset of group query Q with size $\phi|Q|$, for some target proportion $0 < \phi \leq 1$. When $\phi = 1$, FANN is equivalent to AggNN.

Compared to AggNN, FANN is not only better suited for finding semantically meaningful results, but it also allows the user to formulate the group query Q with more flexibility. As pointed out in [11], FANN query results may also be more diverse than AggNN results, since every object in the FANN result set may be related to a distinct subset of Q. For these reasons, FANN is a more difficult problem than AggNN, and existing solutions for the AggNN problem cannot be effectively applied for FANN.

In this paper, we propose the FADT (Flexible Aggregation through Dimensional Testing) family of approximation algorithms for the FANN problem in high-dimensional spaces, that possess great efficiency while returning high-quality query results. Our algorithms adopt a multi-step search strategy [12], together with tests for early termination based on a measure of intrinsic dimensionality [13]. The main contributions are:

- Two approximation algorithms for the FANN problem, one for the *sum* and *avg* variants and the other for the *max* variant.
- A theoretical analysis of our methods, showing conditions under which an exact result can be guaranteed. We also show conditions under which approximate result can be guaranteed for a variable distance approximation ratio.
- An extensive experimental evaluation, showing that our algorithms are able to produce query results with both good accuracy and high efficiency in comparison with state-of-the-art competitors, particularly when the data dimensionality is high.

2   Problem Description

Before formally presenting the FANN problem, let us first introduce some needed notation. Let $\mathcal{U}$ be a data domain with a distance metric $d(u, v)$ defined for any two objects $u, v \in \mathcal{U}$. Let $S \subseteq \mathcal{U}$ denote the set of objects in the database. For any object set $X \subseteq \mathcal{U}$ and any object $v \in \mathcal{U}$, let $N_X(v, k)$ denote the set of k nearest neighbors of v within X with respect to distance metric d. The AggNN aggregate distance measure $D(X, v)$ is defined as $D(X, v) = g(d(x_1, v), \ldots, d(x_{|X|}, v))$, where g is an aggregation function (*sum* or *max*), and $x_i \in X$ for $i = 1, \ldots, |X|$.

FANN Problem. Given a group of query objects $Q \subseteq \mathcal{U}$ with size $|Q| = m > 1$, a distance metric d, an aggregation function g, a target proportion $0 < \phi \leq 1$, and a target neighborhood size k, a FANN query returns the k most similar objects to Q from the database S, where the similarity distance is defined as $D_\phi(Q, v) = D(N_Q(v, \lceil \phi m \rceil), v)$. The similarity distance $D_\phi(Q, v)$ denotes the aggregate distance from v to all members of its $\lceil \phi m \rceil$ neighborhood in Q. We call this similarity distance the 'ϕ-aggregate distance'. Since $D_1(Q, v) = D(Q, v)$, the AggNN problem can be viewed as a special case of the FANN problem whenever $\phi = 1$.

3   Related Work

3.1   Flexible Aggregate Similarity Search

As discussed in Section 1, none of the methods proposed for the AggNN problem can be effectively applied to the FANN problem [7–10]. In [11], two exact solutions, R-tree and List, were proposed specifically for FANN. Algorithm R-tree extends the solution for AggNN due to Papadias et al. [8] in low-dimensional settings (up to 10 dimensions), through an adaptation of the pruning bound. The design of the other exact method, List, is based on the well-known *Threshold Algorithm* (TA) [1]. Algorithm List retrieves candidates for the query result from the neighborhoods of every object of group query Q in a round robin fashion, while maintaining a lower bound for the best possible ϕ-aggregate distance of all the unseen objects. The algorithm terminates when at least k of the objects visited have ϕ-aggregate distances no greater than the lower bound. Although the algorithm is capable of handling higher-dimensional data, the execution cost may be prohibitively expensive, as we shall see in Section 5.

In addition to the exact methods R-Tree and List, two simple approximations methods were also proposed in [11]: ASUM, for the *sum* variant of FANN, and AMAX, for the *max* variant. With the ASUM heuristic, a candidate set is generated by taking the union of the k-NN sets for every query object in Q. The k best objects within these km candidates is then returned as the result. The AMAX heuristic is very similar to ASUM, the only difference being that AMAX first finds the center of the minimum enclosing ball (MEB) of the query set neighborhood $N_Q(q, \lceil \phi m \rceil)$, based at each query object $q \in Q$; AMAX then treats these centers as query objects.

3.2   Multi-step Search

Our proposed solutions for the FANN problem make use of the multi-step search strategy originally proposed for the adaptive similarity search problem, which

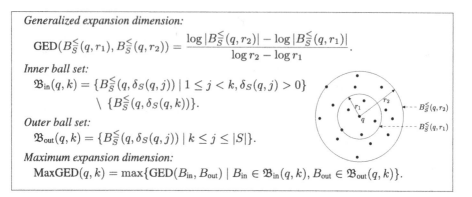

Generalized expansion dimension:

$$\text{GED}(B_{\overline{S}}^{\leq}(q, r_1), B_{\overline{S}}^{\leq}(q, r_2)) = \frac{\log |B_{\overline{S}}^{\leq}(q, r_2)| - \log |B_{\overline{S}}^{\leq}(q, r_1)|}{\log r_2 - \log r_1}.$$

Inner ball set:

$$\mathfrak{B}_{\text{in}}(q, k) = \{B_{\overline{S}}^{\leq}(q, \delta_S(q, j)) \mid 1 \leq j < k, \delta_S(q, j) > 0\}$$
$$\setminus \{B_{\overline{S}}^{\leq}(q, \delta_S(q, k))\}.$$

Outer ball set:

$$\mathfrak{B}_{\text{out}}(q, k) = \{B_{\overline{S}}^{\leq}(q, \delta_S(q, j)) \mid k \leq j \leq |S|\}.$$

Maximum expansion dimension:

$$\text{MaxGED}(q, k) = \max\{\text{GED}(B_{\text{in}}, B_{\text{out}}) \mid B_{\text{in}} \in \mathfrak{B}_{\text{in}}(q, k), B_{\text{out}} \in \mathfrak{B}_{\text{out}}(q, k)\}.$$

Fig. 1. Formal definitions of generalized expansion dimension and maximum expansion dimension (from [14]).

aims to find the most similar objects to a query object with respect to an adaptive similarity measure — one that can be determined by the user at query time [12]. Multi-step search computes a query result using a fixed 'lower-bounding' distance function that is adapted to answer the same query with respect to a user-supplied 'target' distance function. The function d_l is a lower-bounding distance for the target distance d if $d_l(u, v) \leq d(u, v)$ for any two objects u, v drawn from a domain for which both d_l and d are defined.

The proposed FANN solutions build upon our previous work on MAET [13], a heuristic multi-step search algorithm that utilizes, as the basis of an early search termination condition, tests of intrinsic dimensionality ('dimensional tests') according to the *generalized expansion dimension* (GED) [13,14]. For more details on MAET, its theoretical analysis, and its performance in practice, we refer the reader to [13].

3.3 Generalized Expansion Dimension

GED is a relaxation of the *expansion dimension* due to Karger and Ruhl [15], in which the estimation of intrinsic dimension in the vicinity of a query point q is made through the observation of the ranks and distances of pairs of neighbors with respect to q.

For any point $q \in \mathcal{U}$ and any point set $X \subseteq \mathcal{U}$, let $\delta_X(q, k)$ be the k-th smallest distance (with respect to d) from q to the points in X. Given a point $q \in \mathcal{U}$ and a radius $r \geq 0$, let $B_{\overline{S}}^{\leq}(q, r) \subseteq S$ be the closed ball of points centered at q with radius r containing all $v \in S$ satisfying $d(q, v) \leq r$, and let $B_{\overline{S}}^{<}(q, r) \subseteq S$ be the open ball containing all $v \in S$ satisfying $d(q, v) < r$. Note that the definition of the closed ball provided here allows a ball to have radius 0, and to contain no points.

Given two closed balls $B_{\overline{S}}^{\leq}(q, r_1)$ and $B_{\overline{S}}^{\leq}(q, r_2)$ with radii $0 < r_1 < r_2$ and cardinalities $0 < |B_{\overline{S}}^{\leq}(q, r_1)| < |B_{\overline{S}}^{\leq}(q, r_2)|$, their GED is defined as in Fig. 1. As argued in [13–15], the cardinality of the two ball sets is an estimator of their spherical volumes; substituting the true volumes for these cardinalities in the formula would reveal the representational dimensionality of the set. As such, the GED is the dimension of the space within which a uniform point distribution would be expected to produce inner and outer ball cardinalities proportional to the observed values. The GED formula therefore serves to relate an estimate of intrinsic dimensionality with the ranks of two neighbors (the ball cardinalities), and their distances to the query (the ball radii).

Also defined in Fig. 1 are the *inner ball set*, the *outer ball set* and the *maximum expansion dimension* (MaxGED), all relative to a point $q \in \mathcal{U}$ and a neighborhood size $k \geq 2$. Note that the MaxGED(q, k) has no definite value when the inner ball set $\mathcal{B}_{\text{in}}(q, k)$ is empty. In other words, in order for MaxGED(q, k) to have a definite value, the ball $B_S^{\leq}(q, \delta_S(q, j))$ must have radius satisfying $0 < \delta_S(q, j) < \delta_S(q, k)$ for at least one choice of j in the range $[1, k-1]$.

4 The *SUM* and *Avg* Variants of FANN

Note that for the FANN problem, the aggregation functions *sum* and *avg* are equivalent. For this reason, and for the ease of description of our algorithm, we will formulate our solution in terms of the *avg* aggregation function.

Let us first introduce some additional notation. For any object set $X \subseteq \mathcal{U}$ and any group query $Q \subseteq \mathcal{U}$, let $N_X(Q, k)$ denote the set of k nearest neighbors of Q within X with respect to ϕ-aggregate distance D_ϕ. Ties are broken arbitrarily but consistently. Note that when $k = 0$, the neighborhood set $N_X(Q, k)$ would be empty. Let $R_X^{\leq}(Q, r)$ be the closed range set of objects $v \in X$ with distance to Q satisfying $D_\phi(Q, v) \leq r$, and let $R_X^{<}(Q, r)$ be the open range set of objects $v \in X$ with distance to Q satisfying $D_\phi(Q, v) < r$. For any object $u \in \mathcal{U}$, let $R_X^{\leq}(u, r)$ be the closed range set of objects $v \in X$ with distance to u satisfying $d(u, v) \leq r$, and let $R_X^{<}(u, r)$ be the open range set of objects $v \in X$ with distance to u satisfying $d(u, v) < r$. Let $\delta_X(Q, k)$ be the k-th smallest ϕ-aggregate distance from Q to all the objects in X.

4.1 Algorithm

Our algorithm, FADT_AVG, is described in Fig. 2. Whereas traditional search techniques usually progress through the exploration of a neighborhood of a single point in the domain (the query point), our aggregate similarity search method simultaneously explores the neighborhoods of m 'start points' M_i, each of which is formed by aggregating a subset of the group query Q. Iterating through the neighborhoods of the start points in round robin fashion, two pruning strategies are applied to reduce the search space. The first strategy utilizes a distance bound based on an auxiliary distance relative to the ϕ-aggregate distance. The second strategy applies a test of the intrinsic dimensionality (in the vicinity of Q) to determine whether early termination is possible.

Algorithm **FADT** (*database S, group query Q, target proportion ϕ, aggregation function* avg *or* max*, neighborhood size k, lower-bounding ratios $\lambda_1, \ldots, \lambda_m$, termination parameter t*)

1: Assume the existence of an index I created with respect to d, together with a method $getnext$ that uses I to iterate through the nearest neighbor list of a target object.
2: Let P be an object set. $P \leftarrow \emptyset$.
3: **AVG**: For each $q_i \in Q$, find the geometric median M_i of $N_Q(q_i, \lceil \phi m \rceil)$.
   MAX: For each $q_i \in Q$, find the center M_i of the minimum enclosing ball of $N_Q(q_i, \lceil \phi m \rceil)$.
   // $m = |Q|$
4: **AVG**: Let α_i be the average of the largest $\lceil \phi m \rceil$ distances from M_i to all the query points in Q, for $i = 1, \ldots, m$.
   MAX: Let α_i be the $\lceil \phi m \rceil$-th largest distance from M_i to all the query points in Q, for $i = 1, \ldots, m$.
5: Let q^* and r^* be the center and radius of MEB(Q), respectively.
6: Initialize $\gamma_i \leftarrow -\infty$ for $i = 1, \ldots, m$.
7: Initialize $k_{in} \leftarrow 0$ and $k_{out} \leftarrow k$.
8: **while** TRUE **do**
9:    **for** $i = 1 \rightarrow m$ **do**
10:       $(v_i, \beta_i) \leftarrow I.getnext(M_i)$.   // $\beta_i = d(M_i, v_i)$
11:       $\gamma_i \leftarrow \lambda_i \cdot \beta_i - \alpha_i$.   // lower distance bound
12:       $P \leftarrow P \cup \{v_i\}$.
13:       **if** $|P| < k$ **then**
14:          Continue to Line 9.
15:       **else if** $|P| = |S|$ **then**
16:          **return** $N_P(Q, k)$.
17:       **end if**
          // prune using distance bound
18:       $\gamma \leftarrow \max\{\gamma_1, \ldots, \gamma_m\}$.
19:       $k' \leftarrow |R_P^{\leq}(Q, \gamma)|$.
20:       **if** $k' \geq k$ **then**
21:          **return** $N_P(Q, k)$.
22:       **else**
          // prune using dimensional test
23:          $k_{in} \leftarrow |R_P^{\leq}(q^*, \gamma - r^*)|$.
24:          **if** $k_{in} > 0$ **then**
25:             $r_{in} \leftarrow \delta_P(q^*, k_{in})$.
26:             $r_{out} \leftarrow \delta_P(Q, k) + r^*$.
27:             $k_{out} \leftarrow |R_P^{\leq}(q^*, r_{out})|$.
28:             **if** $r_{in} > 0$ and $k_{in} \cdot (r_{out}/r_{in})^t < k_{out} + 1$ **then**
29:                **return** $N_P(Q, k)$.
30:             **end if**
31:          **end if**
32:       **end if**
33:    **end for**
34: **end while**

Fig. 2. The description of the FADT_AVG and FADT_MAX variants of Algorithm FADT.

For each query point $q_i \in Q$, the algorithm constructs one start point M_i. The point M_i is constructed as the geometric median of the neighborhood $N_Q(q_i, \lceil \phi m \rceil)$ of q_i within group query Q. The geometric median M_i is the point which minimizes the average of its distances to all points in the full query set neighborhood $N_Q(q_i, \lceil \phi m \rceil)$. Note that M_i is not required to be one of the points of this neighborhood. Algorithm FADT_AVG then searches for query result candidates by sequentially scanning the neighborhood of every M_i (with respect to d) in a round robin fashion. Throughout the search process, the algorithm maintains an object set P, which stores all objects that have been visited so far. The algorithm also maintains m lower bounds $(\gamma_1, \ldots, \gamma_m)$ for the ϕ-aggregate distances of unseen objects to Q, with each γ_i relating to the neighborhood of the geometric median M_i, for $1 \leq i \leq m$. We let γ denote the maximum of these m lower bounds.

The algorithm terminates when one of the following three conditions holds:

- all objects of S have been visited (Line 15); or
- the exact multi-step search termination condition is fulfilled, in which at least k visited objects have ϕ-aggregate distances to Q that are no greater than the maximum lower bound γ (Line 20); or
- a termination condition relating to the intrinsic dimensionality in the vicinity of Q is fulfilled (Line 28).

The test of intrinsic dimensionality at Line 28 (the dimensional test) is based on a user-supplied estimate, in the form of a termination parameter $t > 0$. The test is performed on two closed balls centered at q^*, the center of minimum enclosing ball MEB(Q) of Q. The inner ball has r_{in} as its radius and contains k_{in} points of S, all of which are members of the aggregate query result; the outer ball has r_{out} as its radius and contains at least k_{out} points of S, where $k_{\text{out}} \geq k$. The aim of the dimensional test is to verify that no unvisited points could possibly be included in the outer ball, and therefore that the correct query result has been found.

Lower-Bounding Relationship. The assumption that d is a distance metric implies that for every geometric median M_i $(1 \leq i \leq m)$, every $q_j \in Q$ $(1 \leq j \leq m)$ and any object $v \in \mathcal{U}$, a triangle inequality holds: $d(M_i, v) \leq d(q_j, v) + d(M_i, q_j)$. Together with the definition of ϕ-aggregate distance D_ϕ for the avg variant, we can derive that:

$$\lceil \phi m \rceil \cdot d(M_i, v) \leq \lceil \phi m \rceil \cdot D_\phi(Q, v) + \sum_{q \in N_Q(v, \lceil \phi m \rceil)} d(M_i, q).$$

Let α_i be the average of the largest $\lceil \phi m \rceil$ distances from M_i to all the points in Q (as defined in Line 4 of the description of FADT_AVG), we can derive the following lower-bounding relationship: $d(M_i, v) \leq D_\phi(Q, v) + \alpha_i$. Together with the dimensional test, Algorithm FADT_AVG utilizes this lower-bounding relationship to filter out candidate query result objects. The distance $d(M_i, v)$ from the candidate to geometric median M_i can be regarded as an approximation of $D_\phi(Q, v) + \alpha_i$ For the approximation to be as tight as possible, we introduce a lower-bounding ratio $\lambda_i \geq 1$ such that

$$\lambda_i d(M_i, v) \leq D_\phi(Q, v) + \alpha_i. \tag{1}$$

For each individual object $v \in \mathcal{U}$, consider the maximum possible value of the lower-bounding ratio $\lambda_i \geq 1$ for which the inequality holds. Ideally, λ_i should be set to the smallest such maximum so that the inequality holds for every object v.

4.2   Analysis

Due to space limitations, in this version of the paper we provide only a brief overview of a formal theoretical analysis of FADT_AVG that establishes the accuracy of the method in terms of the choice of termination parameter t.

The proof strategy involves an assumption on the maximum expansion rate of the items encountered in an expanding search from the center q^* of the minimum enclosing ball $\text{MEB}(Q)$ of the query objects — this assumption is embodied in the choice of t. At any stage in the search, if the presence of unvisited query result objects would indicate an expansion rate that would exceed the limit implied by $\text{MaxGED}(q^*, k+1)$, we can conclude that no such unvisited result objects can exist.

Theorem 1. *If $\text{MaxGED}(q^*, k+1)$ has no definite value, then FADT_AVG returns the correct query result regardless of the value of t. Otherwise, if $t \geq \text{MaxGED}(q^*, k+1)$, FADT_AVG returns the correct query result; if $0 < t < \text{MaxGED}(q^*, k+1)$, FADT_AVG returns a $\sqrt[4]{\wp}$-approximate query result with $\wp = k_{\text{out}} + 1$ at termination.*

4.3   Variants

The algorithm for the *max* variant, which we will refer to as FADT_MAX, is very similar to FADT_AVG. Instead of expanding the search from geometric medians of subsets of Q (as in FADT_AVG), Algorithm FADT_MAX initiates its searches from the center C_i of the minimum enclosing ball of query set neighborhood $N_Q(q_i, \lceil \phi m \rceil)$, for each $q_i \in Q$. Due to space limitations, however, further details are omitted from this version.

In addition to the basic versions of FADT_AVG and FADT_MAX, several heuristic variants were proposed, analyzed, and tested. Here, we briefly state only those variants with the best trade-offs between accuracy and efficiency, which we refer to as FADT_AVG5 and FADT_MAX5. Both are obtained by applying the following three heuristic modifications.

 - The radius term r^* in the algorithmic description was substituted by 0.
 - A limited-capacity object buffer of at most zk objects was maintained for the determination of the outer ball.

– The lower-bounding ratios λ_i were dynamically estimated. More precisely, after each neighborhood v_i of the geometric median M_i and distance $\beta_i = d(M_i, v_i)$ are retrieved from the underlying index, the sum of the ϕ-aggregate distance $D_\phi(Q, v_i)$ and the constant α_i is computed first, followed by the ratio of the sum over β_i. The smallest such ratio encountered is stored in λ_{e_i} as the current estimate of λ_i.

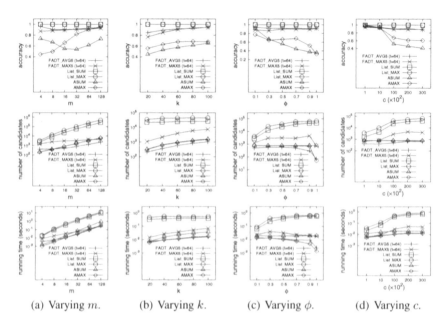

(a) Varying m.   (b) Varying k.   (c) Varying ϕ.   (d) Varying c.

Fig. 3. The results of varying each of the four parameters for generating FANN queries on dataset MNIST. Termination parameter t was set to 64, for which the maximum approximation ratio $\sqrt[t]{|S|}$ was guaranteed to be less than 1.2, given that $|S| = 70,000$.

5   Experimental Results

5.1   Experimental Framework

Data Sets. Four publicly-available data sets were considered for the experimentation, so as to compare across a variety of set sizes, representational dimensions and domains.

– The MNIST data set [16] consists of 70,000 images of handwritten digits from 500 different writers, with each image represented by 784 gray-scale texture values.

– The Amsterdam Library of Object Images (ALOI) [17] consists of 110, 250 images of 1000 small objects taken under different conditions, such as differing viewpoints and illumination directions. The images are represented by 641-dimensional feature vectors based on color and texture histograms (for a detailed description of how the vectors were produced, see [18]).
– The Cortina data set [19] consists of 1, 088, 864 images gathered from the World Wide Web. Each image is represented by a 74-dimensional feature vector based on homogeneous texture, dominant color and edge histograms.
– The Reuters Corpus Vol. 2 news article data set (RCV2) [20] consists of 554, 651 sparse document vectors spanning 320, 647 keyword dimensions, with TF-IDF weighting followed by vector normalization.

FANN Queries. For the experimentation, several factors should be taken into account when generating a set of FANN queries, including the position of the center of $MEB(Q)$ (minimum enclosing ball of group query Q), the radius of $MEB(Q)$, the size $m = |Q|$, the target neighborhood size k, and the target proportion ϕ. For our experiments, we generated a group query Q in the following way. We first determined a ball with a randomly-chosen center covering a certain number c of points in the data space. Then, within this ball, we randomly selected m data points as the group query. Parameter c can thus be regarded as governing the dispersion of the query object sets generated for the experimentation. Unless stated otherwise, the default parameters $m = 16$, $k = 50$, $\phi = 0.5$ and $c = 10, 000$ were used in generating FANN queries.

Methodology. For each test, 100 random FANN queries were generated. Three quantities were measured for the evaluation: query result accuracy, number of candidates visited, and execution time. The results were reported as averages over the 100 queries performed. The number of candidates was calculated as the number of data objects (including duplicates) retrieved from the underlying

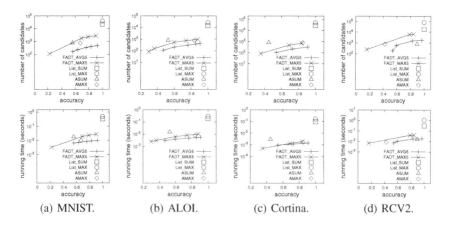

(a) MNIST.   (b) ALOI.   (c) Cortina.   (d) RCV2.

Fig. 4. The results of using default parameters on all the datasets.

index in the course of computing the final query result. For one group query Q, the accuracy of its k-NN query result is defined as the proportion of the result falling within the true k-NN (ϕ-aggregate) distance:

$$| \{v \in Y \mid D_\phi(Q, v) \leq \delta_S(Q, k)\} | / k,$$

where Y denotes the k-NN query result of Q ($|Y| = k$). For ease of comparison among the competing methods, we ignored the costs associated with producing candidate elements from the underlying index structure I, and instead precomputed all required neighborhoods. However, in order to gauge the practicality of the methods tested, in the last set of experiments (Section 5.2), we report the total query cost, including the costs associated with the production of neighborhood lists. For these experiments, we used SASH [21] as the underlying index I, for its ability to answer approximate k-NN queries in very high-dimensional settings with good accuracy and efficiency. The Euclidean distance was used for all experiments.

5.2   Comparison with Other Methods

The following extensive experimental study shows that our algorithms are able to produce query results with good accuracy, while at the same time being very efficient, compared to List, ASUM and AMAX (R-Tree being omitted due to its inability to cope with high-dimensional datasets).

Using Precomputed Neighborhoods. On all the considered datasets, we conducted 4 sets of experiments for the comparison, varying one of the 4 parameters for generating FANN queries respectively, while fixing the rest at their default values ($m = 16$, $k = 50$, $\phi = 0.5$ and $c = 10,000$). Specifically, we varied m from 4 to 128, k from 20 to 100, ϕ from 0.1 to 1, and c from 100 to 30,000. Since similar conclusions can be drawn from all the results, due to space limitations we report only the results for the MNIST dataset. In addition, for all the datasets, we report the results obtained when all parameters are set to the default values stated above.

Variation in the Query Set Size m. The results of varying m are shown in Fig. 3(a). We observe that in all cases, for both the *sum* (or *avg*) FANN problem and the *max* FANN problem, our algorithms FADT_AVG5 and FADT_MAX5 are able to produce query results with reasonable accuracy, while at the same time being very efficient. On average, our algorithms outperform List by approximately one order of magnitude, in terms of both the number of candidates and the execution time. This can be explained by their minimization of aggregate distances through the use of geometric medians or the centers of minimum enclosing balls. Although ASUM and AMAX are also very efficient, for some cases they failed to produce query results with reasonable accuracy. Their failure is likely due to their strategy of always limiting the search to at most km candidates. Such aggressive search strategies may not be able to produce the query result with reasonable accuracy, as justified here and in the following experiments.

Variation in the Neighborhood Size k. Fig. 3(b) shows the results of varying k. We again observe that for all choices of k, our algorithms FADT_AVG5 and FADT_MAX5 maintain their superiority over List by roughly 1 to 2 orders of magnitude, in terms of both the number of candidates and the execution time. We also note that ASUM and AMAX may not be able to produce query results with reasonably good accuracy.

Variation in the Target Proportion ϕ. The results of varying ϕ are shown in Fig. 3(c). Again, FADT_AVG5 and FADT_MAX5 outperform List in terms of both number of candidates and running time for all choices of ϕ, and again, ASUM and AMAX cannot guarantee query results with reasonable accuracy. The superiority of our algorithms over List becomes more and more evident as ϕ increases. In particular, when $\phi = 1$, our algorithms outperform List by approximately 2 to 3 orders of magnitude, in terms of both the number of candidates and the execution time. This shows that our algorithms benefit more from utilizing geometric medians or centers of MEB as ϕ grows.

Variation in the Query Set Dispersion Parameter c. Fig. 3(d) shows the results of varying c. Similarly, algorithms FADT_AVG5 and FADT_MAX5 show their superiority over List in terms of both number of candidates and running time, and ASUM and AMAX may not be able to produce query results with good accuracy. The superiority of our algorithms over List becomes more and more evident as c increases. Especially, when $c = 30,000$, our algorithms outperform List by approximately 1.5 orders of magnitude in terms of both number of candidates and running time. The behaviors of our algorithms are relatively stable with respect to c, again due to the benefits of utilizing geometric medians or centers of MEB.

Default Parameters. Here, we show the results for all datasets using the default parameters in generating and processing FANN queries. For our algorithms, parameter t was chosen to be 2^i for all integer choices of i in the range $[-6, 6]$, so as to cover a reasonably wide range of result accuracies. Note that some of the plots have been cropped for the sake of readability.

The results are shown in Fig. 4. Again, we find that FADT_AVG5 and FADT_MAX5 outperform List by 1 to 2 orders of magnitude, in terms of both the number of candidates and the execution time, and again we observe the inability of ASUM and AMAX to achieve reasonably high accuracies.

Online Neighborhood Generation. For our last set of experiments, for all algorithm-dataset combinations, we measured the execution costs of queries using the SASH as the underlying index for online neighborhood generation. As a baseline comparison, we also computed the query results using a 'brute force' search (BF) in which the ϕ-aggregate distance from Q is explicitly computed for each object in the database. In the plots, the running time is presented as a proportion of the time required by BF. Again, for our algorithms, parameter t was chosen to be 2^i for each choice of $i \in [-6, 6]$, and again, some of the performance plots have been cropped for the sake of readability.

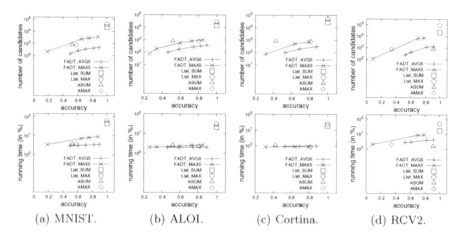

(a) MNIST.   (b) ALOI.   (c) Cortina.   (d) RCV2.

Fig. 5. The results of using default parameters on all the datasets with SASH as the underlying index. The average running times of BF for MNIST, ALOI, Cortina and RCV2 are approximately 1.67, 2.16, 3.92 and 10.24 seconds, respectively.

The results are shown in Fig. 5. Due to the high computational cost associated with online neighborhood generation, in terms of the execution time, the superiority of our algorithms over List is less significant than when neighbor lists have been precomputed. Nevertheless, FADT_AVG5 and FADT_MAX5 still outperform List by approximately 0.5 to 1.5 orders of magnitude. Compared to the baseline method BF, our algorithms maintain their superiority in execution cost by roughly 1 to 2 orders of magnitude, whereas the cost of Algorithm List may approach that of BF.

Acknowledgments. Michael Houle acknowledges the financial support of JSPS Kakenhi Kiban (A) Research Grant 25240036, JSPS Kakenhi Kiban (B) Research Grant 15H02753, and JSPS Kakenhi Kiban (C) Research Grant 24500135. Vincent Oria acknowledges the financial support of NSF under Grant 1241976.

References

1. Fagin, R., Lotem, A., Naor, M.: Optimal aggregation algorithms for middleware. In: Proc. Symp. on Principles of Database Systems (PODS), pp. 102–113 (2001)
2. Marian, A., Bruno, N., Gravano, L.: Evaluating top-k queries over web-accessible databases. ACM Transactions on Database Systems (TODS) **29**(2), 319–362 (2004)
3. Bernecker, T., Emrich, T., Graf, F., Kriegel, H.P., Kröger, P., Renz, M., Schubert, E., Zimek, A.: Subspace similarity search using the ideas of ranking and top-k retrieval. In: Proc. ICDE Workshop DBRank, pp. 4–9 (2010)
4. Razente, H.L., Barioni, M.C.N., Traina, A.J.M., Traina, Jr., C.: Aggregate similarity queries in relevance feedback methods for content-based image retrieval. In: Proc. Symp. on Applied Computing, pp. 869–874 (2008)

5. Beliakov, G., Calvo, T., James, S.: Aggregation of preferences in recommender systems. In: Ricci, F., Rokach, L., Shapira, B., Kantor, P.B. (eds.) Recommender Systems Handbook. Springer (2010)

6. Yiu, M.L., Mamoulis, N., Papadias, D.: Aggregate nearest neighbor queries in road networks. IEEE Transactions on Knowledge and Data Engineering (TKDE) **17**(6), 820–833 (2005)

7. Li, F., Yao, B., Kumar, P.: Group enclosing queries. IEEE Transactions on Knowledge and Data Engineering (TKDE) **23**(10), 1526–1540 (2011)

8. Papadias, D., Tao, Y., Mouratidis, K., Hui, C.K.: Aggregate nearest neighbor queries in spatial databases. ACM Transactions on Database Systems (TODS) **30**(2), 529–576 (2005)

9. Papadias, D., Shen, Q., Tao, Y., Mouratidis, K.: Group nearest neighbor queries. In: Proc. Intern. Conf. on Data Engineering (ICDE), pp. 301–312 (2004)

10. Razente, H.L., Barioni, M.C.N., Traina, A.J.M., Faloutsos, C., Traina, Jr., C.: A novel optimization approach to efficiently process aggregate similarity queries in metric access methods. In: Proc. Intern. Conf. on Information and Knowledge Management (CIKM), pp. 193–202 (2008)

11. Li, Y., Li, F., Yi, K., Yao, B., Wang, M.: Flexible aggregate similarity search. In: Proc. Intern. Conf. on Management of Data (SIGMOD), pp. 1009–1020 (2011)

12. Seidl, T., Kriegel, H.P.: Optimal multi-step k-nearest neighbor search. In: Proc. Intern. Conf. on Management of Data (SIGMOD), pp. 154–165 (1998)

13. Houle, M., Ma, X., Nett, M., Oria, V.: Dimensional testing for multi-step similarity search. In: Proc. Intern. Conf. on Data Mining (ICDM), pp. 299–308 (2012)

14. Houle, M., Kashima, H., Nett, M.: Generalized expansion dimension. In: IEEE ICDM Workshop on Practical Theories for Exploratory Data Mining (PTDM), pp. 587–594 (2012)

15. Karger, D.R., Ruhl, M.: Finding nearest neighbors in growth-restricted metrics. In: Proc. Symp. on Theory of Computing (STOC), pp. 741–750 (2002)

16. LeCun, Y., Bottou, L., Bengio, Y., Haffner, P.: Gradient-based learning applied to document recognition. Proc. IEEE **86**(11), 2278–2324 (1998)

17. Geusebroek, J.M., Burghouts, G.J., Smeulders, A.W.M.: The Amsterdam library of object images. International Journal of Computer Vision (IJCV) **61**(1), 103–112 (2005)

18. Boujemaa, N., Fauqueur, J., Ferecatu, M., Fleuret, F., Gouet, V., Saux, B.L., Sahbi, H.: IKONA: interactive generic and specific image retrieval. In: Proc. Intern. Workshop on Multimedia Content-Based Indexing and Retrieval (MMCBIR) (2001)

19. Rose, K., Manjunath, B.S.: The Cortina data set. http://www.scl.ece.ucsb.edu/datasets/index.htm

20. Reuters Ltd.: Reuters corpus, vol. 2, multilingual corpus. http://trec.nist.gov/data/reuters/reuters.html

21. Houle, M.E., Sakuma, J.: Fast approximate similarity search in extremely high-dimensional data sets. In: Proc. Intern. Conf. on Data Engineering (ICDE), pp. 619–630 (2005)

Similarity Joins and Beyond: An Extended Set of Binary Operators with Order

Luiz Olmes Carvalho$^{(\boxtimes)}$, Lucio F.D. Santos, Willian D. Oliveira,
Agma Juci Machado Traina, and Caetano Traina Jr.

Institute of Mathematics and Computer Sciences, University of São Paulo,
São Carlos - SP, Brazil
{olmes,luciodb,willian,agma,caetano}@icmc.usp.br

Abstract. Similarity joins are troublesome database operators that often produce results much larger than the user really needs or expects. In order to return the similar elements, similarity joins also require sorting during the retrieval process, although order is a concept not supported in the relational model. This paper proposes a solution to solve those two issues extending the similarity join concept to a broader set of binary operators, which aims at retrieving the most similar pairs and embedding the sorting operation only as an internal processing step, so as to comply with the relational theory. Additionally, our extension allows to explore another useful condition not previously considered in the similarity retrieval: the negation of predicates. Experiments performed on real and synthetic data show that our operators are fast enough to be used in real applications and scale well both for multidimensional and non-dimensional metric data.

Keywords: Similarity search · Similarity joins · Query operators

1 Introduction

Similarity joins are becoming important database operators in several scenarios, such as near-duplicate detection, string matching and data mining support [7,10]. Those operators receive two relations T_1 and T_2 and return pairs of tuples $\langle t[T_1], t[T_2] \rangle$ that meet a similarity predicate. The most common types of similarity joins found in the literature are the range join, the k-nearest neighbor join and the k-distance join [1].

Usually, the results of the range and the k-nearest neighbor joins are not sent directly to the user, as they are mainly applied as *preprocessing operators* [3,15] or as *intermediate operators*, once their result set cardinality is usually very large. Those high-cardinality results are often not intuitive to the user, being sent to another algorithm [1]. In most applications requiring similarity joins, the users are usually interested in the few *most similar* pairs [4,9]. Thus, the k-distance

---

The authors are grateful to FAPESP, CNPQ, CAPES and Rescuer (EU Commission Grant 614154 and CNPQ/MCTI Grant 490084/2013-3) for their financial support.

G. Amato et al. (Eds.): SISAP 2015, LNCS 9371, pp. 29–41, 2015.
DOI: 10.1007/978-3-319-25087-8_3

Fig. 1. The negation of predicates: (a) Combining sensors and candidate location to install a sensor; (b) Answer with the negation of a range condition.

join is the similarity operator most suitable to be a *query operator*, because it results in just the k most similar pairs.

However, in order to choose the k most similar pairs in general, the k-distance join operator demands additional operations, such as sorting, before the final result is obtained. Therefore, k-distance joins internally performs more operations than those defined in the "classical" join: in fact, the non-similarity based joins are defined as a Cartesian product followed by a selection. Thus, in a strict sense, the so called k-distance join is not a join operator but an extended type of binary similarity operator that also requires ordering.

In addition to ordering, the same concept employed to define the k-distance join operator can be generalized to support also the range and the k-nearest neighbor joins by interposing their respective similarity selections. Such more generic operator enables to explore a kind of condition common in relational databases but, to the best of our knowledge, it was still not adequately explored in similarity joins: the *negation* of the similarity operators.

Consider the following example. The São Paulo Brazilian State is an important sugar cane producer, providing ethanol to the country. The expansion of such cultivar demands monitoring the climate measures such as temperature and precipitation. For this purpose, a small number of climate sensors (represented as stars, in Fig. 1(a)) were positioned in the most productive areas, where each sensor covers a radius of about 10 km. To improve monitoring, new additional locations sensors should be installed (the diamonds in Fig. 1(a)), but the current budget allows for the installation of just k new sensors. Where should the new sensors be installed, so they are close but not inside the already monitored areas?

This scenario requires to consider not only the locations closest to each sensor, but also those which are outside their covering area. The problem can be solved employing the negation of the range predicate. In addition, the order among the pairs of sensors and the new locations must be considered, as Fig. 1(b) shows, where the stars are combined with the circles (the closest locations not yet covered).

This paper extends the definition of similarity joins to generate a broader set of binary similarity operators, which we call *wide-joins* (Sect. 3). They are defined as a Cartesian product followed by a selection based on order, where the ordering is obtained during the similarity evaluation. Broadening the k-

Table 1. Symbols employed in the paper.

Symbol	Meaning	Symbol	Meaning
d	a metric	$t[A]$	the value of attribute A in tuple t
S_1, S_2	attributes subject to a metric	Π	extended projection [8]
S	metric data domain	θ	a predicate
T_1, T_2, T_R	relations	$\bowtie$	similarity join

distance join, those new operators aims at computing the most similar pairs, being general enough to support range and k-nearest neighbor predicates and also their negation.

We performed an extensive scalability evaluation of the proposed operators using real and synthetic datasets (Sect. 4). The results obtained show that the "ordered similarity joins" present computational cost equivalent to the existing similarity joins whereas returning a smaller and more significant result set.

The major contributions of this paper are: (i) we present the k-distance join as an extended operator and provide a theoretical ground to support it; (ii) we embed the sorting concept into a similarity operator in a way compatible with the relational theory; (iii) we explore the negation of a similarity predicate; (iv) we enable any similarity join to be used as a final *query operator*. Finally, we outline the main ideas and devise future improvements (Sect. 5).

2   Related Work

Similarity search is the information retrieval process where the answer consists of a set of elements recognized, in some sense, as similar to others. The basic query operators that perform similarity retrieval are the similarity selections [2,14] and the similarity joins. They are typically applied over data in a metric space [13]. Formally, let S be a metric data domain subjected to a distance function d, that is $d : S \times S \mapsto \mathbb{R}^+$, T_1 and T_2 be two relations containing attributes $S_1 \in T_1$ and $S_2 \in T_2$ with values sampled from S, $\xi \in \mathbb{R}^+$ be a similarity threshold and $k \in \mathbb{N}^*$ be a constant (Table 1 summarizes the symbols). The similarity range join combines the tuples $t[T_1]$ and $t[T_2]$ such that $d(t[S_1], t[S_2]) \leq \xi$. The k-nearest neighbor join combines each tuple whose attribute $t[S_1]$ is one of the k most similar value to $t[S_2]$, totaling $k * |T_1|$ pairs. Finally, the k-distance join retrieves the k most similar pairs $\langle t[S_1], t[S_2] \rangle$.

Similarity joins have been extensively investigated in the literature. The study introduced in [14] defines algebraic equivalence rules holding in similarity selection, grouping and join operators. It also presents another join that combines range and *one*-nearest neighbor into a conjunctive predicate. It departs from ours as we aim at retrieving the most similar pairs, extending the k-distance join, whereas a range predicate does not ensure such cardinality control.

Several studies process similarity joins using metric structures. The basic idea is to insert the elements from one or both relations into a data structure that speeds up retrieval. Previous studies like [5,12] employ the eD-Index, but they focus only in range joins. Another study [7] proposed pruning techniques on the M-Tree to improve the detection of closest pairs. Our proposal departs from that once we detect closest pairs extending similarity join operators. The List of Twin Clusters (LTC) [11] was specially designed to process similarity joins, but LTC does not ensure the closest pairs, which we are interested in.

A non-indexing approach to process similarity joins in metric spaces is the Quickjoin [10]. It divides the search space into smaller regions to reduce the complexity of nested-loops. However, Quickjoin was designed to compute only range joins, whereas we focus on the k-distance join type. An enhanced version of Quickjoin to process k-nearest neighbor join was introduced in [6]. However, that version computes approximate nearest neighbors, in which some relevant pairs may be lost and replaced by less similar ones, while we are interested in an accurate version of the similarity join operator.

The theory of wide-joins was introduced in [2]. We extend that study including: (i) a reformulation of the basic similarity join definition in a generalized concept which allows to perform any kind of comparison involving similarity; (ii) the support to handle negation; and (iii) experimental performance comparison of similarity joins, wide-joins and their algebraic expressions.

3   Proposal

Considering the similarity join operators presented in Sect. 2, the range join can be expressed as the Cartesian product followed by a range selection. Likewise, the k-nearest neighbor join is equivalent to a Cartesian product followed by a k-nearest neighbor selection. Therefore, both join operators have corresponding similarity selection operators supporting them. Both join operators can also be seen as equivalent to similarity selection operators when T_1 has a single tuple (the query center) and T_2 contains the elements to be queried. However, the k-distance join does not have a corresponding similarity selection operator to produce its result when attached to the Cartesian product.

In order to find the global k most similar pairs, the k-distance join operator must compute the candidate pairs to compose the result (Cartesian product), evaluate their similarity (distance computation) and *order* the pairs following the similarity criterion (distance), allowing to finally filter only the most similar pairs (final selection) in a general context. Thus, the k-distance join is expressed in relational algebra as in (1).

$$\sigma_{(ord \leq \kappa_f)} \left(\Pi_{\{T_1, T_2, \mathcal{F}(dist) \to ord\}} \left(\Pi_{\{T_1, T_2, d(t[S_1], t[S_2]) \to dist\}} \left(T_1 \times T_2 \right) \right) \right) . \quad (1)$$

In (1), $\mathcal{F}$ is a function that projects the ordinal value of each value computed by the metric d into the extended attribute ord. We employ $\mathcal{F}$ such as a function that receives the distances between values $t[S_1], t[S_2]$ and returns the ordinal classification of each of those dissimilarity values. In such way, although the k-distance join demands to order the elements, the ordering concept is contained inside the operator remaining compatible with the relational theory, once each extended projection Π operator receive a relation and return a relation. In fact, this definition follows the same concept that allows aggregate functions such as *sum, average*, etc. to be employed in the relational model. Finally, the elements are not physically nor logically sorted, as the ordering is embodied in the attribute *ord*, which is projected out in the end, but enables to select the k most similar pairs in general.

3.1   Similarity Joins with Order: The Theory of Wide-Joins

Although (1) provides the algebraic definition of the k-distance join, it also defines the conceptual basis to create an extended set of similarity-based operators that includes the order concept as an internal part of the processing. In (1), it is possible to interpose a similarity selection operator between the Cartesian product and the extended projection in order to embrace joins. Once those operators are composed of additional operations, beyond just a Cartesian product followed by a selection (as in the join definition), we refer to them as the "similarity wide-joins" and formalize them in Definition 1.

Definition 1 (Similarity Wide-Join: $\bowtie$). *Let S be a metric data domain subjected to a distance function d, that is, $d : S \times S \mapsto \mathbb{R}^+$, T_1 and T_2 be two relations containing attributes $S_1 \in T_1$ and $S_2 \in T_2$ with values sampled from S, κ_f be an upper bound parameter and θ be a similarity-based predicate. Then, a similarity wide-join $T_1 \overset{(S_1\,\theta\,S_2),\kappa_f}{\bowtie} T_2$ is a binary operator that performs an inner similarity join using the predicate θ, order the intermediate result by the dissimilarity among values $t[S_1]$ and $t[S_2]$ and returns the κ_f tuples having $t[S_1]$ and $t[S_2]$ most similar in general. The wide-join is expressed in relational algebra according to (2).*

$$T_1 \overset{(S_1\,\theta\,S_2),\kappa_f}{\bowtie} T_2 \equiv \sigma_{(ord \leq \kappa_f)}\left(\Pi_{\{T_1,T_2,\mathcal{F}(d(t[S_1],t[S_2])) \to ord\}}\left(T_1 \overset{(t[S_1]\,\theta\,t[S_2])}{\bowtie} T_2 \right) \right) . \quad (2)$$

Similarity wide-joins follow a k-distance join-like definition, with function $\mathcal{F}$ executing as aforesaid. They also employ the ordering concept internally, compatible with the relational theory. In addition, similarity wide-joins make flexible to express both the desired cardinality of the result set and the similarity-based predicate that composes the inner similarity join, where each variation of the similarity condition generates a distinct type of wide-join. Usually, the similarity predicate is expressed by a single-term with comparisons based on range or k-nearest neighbors, but those conditions can be combined to obtain results from more elaborated predicates.

3.2   Single-term Predicates

The most straightforward type of wide-join does not employ a comparison based on similarity, but it corresponds to the predicate $\theta = \text{true}$. In such case, the inner join appearing in (2) becomes a Cartesian product, like in (1). Thus, a wide-join employing a true predicate results in the k-distance join operator.

The other types of wide-joins are directly obtained from using range and nearest neighbor conditions. When $\theta = (t[S_1] \, Range(d, \xi) \, t[S_2])$, the resulting operator is the range wide-join. Range wide-joins enable to combine a range join and k-distance join in a single operator. It selects a restricted number of the κ_f most similar elements among pairs $\langle t[T_1], t[T_2] \rangle$ such that $d(t[S_1], t[S_2]) \leq \xi$. Either range or k-distance joins alone do not produce the same result. Naturally, the former fails in restricting only the most similar pairs and the latter can select pairs whose dissimilarity exceeds the value ξ. Moreover, it is slower to retrieve a composition of a range join followed by a selection of the most similar pairs, like the right-hand side of (2), than to embed those operations into a single operator, as discussed in the experimental section.

A predicate $\theta = (t[S_1] \, kNN(d, k) \, t[S_2])$ produces the k-nearest neighbor wide-join. Like the range wide-join, the k-nearest neighbor wide-join reduces the cardinality of the k-nearest neighbor join from $k * |T_1|$ to κ_f. However, it requires two parameters related to quantities: k for the k initial nearest neighbor join and κ_f for the final number of pairs.

Values of $\kappa_f \leq k$ filter out a subset of the k-nearest neighbor join result. As the wide-join operator retrieves the most similar pairs, the intermediary result eventually corresponds to a k-distance join operator, when $\theta = true$. If k ensures a large selectivity, such setup allows using function $\mathcal{F}$ to optimize the operator, once it allows selecting a reduced number of pairs.

In metric spaces containing a number of denser regions, setting $\kappa_f \leq k$ may lead to many pairs too much similar among themselves, which may not add valuable information to the query answer. However, when $\kappa_f > k$, the k-nearest neighbor wide-join assumes a more exploratory behavior, and returns not only the most similar pairs from the subset of the k-nearest neighbor join in general, but also pairs distributed along the entire search space, despite the existence of denser regions.

Although similarity queries including range and nearest neighbor operators are the most frequent ones, the complement of those single condition is supported in the predicate logic and can be employed for similarity retrieval, producing not-in-range and not-the-nearest neighbors similarity comparators.

3.3 Negation of Single-term Predicates

The predicate negation of a term exists in the relational algebra and is well explored in relational databases. For instance, queries employing operators like $\neq$ (negation of a '='), $\geq$ (negation of a '<') or < (negation of a '≥') are common.

In similarity queries, the negation of a similarity operator becomes troublesome in two main aspects. First, retrieving similar elements corresponds to a "direct" predicate, not to its negation. For example, returning the "10-farthest neighbors" is distinct from returning the "not 10-nearest neighbor". Second, the negation of similarity predicates return a set with very large cardinality. For example, a "not 10-nearest neighbor" query returns all elements of the database, except for the "10-nearest neighbors". With respect to similarity joins, negation retrieves even more pairs. Wide join is well suited to help taming such problems.

Negating a predicate has an interesting motivation in similarity retrieval: once the most similar elements are already known, but the user wants to know *something else*, how to obtain the elements beyond the ones already known? This is a distinct problem from the incremental k-NN selection, as here the user is not interested in the k-NN ones. An example of negating the predicate in a similarity join was provided in Sect. 1 and such situations are treated also negating the predicate term in the wide-join. Following the definitions introduced in Sect. 3.2, negating the case where $\theta = \mathtt{true}$ does not apply to similarity joins nor to wide-joins, as its negation $\theta = \mathtt{false}$ always returns an empty set.

For a θ composed of a range negation $\neg(t[T_1] \, Range(d, \xi) \, t[T_2])$, the not-range comparator selects the pairs where $d(t[S_1], t[S_2]) > \xi$. However, such condition presents the drawback of returning a result with huge cardinality. A common way to solve such shortcoming in similarity range join operators is to increase the similarity threshold ξ to obtain a reduced set of elements. Nevertheless, as ξ increases, the retrieved elements lie more and more in the farthest regions of the space, which disrupts the similarity and the negation concepts, once the obtained elements are not similar to the query centers. Thus, predicates based on the not-range conditions are computed using wide-joins, where the upper bound limit κ_f ensures retrieving the similar elements that are beyond the threshold ξ and also prevents the operator from returning too many elements.

Algorithm 1 introduces a nested-loop procedure to compute not-range wide-joins. Line 4 performs the *not-range* comparison. While the result T_R contains less than κ_f pairs, the tuples whose similarity exceed the threshold ξ are temporarily included in the answer (lines 5–6). As soon as T_R has more than κ_f pairs, lines 9–11 replaces the most dissimilar one with the currently analyzed pair. The condition in line 5 ensures a result set with at most κ_f pairs and

---

Algorithm 1. Wide-join with $\neg Range$ comparison

1: **for** $t_1 \in T_1$ **do**
2:       **for** $t_2 \in T_2$ **do**
3:           $dist \leftarrow d(t_1[S_1], t_2[S_2])$;
4:           **if** $(dist > \xi)$ **then**
5:               **if** $|T_R| < \kappa_f$ **then**
6:                   $T_R \leftarrow T_R \cup \{\langle t_1[S_1], t_2[S_2], dist \rangle\}$;
7:               **else**
8:                   Let $w \in T_R$ be the tuple with the greater $d(t_2[S_1], t_2[S_2])$ value;
9:                   **if** $dist < w.dist$ **then**
10:                      $T_R \leftarrow T_R \cup \{\langle t_1[S_1], t_2[S_2], dist \rangle\}$;
11:                      $T_R \leftarrow T_R - \{w\}$;

12: **return** T_R;

---

avoids the $\neg Rng$ predicate to return unnecessary elements. As it requires the maximum upper bound κ_f, this solution is specific for wide-joins.

Likewise, when $\theta = \neg(t[S_1] \, kNN(d,k)t[S_2])$, the not-$k$-nearest neighbor wide-join retrieves the most similar pairs beyond the result of a k-nearest neighbor join, but restricted to the upper bound κ_f. Its implementation is similar to Alg. 1, just varying three key points: (i) the condition in line 4 is suppressed; (ii) the cardinality in line 5 is checked to be less than $(\kappa_f + k)$; and (iii) for each $t_1 \in T_1$, T_R becomes a temporary result T_{temp}, where the procedure returns a result $T_R = T_{temp_1} \cup T_{temp_2} \cup \ldots \cup T_{temp_n}$. The remainder operations defined presented in the right-hand side in (2) (projection and selection) are performed subsequently to the processing of those unary not-range and not-k-nearest neighbor conditions. Following, we explore predicates composed of more than one term.

3.4   Multiple-Term Predicates

Sections 3.2 and 3.3 introduced four similarity-based comparators: range, k-nearest neighbor and their respective complements. However, the similarity predicate can be assumed to be in the form as $\theta = \tau_1 \, \varphi \ldots \varphi \, \tau_n$, where φ is a logical connective and the term τ is one of the four comparators previously defined.

It is not straightforward to enumerate the number of distinct types of wide-join instances generated by combining similarity terms, as some of them are equivalent to others. However, such discussion is beyond the scope and space of this paper. As multiple-term predicates connect terms τ in a conjunctive ($\wedge$) or disjunctive ($\vee$) way, each term τ can be processed separately and each individual result combined to others respectively executing intersection and union set operations in place of $\wedge$ and $\vee$.

Following, we present some optimization options to compute the wide-joins either for single or multiple-term predicates.

3.5   Optimizing Wide-Joins Processing

Similarity wide-joins are usually processed using nested-loops like in Alg. 1, i.e., performing $|T_1| * |T_2|$ distance computations to obtain the result set. This approach presents a high computational cost, but it enables to compute any type of wide-join and to combine any kind of data, either multidimensional or purely metric. However, some improvements can be applied to speed up its processing.

Indexing the elements in T_2 is an effective technique often employed in the literature and can also be applied to wide-joins. Once metric structures can benefit from properties such as the triangle inequality to prune elements, they usually reduce the number of I/O operations when processing the operator.

Aiming at decreasing the CPU time when processing the inner join, it is possible to employ an extended version of the Cartesian product operator that returns a triple $\langle t[T_1], t[T_2], d(t[S_1], t[S_2]) \rangle$, as is shown in lines 6 and 10 of Alg. 1. Thus, the similarity distance between $t[S_1]$ and $t[S_2]$ does not need to be recomputed in the subsequent operations, such as in the function $\mathcal{F}$ in (2).

Finally, the operations performed by the extended projection and the final selection in the right-hand side of (2) can be directly executed if the result of the inner similarity join is inserted into a priority queue, where the more similar $t[S_1]$ and $t[S_2]$ are, the greater is the priority in the list. Thereafter, the final selection only removes κ_f elements from the priority queue when composing the final answer.

4   Experiments

This paper reports on the enlargement of similarity joins in a broader set of binary operators that employs the sorting concept internally, allowing extending the applicability of joins in similarity queries. We conducted our experimental studies to evaluate the scalability of the proposed operators by varying the cardinality of the joined relations, the data distribution and its dimensionality, and performed an analysis on how setting the parameters $(\xi,\ k,\ \kappa_f)$ influences the performance.

We describe the results of several synthetic datasets (Synth) with distinct dimensionality and cardinality, and a real one (Protein). The Synth data sets vary from 1,000 to 100,000 points in 2, 4, 8, 16, 32 and 64 dimensions each set, generated according to Uniform and Normal distributions. For those datasets, we used the L_2 metric. The Protein[1] data set consists of 12,866 chains of amino-acids represented by characters. This is a purely metric data set and allows to evaluate the operators on data that cannot be represented in the multidimensional space model. We retained proteins whose length varies between 2 and 15 amino-acids and employed the well-known Levenshtein edit distance.

The experiments ran in a computer with an Intel® Core™ i7-4770 processor, running at 3.4 GHz, with 16 GB of RAM on the operating system Ubuntu 14.04. We implemented all distinct operators in the same framework, written in C++, and the elements of both relations remain in disk, that is, tuples are loaded in memory only when they are required. The results obtained are depicted in Fig. 2 and the default parameter setup of each experiment can be found in Table 2.

Figure 2(a) shows the performance of the range wide-join and its algebraic version when the cardinality of the Synth dataset increases. The algebraic version corresponds to the combination of operators expressed in the right-hand side in (2). As it can be seen, the range wide-join operator (Rng) is at least 6.62% faster ($|T_1|, |T_2| = 1,000$) than the algebraic operator composition (AlgRng), but the largest performance gain (71.57%) was observed when both cardinalities are larger, as when they are equal to 10,000. The figure also compares the range wide-join and the similarity range join (SimRng) present in the literature. As expected, SimRng executes faster than the Rng, because the latter performs all the processing of the former plus some additional operations. Even so, the SimRng was in average only 6.87% faster than the range wide-join.

Figure 2(b) presents the previous comparison regarding the operators based on the nearest neighbor predicate. The k-nearest neighbor wide-join (KNN) was

---

[1] Proteins: http://www.uniprot.org/uniprot Access: Apr 27, 2015

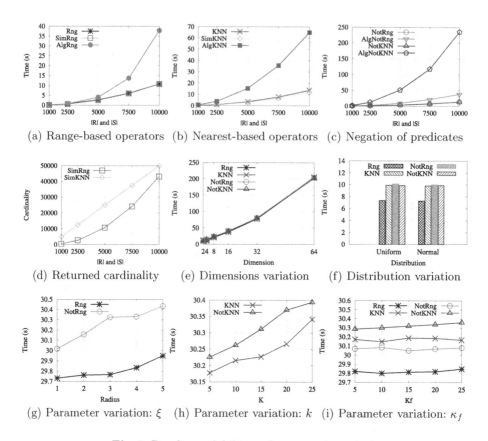

(a) Range-based operators (b) Nearest-based operators (c) Negation of predicates

(d) Returned cardinality (e) Dimensions variation (f) Distribution variation

(g) Parameter variation: ξ (h) Parameter variation: k (i) Parameter variation: κ_f

Fig. 2. Results: scalability and parametric analysis

in average 76.78% faster than its algebraic version (AlgKNN) regarding increasing cardinality. The KNN wide-join was also compared to the similarity k-NN join (SimKNN). Analogous to Fig. 2(a), the wide version is expected to be slower than the SimKNN, but the latter was in average 14.25% faster than the former.

Figures 2(a) and 2(b) show that although our proposed operators are not faster than the corresponding plain similarity joins found in the literature, the additional processing introduced to compute the most similar pairs does not influence the computational complexity of both types of joins.

Figure 2(c) shows the performance of the wide-joins employing the negation of the range and k-nearest neighbor conditions. The new proposed NotRng and NotKNN operators executed in a time quite similar, once they share the same implementation structure (Section 3.3), but the NotRng was in average 9.45% faster than the NotKNN. Notice in the same figure that NotRng and NotKNN are in average 63.86% and 92.62% faster than their algebraic versions, respectively.

Additionally, the negation of a similarity predicate (Fig. 2(c)) follows the same behavior of its traditional versions, where range is faster than the nearest

Table 2. Parameter configuration: default values in bold

Parameter	Dataset	Values
$|T_1|$	Protein	861 (proteins present in humans)
	Synth	**1,000**; 2,500; 5,000; 7,500; 10,000
$|T_2|$	Protein	12005 (proteins not present in humans)
	Synth	1,000; 2,500; 5,000; 7,500; 10,000; **100,000**
ξ	Protein	1; 2; 3; 4; 5
	Synth	0.25; 0.50; 0.75; **1.00**; 1.25
k	All	**5**; 10; 15; 20; 25
κ_f	All	5; **10**; 15; 20; 25
Dimension	Synth	2; **4**; 8; 16; 32; 64
Distribution	Synth	normal; **uniform**

neighbors, as can also be seen comparing Figs. 2(a) and 2(b). Comparing the results in Figs. 2(a), 2(b) and 2(c), one can see that the computational complexity of the join operators follow the theoretical $\mathcal{O}(|T_1| * |T_2|)$ prediction.

Figure 2(d) shows the total amount of pairs returned by the inner similarity range and k-nearest neighbor joins before the subsequent operations of the wide-join definition. This figure confirms that both types of similarity joins retrieve more pairs than the user is usually interested in, and, eventually, most of them are discarded, which is a waste of computational resources. Distinctly, the wide-joins retrieved only 10 pairs in each run on the Synth set (as κ_f in Table 2).

Figure 2(e) studies the effect of the dimensionality variation in the performance. As it can be noticed, the difference among the proposed operators is small. However, as the number of dimensions increases, the metric becomes more computationally expensive to compute and the performance of the operators reduces.

Figure 2(f) considers the performance in 2D data following the normal and uniform distributions. The execution time is equivalent in both distributions, showing that the wide-joins were not influenced by the data distribution.

Following, Figs. 2(g), 2(h) and 2(i) study the effect of parameter variation (ξ, k, κ_f), using the Protein dataset. Figures 2(g) and 2(h) shows that as the radius or k increases, more elements are combined in the join phase and included in the partial result, which smoothly reduces the performance of the sorting phase. As shown in Fig. 2(i), κ_f restricts more similar pairs from the result sorted in the previous steps. Thus, using κ_f to filter the most similar tuples leads to a linear processing time in the result size. Thus, even when that parameter increases, the answer growth rate is not big enough to influence the overall performance.

5   Conclusion

Similarity join operators present two main drawbacks when applied to the relational environments: their resulting cardinality is usually larger than necessary, requiring post-processing; and they often require an ordering step, a concept that is not acknowledged by the relational theory. In order to address those problems, this paper presented a complete set of binary similarity operators, namely the wide-join, that embraces the join concept and produces a more meaningful result set than the plain similarity joins. Wide-joins enable retrieving the most similar tuple pairs in general, representing the ordering among the elements internally, not requiring to sort the input nor the output data, thus complying to the relational model.

We provided the wide-join definition and specified the distinct kinds of predicates that the operator is able to process, where each distinct similarity condition generates a variant of the wide-join operator. We also presented an algorithm aiming at showing the wide-join usability in real applications and provided guidelines to implement the main instances of the operator.

The experiments performed on synthetic and real datasets, including nondimensional and multidimensional data, showed that wide-joins execute with performance equivalent to the existing similarity joins whereas providing a result set significantly smaller and more meaningful to the user.

As a future work, we are now exploring the algebraic properties on how the wide-join operators interact with the other similarity operators and instances of wide-joins employing multiple-term predicates.

References

1. Böhm, C., Krebs, F.: The k-nearest neighbour join: turbo charging the kdd process. Knowledge and Information Systems **6**(6), 728–749 (2004)
2. Carvalho, L.O., Oliveira, W.D., Pola, I.R.V., Traina, A.J.M., Traina Jr, C.: A 'wider' concept for similarity joins. Journal of Information and Data Management **5**(3), 210–223 (2014)
3. Chaudhuri, S., Ganti, V., Kaushik, R.: A primitive operator for similarity joins in data cleaning. In: Proc. 22nd Int. Conf. on Data Engineering, p. 12 (2006)
4. Cheema, M.A., Lin, X., Wang, H., Wang, J., Zhang, W.: A unified framework for answering k closest pairs queries and variants. IEEE Trans. on Knowledge and Data Engineering **26**(11), 2610–2624 (2014)
5. Dohnal, V., Gennaro, C., Zezula, P.: Similarity join in metric spaces using ed-index. In: Mařík, V., Štěpánková, O., Retschitzegger, W. (eds.) DEXA 2003. LNCS, vol. 2736, pp. 484–493. Springer, Heidelberg (2003)
6. Fredriksson, K., Braithwaite, B.: Quicker range- and k-NN joins in metric spaces. Information Systems **52**, 189–204 (2014). doi:10.1016/j.is.2014.09.006
7. Gao, Y., Chen, L., Li, X., Yao, B., Chen, G.: Efficient k-closest pair queries in general metric spaces. The VLDB Journal **24**(3), 415–439 (2015)
8. Garcia-Molina, H., Ullman, J.D., Widom, J.: Database systems: the complete book. Pearson (2009)
9. Ilyas, I.F., Beskales, G., Soliman, M.A.: A survey of top-k query processing techniques in relational database systems. Computing Surveys **40**(4), 395–420 (2008)
10. Jacox, E.H., Samet, H.: Metric space similarity joins. ACM Trans. on Database Systems **33**(2), 7:1–7:38 (2008)
11. Paredes, R., Reyes, N.: Solving similarity joins and range queries in metric spaces with the list of twin clusters. Journal of Discrete Algorithms **7**(1), 18–35 (2009)

12. Pearson, S.S., Silva, Y.N.: Index-based R-S similarity joins. In: Traina, A.J.M., Traina Jr, C., Cordeiro, R.L.F. (eds.) SISAP 2014. LNCS, vol. 8821, pp. 106–112. Springer, Heidelberg (2014)
13. Searcóid, M.Ó.: Metric spaces. Springer (2007)
14. Silva, Y.N., Aref, W.G., Larson, P.A., Pearson, S., Ali, M.H.: Similarity queries: their conceptual evaluation, transformations, and processing. The VLDB Journal **22**(3), 395–420 (2013)
15. Xiao, C., Wang, W., Lin, X., Yu, J.X., Wang, G.: Efficient similarity joins for near-duplicate detection. ACM Trans. on Database Systems **36**(3), 15:1–15:41 (2011)

Diversity in Similarity Joins

Lucio F.D. Santos$^{(\boxtimes)}$, Luiz Olmes Carvalho, Willian D. Oliveira,
Agma J.M. Traina, and Caetano Traina Jr.

Institute of Mathematics and Computer Sciences, University of São Paulo,
São Carlos, SP, Brazil
{luciodb,olmes,willian,agma,caetano}@icmc.usp.br

Abstract. With the increasing ability of current applications to produce
and consume more complex data, such as images and geographic information, the similarity join has attracted considerable attention. However,
this operator does not consider the relationship among the elements in
the answer, generating results with many pairs similar among themselves,
which does not add value to the final answer. Result diversification methods are intended to retrieve elements similar enough to satisfy the similarity conditions, but also considering the diversity among the elements
in the answer, producing a more heterogeneous result with smaller cardinality, which improves the meaning of the answer. Still, diversity have
been studied only when applied to unary operations. In this paper, we
introduce the concept of *diverse similarity joins*: a similarity join operator that ensures a smaller, more diversified and useful answers. The
experiments performed on real and synthetic datasets show that our
proposal allows exploiting diversity in similarity joins without diminish
their performance whereas providing elements that cover the same data
space distribution of the non-diverse answers.

Keywords: Similarity joins · Result diversification · Query processing

1 Introduction

Nowadays, huge amount of information are produced by the applications, and
the modern Relational Database Management Systems (RDBMS) must handle
more complex data types, such as images, videos, genetic sequences, geographic
information. Unlike scalar data types (numbers, dates and strings), it makes
no sense to compare complex data by equality or by order relationships: they
are better compared by similarity. Similarity-based variations of the classical
relational operators are being investigated to support similarity and included in
RDBMS, such as similarity selections [13] and similarity joins [7,8].

There exist several types of similarity joins, but the similarity range one
(often called just *similarity join*) is the most discussed in the literature
[4,8,11,13]. It retrieves pairs of elements such that elements in each pair are

The authors are grateful to FAPESP, CNPQ, CAPES and Rescuer (EU Commission
Grant 614154 and CNPQ/MCTI Grant 490084/2013-3) for their financial support.

G. Amato et al. (Eds.): SISAP 2015, LNCS 9371, pp. 42–53, 2015.
DOI: 10.1007/978-3-319-25087-8_4

similar up to a maximum threshold. Range join is useful in several contexts [8], such as string matching, data cleaning, and near-duplicate object detection. However, let us consider an application example where similarity joins may produce results whose intuitiveness and expressiveness can be improved. Assume an emergency scenario where one or more incidents require immediate providence to reduce adverse consequences to life and property. Suppose that an emergency crowd-source-based control system can receive many eyewitnesses reports containing photos with meta information, such as their geolocation coordinates. In this scenario, it is reasonable to consider that the system with several identified incidents can receive a large amount of photos, leading to the following question: *"How to capture a broad vision around the incidents region using a reduced number of photos?"*. There are two possible ways to use similarity join operators to answer the question, computing similarity using incident locations with the geolocation of the photos up to a maximum threshold.

The first way uses a similarity join as a final query generator. However, many pairs can bring the same information, as many photos come from the same point of view (near-duplicate perspectives). The second way employs a similarity join as a *pre-processing* operator, sending the join result to a clustering algorithm that summarizes the answer. Both alternatives have drawbacks. The former only considers similarity among the incident and the photos, not taking into account the similarity among the pairs in the answer. For example, a result composed of pairs in the form $\langle s_1, s_a \rangle$ and $\langle s_2, s_b \rangle$ where s_1 and s_2 are incident locations, $\langle s_1, s_a \rangle$ and $\langle s_2, s_b \rangle$ are pairs similar among themselves and s_a and s_b are close locations photos, leads to a large cardinality answer set, which requires more attentive effort from the emergency control system staff. The second alternative trades the problem of "analyzing too many similar pairs" for "properly tune a clustering algorithm" executed in a follow-up operation. The performance of clustering algorithms is directly affected by the cardinality of the input, thus it typically needs to be executed more than once to find proper results. This alternative is computationally costlier, but eases and improves the staff's job.

Surpassing those drawbacks, a more interesting answer is to capture a whole perspective about each incident region, taking advantage of the relationship among the reports to obtain a more diversified view of the incidents. In order to obtain more relevant photos with such a holistic vision about the search space, several studies introduced a diversity factor in similarity queries [6, 12, 17]. Query result diversification aims at computing not only a result set with elements similar enough to satisfy the similarity conditions, but also to get a set of elements diverse among themselves to produce a more heterogeneous result. However, to the best of our knowledge, the result diversification definitions found in the literature [2, 5, 6, 12, 14, 16, 17] were always applied to unary operations, and none of them explored combining two relations, such as in a join operation.

This paper introduces the concept of *diverse similarity joins*: a binary operation that receives two relations and combines their elements meeting a similarity predicate but that also ensures a smaller, diversified answer. Our proposal was evaluated using real and synthetic datasets. The results show that diverse similarity joins have equivalent computational costs of the similarity joins, yet retrieving a broader distribution of the elements from both input sets. The main contributions of this paper are summarized as follows:

- The definition of a theoretical basis to combine the concepts of diversity queries and similarity joins.
- Introducing an operator that improves the usefulness of similarity joins.
- An efficient algorithm to compute diverse similarity joins.

The remainder of this paper is organized as follows: Section 2 reviews related works. Section 3 introduces the diversity join concept. Section 4 presents experimental evaluation and results. Section 5 concludes the paper and discusses future research directions.

2   Related Work

Similarity Joins: Let us assume that it exists two relations T_1 and T_2, each one having one attribute (or a set of attributes) $S_1 \in T_1$ and $S_2 \in T_2$ sharing the same complex data domain $\mathbb{S}$, that is $Dom(S_1) \subseteq \mathbb{S}$ and $Dom(S_2) \subseteq \mathbb{S}$. The similarity range join of T_1 and T_2 retrieves all tuple pairs $\langle t_i, t_j \rangle \,|t_i \in T_1, t_j \in T_2$ such that the distance between the values of the corresponding attributes do not exceed a maximum similarity threshold ξ, that is $d(t_i[S_1], t_j[S_2]) \leq \xi$. It also exists a similarity join whose limit is not a maximum similarity threshold, but a maximum number of elements, the *k-nearest neighbor join*. It retrieves the pairs $\langle t_i, t_j \rangle \,|t_i \in T_1, t_j \in T_2$ such that the value of the complex attribute $t[S_1]$ in the right relation is one of the k most similar to the value of the complex attribute $t[S_2]$ in the left relation. The range join is the fastest and most common type of similarity join [8].

Similarity joins can be processed by nested-loops, which implies to perform $|R| * |S|$ distance computations [10]. Despite its high computational cost, this approach enables to perform any type of similarity join and to combine any kind of data. However, it is possible to employ an index data structure in order to improve performance. The main idea is to store the elements of one or both joined relations into a data structure that speeds up accesses. Examples of such structures are the eD-Index, employed in studies such as [4,11], and the List of Twin Clusters (LTC), introduced in [10]. Whereas eD-Index is well-suitable to compute range joins, LTC also processes the k-nearest neighbor variant.

Similarity joins can also be computed by non-indexing approaches based on the divide-and-combine strategy. These techniques intend to partition the search space and to group the elements. Examples of such techniques include the Quickjoin [8], Epsilon Grid Order (EGO) [1] and its extensions [9], and the Generic External Space Sweep (GESS) [3]. Quickjoin improves similarity joins in multidimensional spaces by dividing the elements into small groups, in a way that enables them to be efficiently joined by nested-loops. However, Quickjoin only processes range joins whereas variations aimed at processing the k-nearest neighbors join [7] computes only approximate answers. EGO and GESS are specific to handle dimensional data and cannot be applied in metric spaces in general.

Search Result Diversification: The main idea of adding a diversity factor into similarity queries is to bring better information than a result based only on similarity, as diversity allows the user to have a broader perspective of the possible results over larger portions of the data space distribution. Diversity has been exploited in several areas, such as information retrieval [5], recommendation systems [2] and similarity queries [12,14]. Most approaches use metadata associated to the elements, such as taxonomies in document sets [5], cluster attributes in annotated data [2,16], and distances among the elements in dimensional and metric spaces. However, processing external information is often computationally expensive and restricts its use to datasets that have this information [17].

Other approaches have pursued diversity without using extra information. Called distance-based approaches, they can be classified as two main groups: Optimization and Separation Distance. The optimization approach considers that similarity and diversity must compete to each other, taking a user-defined diversity preference as input, so that the results of pure-similarity algorithms (configured to retrieve more than the k elements requested by the user) can be re-ranked, inducing diversity among elements based on a bi-criteria objective function. However, this diversity definition results in an NP-hard problem [12,17] and restricts their usage to k-nearest neighbor queries.

The separation distance approach considers that there is a minimum distance ξ_{min} among elements in the answer, such that pairs of elements closer than ξ_{min} are considered too much similar to each other and only one is included in the final result [6,14]. An example of using this approach is the k-Distinct Nearest Neighbors (kDNN) query [14]. The kDNN query builds on the classic k-NN query, but excluding all elements that are too similar by restricting the result relation $T_R = \{\forall s_x, s_z \in T_R, \forall s_y \in T - T_R : d(s_x, s_z) \geq \xi_{min} \wedge (d(s_x, s_q) \leq d(s_y, s_q) \vee \exists s_w \in T_R : d(s_y, s_w) < \xi_{min})\}$, where T is a relation, ξ_{min} is a fixed user-defined separation distance and s_q is the query center. Although the separation distance approaches have a reduced computational cost compared to the optimization ones, setting up the separation distance for each element in a relation T_1 to join to another relation T_2 makes such diversity definition less intuitive to the user, as it requires defining a fixed separation distance for all elements in each relation.

Other recent approach defines diversity without requiring more information from the user about the separation distance, called the Result Diversification based on Influence (RDI) [12]. This technique is based on a minimum distance that can be automatically estimated using the concept of "influence" intensity (I), which is defined as the inverse of the similarity distance between two elements. Let s_i, s_j and s_q be elements in a relation T. Then s_j is more influenced by s_i than by s_q iff $I(s_i, s_j) \geq I(s_j, s_q)$. For a query centered at s_q, the RDI goal is to retrieve a diversity result set $T_R \subset T$ by selecting elements in T that are similar to s_q (nearest or in the range), but also considering the minimum distance between two elements $s_i, s_j \in T_R$ by the influence intensity I.

The diversity property was defined only for the selection operator, thus considering only one relation. The proposal in this paper extends those aforestated studies to compute the diversity of similarity joins of two relations T_1 and T_2. Thus, we intend to compute a result composed of pairs $\langle s_i, s_2 \rangle$ where $s_i \in T_1$ and $s_2 \in T_2$ and each s_i is similar enough to elements s_2 to satisfy the similarity request but, increasing the diversity among the elements s_i in the result.

3   Diversified Similarity Joins

Let us first hone our intuition about how the combination of antagonistic concepts as similarity and diversity improves the applicability of the similarity joins. Typically, queries are expressed by combining searching operators, as for example, the join operator is composed of the Cartesian product followed by a selection. Thus, the answer of a similarity range join has properties related to the "range selection" operator, an operator that selects the elements within a similarity threshold from a query center.

As already discussed in Sect. 1, a pure similarity criterium (without considering diversity) results in a relation including elements too much similar to each other, which increases the result cardinality without increasing the information content. Figure 1 (a) represents elements of two relations T_1 and T_2 in an Euclidean bi-dimensional space, where stars represent elements in T_1, triangles represent elements in T_2 and circles delimits elements of T_2 combined with elements T_1. However, in the same way that a diversified similarity selection operator can improve the result evaluating diversity, a diverse similarity join may provide a better perspective too, as shown in Fig. 1 (b). Notice that each element in T_1 can be associated to a varying number of elements in T_2, depending on the data space distribution. In Fig. 1 (b), the squares represent elements in T_2 skipped to be paired with elementos from T_1, as they probably only repeat the information already presented by another pair already in the result, meeting what we want in the emergency scenario presented in Sect. 1.

We now state the problem of the diversified similarity join. Let T_1 and T_2 be two relations, each one having an attribute sharing the same complex data domain $\mathbb{S}$. Our objective is to retrieve all pairs of tuples from both relations such that the distance between the corresponding complex attributes does not exceed a maximum similarity threshold, at the same time ensuring that the resulting pairs are diverse among each other. The **Diversified Similarity Join** or just **DS-join** is defined as follows:

Definition 1. *Diversified Similarity Join Operator (DS-join): Let T_1 and T_2 be two relations, each one having one attribute (or a set of attributes) $S_1 \in T_1$ and $S_2 \in T_2$ sharing the same complex data domain $\mathbb{S}$. Let also d be a metric defined on $\mathbb{S}$, ξ be a distance threshold and $RngDiv(d, \xi)$ be a similarity with diversity range comparison operator. The diversified similarity join $T_1 \overset{S_1 RngDiv(d,\xi) S_2}{\bowtie} T_2$ combines tuples of T_1 and T_2 whose distance between the pair of elements $d(t[S_1], t_i[S_2])$ is less than or equal to the given threshold ξ and*

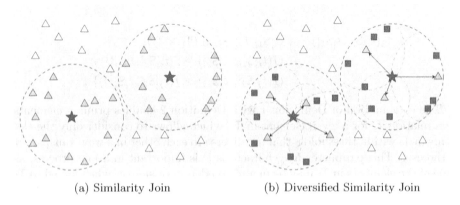

(a) Similarity Join   (b) Diversified Similarity Join

Fig. 1. Similarity joins in an Euclidean bi-dimensional space. Stars are the elements of the first relation T_1, triangles represent the elements of the second relation T_2. Dashed circles groups the elements of T_1 paired with the elements of T_2. Squares are the elements of the T_1 not paired with T_2 in a diversified similarity join.

ensures that the pairs in the result relation $\langle t[S_1], t_i[S_2] \rangle$, $\langle t[S_1], t_j[S_2] \rangle \in (T_1 \times T_2)$ *are separated by a minimum distance based on the relative position of* $t_i[S_2]$ *and* $t_j[S_2]$ *to* $t[S_1]$*, that is:*

$$T_1 \overset{S_1\,RngDiv(d,\xi)S_2}{\bowtie} T_2 = T_R = \{\langle t[S_1], t_i[S_2] \rangle \in (T_1 \times T_2) \mid d(t[S_1], t_i[S_2]) \leq \xi \,\wedge$$
$$\forall \langle t[S_1], t_i[S_2] \rangle \in T_R : \langle t[S_1], t_i[S_2] \rangle \notin \overset{\curvearrowright}{T}_2(t_j[S_2], t_i[S_2])\} \ .$$

Let us interpret Definition 1 and see how it formalizes our intuition. The first part of that definition $\langle t[S_1], t_i[S_2] \rangle \in (T_1 \times T_2) \mid d(t[S_1], t_i[S_2]) \leq \xi$ ensures that only the tuples holding the most similar complex attribute values (restricted by the user threshold ξ) in T_1 will be paired to the tuples in T_2. The second part, $\forall \langle t[S_1], t_i[S_2] \rangle \in T_R : \langle t[S_1], t_i[S_2] \rangle \notin \overset{\curvearrowright}{T}_2(t_j[S_2], t_i[S_2])$ ensures the diversity in the final result, selecting only the pairs whose complex attributes are separated by a minimum distance. We consider that only the tuples $t_i, t_j \in T_2$ farther than a minimum distance to the others in T_R (estimated in an automatic way, using the concept of influence Intensity described in Sect. 2) will be selected. The tuples influenced by $t_i[S_2]$ is represented by $\overset{\curvearrowright}{T}_2(t_j[S_2], t_i[S_2])$. The intuition is that $t_i[S_2]$ provides more (or equivalent) information as $t_j[S_2]$ to each respective element $t[S_1]$. Thus, we only need to pair up one of them in the final result. Definition 2 shows how to generate a set of elements $t_j[S_2]$ that can be surely skipped as candidates from relation T_2 based on the element $t_i[S_2]$ for each corresponding element $t[S_1] \in T_1$.

Definition 2. Strong Influence Set – $\overset{\curvearrowright}{T}_2$: *Given an value* $t[S_1]$ *in relation* T_1 *and another* $t_i[S_2]$ *in relation* T_2*, the strong influence set* $\overset{\curvearrowright}{T}_2$ *of* $t_i[S_2]$ *for each* $t[S_1]$ *is:*

$$\tilde{T_2}(t_i[S_2], t[S_1]) = \{ \langle t[S_1], t_j[S_2] \rangle \in (T_1 \times T_2) |$$
$$(I(t_i[S_2], t_j[S_2]) \geq I(t_i[S_2], t[S_1])) \wedge$$
$$(I(t_j[S_2], t_i[S_2]) \geq I(t_j[S_2], t[S_1])) \} \ .$$

The combination of Definition 1 with Definition 2 enables pruning elements from relation T_2 for a given element in T_1, which allows to consider only the elements in T_2 within threshold ξ that are diverse to each other in a way transparent to the user. This parameter-free characteristic is important in a join process, as some of the elements in T_2 may be in a region denser or sparser when joined to T_1 and the influence intensity is automatically estimated during the join execution, based solely on the data space distribution. .

An Algorithm for the Diversified Similarity Joins: The traditional way to compute similarity joins is through a nested-loop approach. Algorithm 1 combines a nested-loop with the operations to prune elements influenced by others in the second relation to support diversity in similarity joins. Lines 2–8 execute the inner similarity join. In line 6, the two elements that are in the threshold ξ are concatenated and included in the result, together their the similarity. The element pairs are kept ordered in T_{temp} with respect to their similarity to $t[S_1]$. The use of a priority queue eases storing the elements in that part of the algorithm. As the diversity concept is applied for each element of the first relation, the function Diverse (line 9) is applied after all the elements in the threshold ξ are selected.

In Algorithm 2 (Diverse function), the intuition is to prune elements too similar by keeping the maximum threshold requirement, such that the answer is a subset of the traditional similarity join, reducing the cardinality of the answer using only the elements in the relation T_2 that can provide a broader vision around each element in T_1. Thus, the distance from a diverse candidate

---

Algorithm 1. Nested-loop DS-join

Input : Relations T_1 and T_2;
Output: The relation T with diverse elements of T_1 joined to T_2.

1: $T \leftarrow \varnothing$;
2: **for** $t[S_1] \in T_1$ **do**
3:     $T_{temp} \leftarrow \varnothing$;
4:     $T_{div} \leftarrow \varnothing$;
5:     **for** $s \in T_2$ **do**
6:        **if** $(dist(t[S_1], s) \leq \xi)$ **then**
7:           $dist \leftarrow d(t[S_1], s)$;
8:           $T_{temp} \leftarrow T_{temp} \cup \{\langle t[S_1], s, dist \rangle\}$;
9:     $T_{div} = \text{Diverse}(T_{temp})$;
10:    $T \leftarrow T \cup T_{div}$;

Algorithm 2. Diversifying Result Sets

Input : Relation T_{temp};
Output: The relation T_{div} with the diverse elements of T_{temp}.

1:   $T_{div} \leftarrow \varnothing$;
2:   $divCand \leftarrow$ The element s with the lower value of dist;
3:   **while** T_{temp} *is not empty* **do**
4:      set $divCand$ as diverse candidate;
5:      **for** *each* $t \in T_{div}$ **do**
6:         **if** $divCand \in \check{T}_2(t, t[S_1])$ **then**
7:            set $divCand$ as non-diverse candidate.;
8:            **break**
9:      **if** *$divCand$ is a diverse candidate* **then**
10:        insert $\langle t[S_1], divCand, dist \rangle$ in T_{div} ;
11:      $divCand \leftarrow$ next element s with the lower value of dist ;

---

in relation T_2 to element $t[S_1] \in T_1$ must be minimal among all the elements in T_2. In such way, we start assuming that the minimum distance between the elements T_2 is zero. At each iteration, the closest element ($divCand$) is considered a diverse candidate (Lines 3–4). Thereafter, that element is evaluated if it belongs to the strong influenced set already selected. If so, then $divCand$ is influenced by t and it is tagged as non-diverse (lines 5–7). Moreover, if $divCand$ is not influenced by any element in T_{div}, then $divCand$ is inserted into the result set (lines 9–10). Notice that the number of diverse candidates depends on the data space distribution around the element in T_1. This process repeats until no other element exists in T_{temp} to be analyzed (line 3).

4   Experiments

In this section we compare the proposed DS-join operator to the traditional non-diverse similarity range join (Sim-join) and to a diversity algorithm based on the distinct nearest neighbors (Dist-join) [14], as the other diversity algorithms from the literature can be only applied to k-nearest neighbor query operators. Dist-join employs a concept similar to that used in DS-join, as it considers elements diverse based on a separation distance (ξ_{min}). However, Dist-join requires the definition of a ξ_{min}, which is fixed for every element in relation T_1. All the compared algorithms follow the nested-loop join strategy to enable fair comparisons. The objective here is to compare the impact of using different diversity definitions on similarity joins.

We follow two strategies to evaluate our proposal: the first (Sect. 4.1) evaluates the impact of varying parameter ξ on the performance and on the cardinality of the result using real datasets; the second strategy (Sect. 4.2) performs a scalability analysis of the DS-join operator varying the cardinality of the joined relations and the data dimensionality, using synthetic datasets.

Table 1. Experimental setup

Parameter	Dataset	Values
	Aloi	100
$|T_1|$	Proteins	861
	Synth	(1,000); 2,500; 5,000; 7,500; 10,000
	Aloi	72,000
$|T_2|$	Proteins	12,005
	Synth	1,000; 2,500; 5,000; 7,500; (10,000)
ξ	Aloi	(1.0); 2.0; 3.0; 4.0; 5.0
	Proteins	(5); 6; 7; 8; 9
Dimension	Synth	2; (4); 8; 16; 32

We evaluated the results by processing two real datasets (Aloi, Proteins) and several synthetic ones (Synth) with distinct dimensionality and cardinality. The Aloi[1] dataset is composed of 1,000 main objects rotated in steps of 5° from 0 to 360°, generating 72 images per object and a total amount of 72,000 distinct images. This dataset has 144 features obtained using the color moment extractor [15]. Additionally, the Manhattan distance was used to compute the similarity between the elements. The Proteins[2] dataset consists of 12,866 chains of amino acids represented by characters. This is a purely metric dataset and allows to evaluate DS-joins over data that cannot be represented in a multidimensional space model. We retain proteins whose length varies between 2 and 15 amino acids. The metric employed in this dataset is the well-known Levenshtein Edit distance. The Synth datasets vary from 1,000 to 10,000 points in 2, 4, 8, 16 and 32 dimensions each set, generated at random (uniform). Every Synth dataset used the Euclidean distance to evaluate the similarity among the elements. Table 1 summarizes the parameter variations and indicates the default values in parenthesis when they are not specified in the test description.

The experiments were executed in a computer with an Intel® Core™ i7-4770 processor, running at 3.4 GHz, with 16 GB of RAM on the operating system Ubuntu 14.04. All the algorithms were implemented in C++, using the same programming framework and both joined search spaces remains in disk, that is, elements are loaded in memory only when they are required to be joined.

4.1 Performance and Result Size Evaluation

In order to evaluate the retrieval performance of our proposal, we measured the running time and the number of elements in the final result obtained by DS-join, Dist-join and the traditional Sim-join. We present the behavior analysis in a high-dimensional (Aloi) and in a purely metric dataset (Proteins), since

---

[1] Aloi: http://aloi.science.uva.nl Access: April 19, 2015
[2] Proteins: http://www.uniprot.org/uniprot Access: April 19, 2015

they encompass two representative cases regarding complex data. The maximum threshold ξ was chosen so as its smaller value retrieves about 1% of the amount of elements of the Cartesian product and the larger value retrieves about 10% of that total. Naturally, each distinct dataset has a different range of values ξ.

Figure 2(a) shows the running time for the Aloi dataset. For this experiment, we used two values for the separation distance parameter of Dist-join: 1.0 and 1.3. They accomplish respectively the best diversification and the fastest performance according to the authors [14]. As it can be seen, all the algorithms have almost the same execution time when the maximum threshold is small, as the final result has only few elements. However, as the maximum threshold increases, both configurations of Dist-join have their performance degraded, being on average 10 times slower than the Sim-join. This happens as the separation distance parameter is fixed to each element for relation T_1, without considering the distance distribution of the elements around it in the relation T_2. However, the experiments showed that DS-join is much faster than the Dist-join, and when compared to the (non-diverse) Sim-join, it was on average 20% slower. Figure 2(d) shows the result set sizes for the Aloi dataset. For this experiment, the intuition is that a good diversity join algorithm will select a reduced number of elements that cover the same data space distribution of the non-diverse Sim-join. Both configurations of the Dist-join was outperformed by DS-join, covering the same data space, but using only 10% of elements used by Sim-join.

Figure 2(b) shows the running time for the Proteins dataset. In this experiment, we defined the separation distance parameter of Dist-join as 6. DS-join outperformed Dist-join been at least 2 times faster for smaller values of ξ and up to one order of magnitude faster when 10% of the elements are retrieved by the join. In addition, the large amount of time spent by Dist-join was not enough to reduce the number of elements as DS-join does, since it retrieved, in average, 3 times more elements, as presented in Fig. 2(e).

The presented results highlights that our DS-join executes diversity in similarity joins in a equivalent time to the "pure" similarity join (Sim-join), while the closest diversity algorithm compared can be 10 times slower. Moreover, DS-join does not require any new parameters. Allowing to make the use of diversity in joins transparent and intuitive.

4.2 Scalability

In order to evaluate the scalability of the proposed DS-join algorithm, we performed two experiments over synthetic datasets. We first employed the new algorithm to a cardinality-behavior analysis regarding the running time. Figure 2(c) shows the effect of increasing the cardinality of both relations T_1 and T_2. As it can be seen, the DS-join follows the same behavior of the non-diverse algorithm Sim-join with a slightly difference (less than 7%). This result shows that the inclusion of diversity in the similarity join operator does not degrade the overall performance regarding both relation variations.

The second scalability experiment evaluates the effect of the dimensionality variation in the DS-join performance. Figure 2(f) shows that the difference

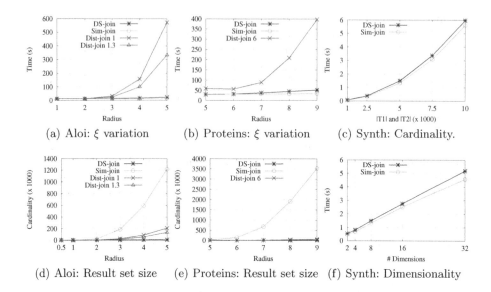

Fig. 2. Performance and result set size graphs showing the impact of diversity in join operators. In all graphs, lower values correspond to better algorithms.

between DS-join and Sim-join is always very close (less than 10%) regarding the dimensionality of the relations. As already expected, the cost of DS-join increase with the dimensionality, once the distance functions become more computationally expensive to compute and DS-join must consider both similarity and the diversity among the elements. However, the slim cost increase is surpassed by the gain in the response meaning obtained by including the diversity.

5 Conclusion

The similarity operators are attracting considerable attention to process complex data. However, similarity-based operators often retrieve result with elements too much similar among themselves, which does not add valuable information to the final answer.

Result diversification provides a promising solution, making it possible to retrieve elements similar enough to satisfy similarity conditions but also considering the diversity among them. Until now, diversity have been applied only to unary selection operators. In this paper we introduced the concept of diversity in similarity joins, ensuring a diversified and more useful answer. We applied the diversity over the range join operator to prune the elements that are too similar to each other, reducing the cardinality of the answer. Our experiments showed that it is possible to consider the diversity among the elements in the result of the similarity join operator without significant impact on their performance.

As a future work, we are exploring the benefits of including diversity also in the k-nearest neighbor joins. For this kind of join, the diversity concept provides a different vision, as it will prune too similar elements in T_1, exchanging them with others less similar to keep k elements in the answer, meeting the requirement of pairing up each element of T_2 to k elements in T_1, but where there is no commitment to represent the same data space distribution of the non-diverse answers, which is more suitable for exploratory queries.

References

1. Böhm, C., Braunmüller, B., Krebs, F., Kriegel, H.P.: Epsilon grid order: an algorithm for the similarity join on massive high-dimensional data. In: ACM SIGMOD Record, vol. 30(2), pp. 379–388 (2001)
2. Boim, R., Milo, T., Novgorodov, S.: Diversification and refinement in collaborative filtering recommender. In: Proc. 20th CIKM, pp. 739–744 (2011)
3. Dittrich, J.P., Seeger, B.: Gess: a scalable similarity-join algorithm for mining large data sets in high dimensional spaces. In: Proc. 7th ACM SIGKDD, pp. 47–56 (2001)
4. Dohnal, V., Gennaro, C., Zezula, P.: Similarity join in metric spaces using eD-index. In: Mařík, V., Štěpánková, O., Retschitzegger, W. (eds.) DEXA 2003. LNCS, vol. 2736, pp. 484–493. Springer, Heidelberg (2003)
5. Dou, Z., Hu, S., Chen, K., Song, R., Wen, J.: Multi-dimensional search result diversification. In: Proc. 4th WSDM, pp. 475–484 (2011)
6. Drosou, M., Pitoura, E.: Disc diversity: result diversification based on dissimilarity and coverage. Proc. VLDB Endowment 6(1), 13–24 (2012)
7. Fredriksson, K., Braithwaite, B.: Quicker range- and k-NN joins in metric spaces. Information Systems 52, 189–204 (2015)
8. Jacox, E.H., Samet, H.: Metric space similarity joins. ACM TODS 33(2), 7:1–7:38 (2008)
9. Kalashnikov, D.V.: Super-ego: fast multidimensional similarity join. The VLDB Journal 22(4), 395–420 (2013)
10. Paredes, R., Reyes, N.: Solving similarity joins and range queries in metric spaces with the list of twin clusters. Journal of Discrete Algorithms 7(1), 18–35 (2009)
11. Pearson, S.S., Silva, Y.N.: Index-based R-S similarity joins. In: Traina, A.J.M., Traina Jr, C., Cordeiro, R.L.F. (eds.) SISAP 2014. LNCS, vol. 8821, pp. 106–112. Springer, Heidelberg (2014)
12. Santos, L.F.D., Oliveira, W.D., Ferreira, M.R.P., Traina, A.J.M., Traina Jr., C.: Parameter-free and domain-independent similarity search with diversity. In: Proc. 25th SSDBM, pp. 5:1–5:12 (2013)
13. Silva, Y.N., Aref, W.G., Larson, P.A., Pearson, S., Ali, M.H.: Similarity queries: their conceptual evaluation, transformations, and processing. The VLDB Journal 22(3), 395–420 (2013)
14. Skopal, T., Dohnal, V., Batko, M., Zezula, P.: Distinct nearest neighbors queries for similarity search in very large multimedia databases. In: Proc. 11th WIDM, pp. 11–14 (2009)
15. Stricker, M., Orengo, M.: Similarity of color images. In: Proc. 3rd SPIE, pp. 381–392 (1995)
16. Van Leuken, R.H., Garcia, L., Olivares, X., Van Zwol, R.: Visual diversification of image search results. In: Proc. 18th Int. Conf. on WWW, pp. 341–350 (2009)
17. Vieira, M.R., Razente, H.L., Barioni, M.C.N., Hadjieleftheriou, M., Srivastava, D., Traina Jr., C., Tsotras, V.J.: On query result diversification. In: Proc. 27th ICDE, pp. 1163–1174 (2011)

CDA: Succinct Spaghetti

Edgar Chávez[1]([⊠]), Ubaldo Ruiz[1,3], and Eric Téllez[2,3]

[1] Department of Computer Science, CICESE, Ensenada, Mexico
{elchavez,uruiz}@cicese.mx
[2] INFOTEC, Mexico City, Mexico
eric.tellez@infotec.com.mx
[3] Cátedra CONACYT, CONACYT, Mexico City, Mexico

Abstract. A pivot table is a popular mechanism for building indexes for similarity queries. Precomputed distances to a set of references are used to filter non-relevant candidates. Every pivot serves as a reference for all, or a proper subset of, the objects in the database.Each pivot filters its share of the database and the candidate list for a query is the intersection of all the partial lists.The *spaghetti* data structure is a mechanism to compute the above intersection without performing a sequential scan over the database, and consist of a collection of circular linked lists.

In this paper, we present a succinct version of the spaghetti. The proposed data structure uses less memory and, unlike the original spaghetti, it can compute the intersection using an arbitrary order of the component sets. This later property enables more sophisticated evaluation heuristics leading to faster intersection computation.

We present the analysis of the performance, as well as a comprehensive set of experiments where the new approach is proven to be faster in practice.

1 Introduction

One of the simplest methods to speed up proximity search is a pivot table. The method consists in precomputing the distances of the elements of the database to a fixed set of pivots. Those distances are used to filter out the database objects that are unlikely to be in the query outcome. There are several variations of this simple idea presented in the literature. Different pivot-based indexes compute the intersection with different methods. The different techniques for computing intersections are surveyed in the extended version of the paper, in this manuscript we will only compare the standard spaghetti (SPA) against the succinct version (CDA).

2 Spaghetti

The spaghetti [1] computes the intersection of a collection of k intervals from k ordered sets. We will call *pivots* to the ordered sets, because they are obtained from a pivot table. The algorithm is described briefly below.

© Springer International Publishing Switzerland 2015
G. Amato et al. (Eds.): SISAP 2015, LNCS 9371, pp. 54–64, 2015.
DOI: 10.1007/978-3-319-25087-8_5

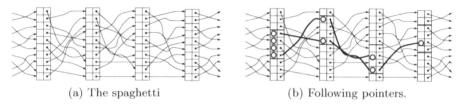

| (a) The spaghetti | (b) Following pointers. |

Fig. 1. Following pointers in the spaghetti. The probability of traversing the entire spaghetti is the product of the "chance" windows in each array.

Preprocessing

1. For each pivot calculate and save the distances to each database's element.
2. Sort each array saving the permutation with respect to the preceding array (as illustrated in Figure 1a)

Querying. Given k intervals, defining k sets, $[a_1, b_1], \cdots, [a_k, b_k]$ (with $a_i = d(p_i, q) - r$ and $b_i = d(p_i, q) + r$)

1. Obtain the *index* intervals $[I_1, J_1], \cdots, [I_k, J_k]$ corresponding to each set.
2. Follow each point through the pointers to find out if it falls inside all the *index* intervals.
3. If a point falls inside *all* the index intervals it is in the intersection.

Figure 1a shows the construction of the data structure. In Figure 1b, the thick pointer represents a successful path, that is, the object followed is in the intersection.

3   Time Complexity Analysis

The spaghetti implements a practical way of finding the intersection of k sets. For simplicity lets consider that each set s_i, associated with the i-th pivot, have exactly m elements; hence the fraction of points captured in each set s_i is $\varepsilon = \frac{m}{n}$ with n the size of the database. The fraction ε depends on the radius of the query, the intrinsic dimension of the database and the distance function used to compare the elements. The goal is to analyze the complexity of the algorithm for isolating the candidate list, i.e. for finding the intersection of the k sets.

3.1   Unsuccessful Search

Consider the probability of paying $1, 2, \ldots, k$ operations for each point (one for each pivot). We will proceed inductively on k. Let $\mathbf{c}$ be the random variable describing the cost of traversing the spaghetti for one individual point, $\mathbf{c}$ will take values in the discrete interval $[1 \ldots k - 1]$. The analysis makes sense for values of $k \geq 2$ (for $k = 1$ we pay 0 in traversing the spaghetti).

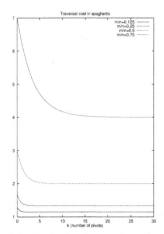

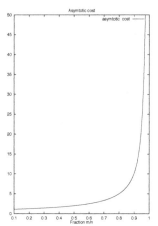

(a) Cost for traversing the spaghetti data structure, as a function of the number of pivots. Plots for different fractions m/n are shown.

(b) Asymptotic cost for traversing the spaghetti, as a function of the fraction $\varepsilon = m/n$.

If $k = 2$ then for any point in the first set we pay exactly 1 for traversing the spaghetti. If $k = 3$ we pay 1 with the probability of finding it in the second set and *not* finding it in the third set e.g. $P(1) = (1 - \varepsilon)$; similarly we pay 2 with the probability of finding it in the first set, the second set and the third set e.g. $P(2) = \varepsilon^2$. For $k > 3$ the reasoning is similar.

The following formula accounts for the probability of an arbitrary k.

$$P(\mathbf{c}) = \left\{ \begin{array}{l} \varepsilon^{\mathbf{c}-1}(1 - \varepsilon)\text{if}\, \mathbf{c} < k - 1 \\ \varepsilon^{k-1}\text{if}\, \mathbf{c} = k - 1 \end{array} \right\} \tag{1}$$

The expected value of $\mathbf{c}$ is

$$\mathbf{E}(\mathbf{c}) = \sum_{i=1}^{k-1} \mathbf{c} P(\mathbf{c}) \tag{2}$$

This sum is telescopic, and the final expression is

$$\mathbf{E}(\mathbf{c}) = (1 - \varepsilon) + 2(\varepsilon(1 - \varepsilon)) + 3(\varepsilon^2(1 - \varepsilon)) + \cdots + (k - 1)\varepsilon^{k-2} \tag{3}$$

$$\mathbf{E}(\mathbf{c}) = 1 + \varepsilon + \varepsilon^2 + \cdots + \varepsilon^{k-2} \tag{4}$$

Hence the average cost or expected value is

$$\sum_{i=0}^{k-2} \varepsilon^i = \frac{1 + \varepsilon^{k-1}}{1 - \varepsilon} \leq k \tag{5}$$

Figure 2a shows the traversal cost for spaghettis for different values of the fraction m/n. From smaller to larger, the curves in the figure correspond to fractions m/n of values 0.125, $0.25, 0.5$ and 0.75. The respective asymptotic costs, in the same order are 0.5, 1.0, 2.0 and 4.0.

The above analysis implies that on the average increasing the number of pivots does not increase the cost of traversing the spaghetti; in other words for each fraction m/n there is a k such that the cost of traversing the spaghetti for any point is constant, as pointed in Figure 2b.

The above discussion proves that the *average cost* of traversing the spaghetti is $O(m)$ and it differs from the worst case complexity by the factor k

3.2 Successful Search

The above analysis is valid if the final intersection (i.e. the outcome of the query) is empty; or in other words if the search is unsuccessful. If the size of the intersection is s, then the total cost is at least ks; since for each element in the intersection we cannot "quit" in examining the spaghetti.

In this case we have to add the fixed cost of ks to the above calculated cost, hence the cost is obtained making $\epsilon = \frac{m-s}{n}$; accounting for the cost of an unsuccessful search in the interval of size $m - s$ plus the fixed cost ks. Observe that $s \leq m$ in any case.

$$\frac{1 - \left(\frac{m-s}{n}\right)^{k-1}}{1 - \left(\frac{m-s}{n}\right)} + ks \tag{6}$$

4   CDA, the Succinct Spaghetti

In the original spaghetti, the distances from each pivot to all elements in the database are stored in arrays (one per each pivot). Those distance arrays are ordered according to their values, and the permutation of the elements in each array with respect to the one preceding it is saved (see Figure 1a). Our contribution is a new representation. Instead of saving the permutation between each distance array and the one preceding it, the permutation of the elements in each array with respect to the identity is saved (as it is shown in Figure 2). In the identity each element has a unique identifier. The new representation can be used to reduce the number of distance computations, as we will show later.

To solve a query in the spaghetti, we find the sets $S_i = \{x : |d(x, P_i) - d(q, P_i)| < r\}$ for each pivot P_i where $i = 1, \ldots, k$. The intersection of them gives the candidate set. To compute it we need to check if each one of the elements in the set S_j is also in each one of the remaining sets S_k where $k \neq j$. Note that for testing the elements, we need to know their positions in each distance array.

Each array is a permutation, we can compose the permutations to find the position of an arbitrary element passing through the identity. We denote as I to the identity (see Figure 2). Let π_i be the permutation of I when the elements are ordered w.r.t. the distance to P_i. Here $\pi_i(k)$ represents the position in I of

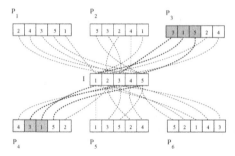

Fig. 2. An example of the new presentation of the spaghetti data structure.

the k-element in the distance array associated to pivot P_i, in Figure 2, $\pi_1(2) = 4$ and $\pi_3(1) = 3$. $\pi_i^{-1}(j)$ represents the relative position to P_i of the j-element in I, in Figure 2, $\pi_1^{-1}(4) = 2$ and $\pi_3^{-1}(3) = 1$. To find the position of an element with respect to a pivot, assuming we know the element's position with respect to another pivot, we just make a composition of the permutations. As a formative example, assume we want to find the position in P_5 of the element in position 3 in P_2. First, we obtain the position of the element in I by computing $\pi_2(3) = 2$. Once we know the element's position in I, we compute its position in P_5 by doing $\pi_5^{-1}(2) = 4$.

4.1   Compact Representation of Permutations

A key component in the new representation is the use of permutations. We are interested in a representation of them that can help to save storage space and also allows to perform fast computations. The problem of succinctly representing a permutation was studied in [2]. Given an integer parameter t, the permutations π_i and π_i^{-1} can be supported by simply representing π_i using an array of n words of $\lceil \lg n \rceil$ bits each, plus an auxiliary array S of at most n/t shortcuts or back pointers. In each cycle of length at least t every t-th element has a pointer t steps back. $\pi_i(k)$ is simply the k-th value in the primary structure, and $\pi_i^{-1}(k)$ is found by moving forward until a back pointer is found and then continuing to follow the cycle to the location that contains the value k. The key idea is in encoding of the locations of the back pointers: this is done with a simple bit vector B of length n, in which a 1 indicates that a back pointer is associated with a given location. B is augmented using $o(n)$ additional bits so that the number of 1's up to a given position and the position of the r-th 1 can be found in constant time using the rank and select operations on binary strings [3]. This gives the location of the appropriate back pointer in the auxiliary array S. For more details, the reader is referred to [2,4].

4.2   Computing the Intersection

In practice, the candidate set for each pivot can contain thousands of elements thus it can be helpful to develop an algorithm to quickly and efficiently evaluate those elements. Using the representation described above it is possible to decide the order in which the candidate sets are going to be compared. In this work, we make use of a strategy called Small vs Small (SVS) which we will describe in the next paragraphs.

Small vs Small (SVS). Let $S_i = \{x : |d(x, p_i) - d(q, p_i) \leq q|\}$ be the candidate set for each pivot p_i. Without lost of generality we consider that $|S_1| \leq |S_2| \leq \ldots \leq |S_k|$. The key idea is to identify the candidate set having the smallest cardinality. This set is intersected with each one of the remaining sets. The sets are visited following an ascending order according to their cardinalities. The final candidates are given by $\cap_{i=1}^{i=k} S_i$. This algorithm guarantees that the number of elements in the intersection is never bigger than the number of elements in the set to be compared. Note that in worst case, the cardinality of the intersection set equals the cardinality of the smallest candidate set.

4.3   Time Complexity Analysis

For simplicity, consider that each interval in each array has m elements. Let s be the size of the intersection set, the minimal cost to compute the intersection is ks since the execution stops until s elements are verified to belong to the k intervals. From subsection 3.2, we have that the cost for computing the intersection of the candidate sets is

$$\frac{1 - \left(\frac{m-s}{n}\right)^{k-1}}{1 - \left(\frac{m-s}{n}\right)} + ks \tag{7}$$

Also we have to consider the cost of computing π_i which is t for each element in each one of the k iterations, thus the cost is

$$t \cdot \left(\frac{1 - \left(\frac{m-s}{n}\right)^{k-1}}{1 - \left(\frac{m-s}{n}\right)} + ks\right) \tag{8}$$

Inverse Permutations. As a consequence of reducing the storage space we have increased the cost of execution, although it is possible to improve the algorithm. The inverse permutation is computed in worst case $m(k - 1)$ times while the direct permutation only m times, thus if during the construction of the index we store π_i^{-1} instead of π_i we reduce the search time. Note that the time only increases when we find the initial candidate set. We have that the cost is

$$\frac{1 - \left(\frac{m-s}{n}\right)^{k-1}}{1 - \left(\frac{m-s}{n}\right)} + ks + tm \tag{9}$$

In this paper, we evaluate the algorithm's performance of the succinct representation saving the inverse permutations.

5   Experimental Results

In this section, we analyze the improvements to the original spaghetti. First, we study the size reduction of the index as a consequence of using a succinct representation for the permutations. Second, we discuss the speed improvement by using the SVS technique.

The experiments were performed on databases containing randomly-generated vectors in six dimensions 4, 8, 12, 16, 20 and 24. The databases have 10^4, 10^5, 2.5×10^5, 10^6, 5×10^6, and 10^7 elements. We also performed experiments with **Nasa**, **Colors**, and **CoPhiR-1M** databases. As usual, vector spaces are indexed without using the coordinates. The query set for each database contains 256 randomly-generated vectors which are not contained in the index.

5.1   Index Size

The goal is to establish a comparison between the original and the succinct representations of the spaghettis. In Figure 3a, we can observe the index size growing for both representations as the number of elements in the database increases. The indexes were built using 120 pivots. In the figure, we can also see how the variation of the parameter t affects the size using the succinct representation.

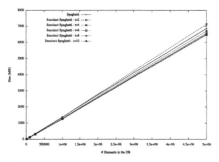
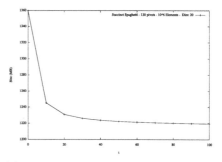

(a) Index size as a function of the number of elements in the database and the value of t.

(b) Index size reduction as a function of t.

Fig. 3. Index sizes for databases containing randomly-generated vectors.

Figure 3b shows a closer look to the index size as a function of t. In this case, a database with 10^6 randomly-generated vectors was used. In the figure, we can note that after some $t = t_m$ is reached, the index size remains almost constant if we continue increase the value of t. Therefore for $t > t_m$ the size reduction does not compensate the time increment to compute the inverse.

In Figures 4a, 4b and 4c we can observe the index size values as function of t for the NASA, Colors and Cophir databases. They present a similar behavior to the one described in the previous experiments.

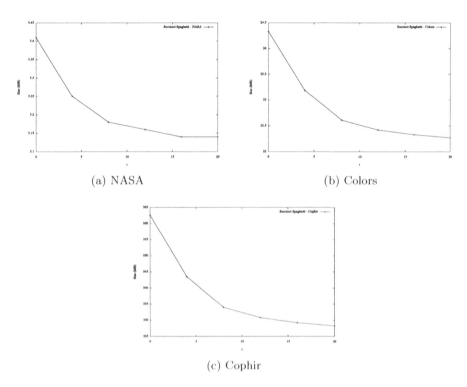

(a) NASA   (b) Colors

(c) Cophir

Fig. 4. Index sizes as a function of t for the NASA, Colors and Cophir databases.

5.2 Computing the Candidate Set

One of the goals of the new representation is reducing the number of distance evaluations to find the candidate set. Figure 5a shows the average size of the candidate sets (after visiting each pivot) in different dimensions using the SVS strategy. Note that the size of the candidate set after visiting a pivot is directly related with the number of distance evaluation that need to be performed in the next pivot. In the figure, we can observe that as the dimension of the database increases also does the size of the candidate set and the number of distance evaluations to find it. Figure 5b shows the results of using a random order (original spaghettis) for the same experiment.

Figures 6a, 6b, and 6c show the size of the approximated candidate sets (after visiting each pivot) for NASA, Colors and Cophir databases. From the previous results, we can observe that in the SVS strategy the initial pivots are the ones discarding the bigger amount of elements, this suggest that we can improve the algorithm's performance using an early stop strategy where only a small number of the ordered pivots are used to compute the candidate set.

Figures 7a, 7b, 7c and 7d show the average time (in secs) to solve a query for the Nasa, Colors, Cophir and Random databases. We test two implementations

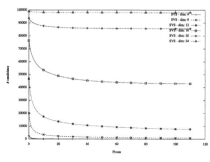
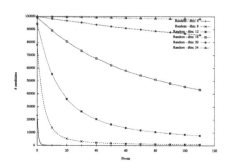

(a) Number of candidates after each comparison using the SVS strategy.

(b) Number of candidates after each comparison using a random strategy (original spaghetti).

Fig. 5. Computing candidate sets in databases with randomly-generated vectors.

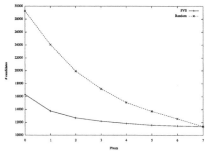
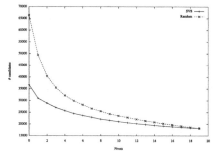

(a) Number of candidates after each comparison for the Nasa database.

(b) Number of candidates after each comparison for the Colors database.

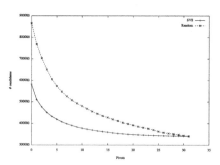

(c) Number of candidates after each comparison for the Cophir database.

Fig. 6. Computing a candidate set in the Nasa, Colors, and Cophir databases.

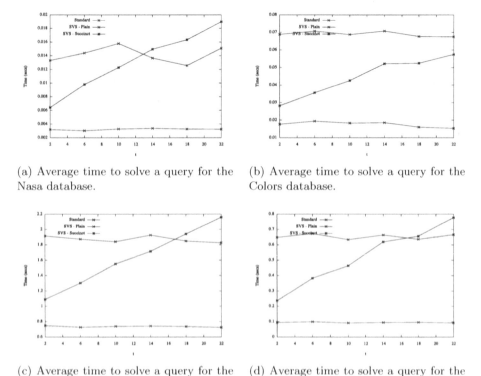

(a) Average time to solve a query for the Nasa database.

(b) Average time to solve a query for the Colors database.

(c) Average time to solve a query for the Cophir database.

(d) Average time to solve a query for the Random database in 12 dimensions.

Fig. 7. Time to solve a query using the standard spaghetti, and the SVS strategy.

of the SVS strategy. In the first one, both the direct and inverse plain permutations are stored. We use this implementation as a ground truth of the best performance for the SVS strategy since no cost is involved to compute the direct permutations. For the succinct version, the inverse permutations are saved in order to improve the algorithm's performance, as it was described in subsection 4.3. For both algorithms, only the first four pivots (early stop condition) where used to compute the candidate set. In the figures, we can observe that the plain version of the SVS strategy is the fastest. The figures also show that for small values of t the succinct version of the SVS strategy is faster than the standard spaghetti. As it is expected, as we increase the value of t (better compresion) it takes more time to compute the direct permutation in the succinct representation and thus the overall time to compute the query also increases. Note that the standard and plain versions of the spaghettis are independent of t and the small variations in time are related to the random process to create the indexes.

6   Conclusions and Future Work

In this paper, we introduce a new representation for the spaghetti data structure based on the permutations of the elements in each pivot to an identity array. Using a succinct representation of those permutations we have shown that it is possible to save memory, that can be used to increase the number of pivots.

We also presented a new method to compute set intersections for using pivot tables. We have shown how the small-vs-small heuristic can be used to quickly trim candidate list, faster than using the original spaghetti.

One key finding is that our method is able to compute almost the final intersection using only a small fraction of the pivots. This fact can be used as an alternate method to perform range queries in multidimensional data, as an alternative to kd-trees. Those results will be reported elsewhere.

References

1. Chávez, E., Marroquin, J.L., Baeza-Yates, R.: Spaghettis: an array based algorithm for similarity queries in metric spaces. In: String Processing and Information Retrieval Symposium, 1999 and International Workshop on Groupware, pp. 38–46. IEEE (1999)
2. Munro, J.I., Raman, R., Raman, V., Rao, S.S.: Succinct representations of permutations. In: Baeten, J.C.M., Lenstra, J.K., Parrow, J., Woeginger, G.J. (eds.) ICALP 2003. LNCS, vol. 2719, pp. 345–356. Springer, Heidelberg (2003)
3. Munro, J.I., Raman, V.: Succinct representation of balanced parentheses and static trees. SIAM Journal on Computing **31**(3), 762–776 (2001)
4. Barbay, J., Munro, J.I.: Succinct encoding of permutations: applications to text indexing. In: Encyclopedia of Algorithms, pp. 915–919. Springer (2008)

Improving Metric Access Methods
with Bucket Files

Ives R.V. Pola[2]([✉]), Agma J.M. Traina[1], Caetano Traina Jr.[1],
and Daniel S. Kaster[2]([✉])

[1] University of São Paulo, São Carlos, Brazil
{agma,caetano}@icmc.usp.br
[2] University of Londrina, Londrina, Brazil
{ives,dskaster}@uel.br

Abstract. Modern applications deal with complex data, where retrieval
by similarity plays an important role in most of them. Complex data
whose primary comparison mechanisms are similarity predicates are usually immersed in metric spaces. Metric Access Methods (MAMs) exploit
the metric space properties to divide the metric space into regions and
conquer efficiency on the processing of similarity queries, like range and
k-nearest neighbor queries.

Existing MAM use homogeneous data structures to improve query
execution, pursuing the same techniques employed by traditional methods developed to retrieve scalar and multidimensional data. In this paper,
we combine hashing and hierarchical ball partitioning approaches to
achieve a hybrid index that is tuned to improve similarity queries targeting complex data sets, with search algorithms that reduce total execution
time by aggressively reducing the number of distance calculations. We
applied our technique in the Slim-tree and performed experiments over
real data sets showing that the proposed technique is able to reduce the
execution time of both range and k-nearest queries to at least half of the
Slim-tree. Moreover, this technique is general to be applied over many
existing MAM.

1   Introduction

The existing Data Base Management Systems (DBMS) were originally developed
to store and retrieve data represented in numeric and short character strings
domains. They are not able to efficiently manage the complex data handled by
current applications, such as multimedia data, georeferenced data, time series,
genetic sequences, scientific simulations, etc. The main reason precluding those
data to be appropriately managed by current DBMSs is because their internal
structures require the data domains to comply with the ordering relationship
(OR) properties, that is, they require that every data element from a domain
can be compared by the $<, \leq, >$ and $\geq$ operators. To manage complex data even
the equality comparison operators $=$ and $\neq$ are almost useless, because identity
seldom occurs (or is not worth pursuing) when retrieving complex data. To query
complex data, comparing by similarity is the most important operation [6].

© Springer International Publishing Switzerland 2015
G. Amato et al. (Eds.): SISAP 2015, LNCS 9371, pp. 65–76, 2015.
DOI: 10.1007/978-3-319-25087-8_6

Similarity search is the most frequent abstraction to compare complex data, based on the concept of proximity to represent similarity embodied in the mathematical concept of metric spaces [8]. The development of the Metric Access Methods (MAMs), also known as distance-based index structures, provides adequate techniques to retrieve complex data, once they are based solely on the distances (similarities) between pairs of elements in a data set. Evaluating (dis)similarity using a distance function is desirable when the data can be represented in metric spaces. Formally, a metric space is a pair $\langle \mathbb{S}, d \rangle$, where $\mathbb{S}$ is the data domain and $d : \mathbb{S} \times \mathbb{S} \rightarrow \mathbb{R}^+$ is the distance function, or metric, that holds the following properties for any $s_1, s_2, s_3 \in \mathbb{S}$:

- Identity $(d(s_1, s_2) = 0 \rightarrow s_1 = s_2)$;
- Symmetry $(d(s_1, s_2) = d(s_2, s_1))$;
- Non-negativity $(0 < d(s_1, s_2) < \infty$, $s_1 \neq s_2)$ and
- Triangular inequality $(d(s_1, s_2) \leq d(s_1, s_3) + d(s_3, s_2))$.

Given a set S in a complex domain $\mathbb{S}$, a similarity query returns a result set $T_R = \{s_i \in S\}$ that meet a given similarity criterion, expressed through a reference element $s_q \in \mathbb{S}$. For example, for image databases one may ask for images that are similar to a given one, according to a specific criterion. There are two main types of similarity queries: the range and the k-nearest neighbor queries.

There are two broad classes of access methods that exploit the properties of metric spaces: those based on the hierarchical division of the space based on ball-shaped regions centered at one element, and those based on pivots sets, which can be implemented as a hierarchy or as hash tables. Dynamic MAMs have had special attention by academy and industry as they do not degrade with updates.

In this paper we propose the Bucket-Slim-Tree (BST), a MAM based both on hash and on ball partitioning, aiming at reducing the number of distance evaluations required to answer a similarity query. BST employs the dynamic MAM Slim-tree as a kind of hash function mapping to buckets delimited by a fixed radius. Such organization allows reducing the overlap among regions improving the search performance in a great extent. Experiments over real data sets reported in the paper show that BST demands less than half of the distance calculations and of the execution to perform similarity queries when compared to the original Slim-tree, in the best results.

The rest of the paper has the following outline. Section 2 discusses the background and existing works related to this one. Section 3 presents the proposed data structure, the Bucket-Slim-Tree. Section 4 shows experiments performed to demonstrate the improvement of the proposed structure. Finally, Section 5 concludes for this work.

2   Background

Metric access methods use only the distances between elements to prune further comparisons in subsets of the elements during search. Pruning techniques require the algorithms to store distances to take advantage of the metric properties and/or of statistics from the distance distribution over the data space. The usual pruning techniques use lower bounds of distances derived from the triangular inequality property. Another approach is to store the minimum and the maximum distances within a group of elements to help discarding entire regions during search algorithm execution.

Many indexing structures were developed exploiting those concepts, such as the Geometric Near Access Tree (GNAT) of Brin [3], leading to the class called Voronoi-based MAMs. The EGNAT [10] is a dynamic variation of GNAT, which provides a mechanism to store elements on disk by creating buckets on the leaf nodes and also enables deletion. Another approach, the ball decomposition scheme, partitions the dataset based on distances from a distinguished element called a Vantage Point (VP), thus creating the so-called VP-tree [14]. The VP-tree construction process is based on finding the median element of a sorted sample list of the elements, which leads to a recursive tree creation. Other disk-based MAM have been proposed based on the VP-tree, such as the MVP-tree [2], where multiple vantage points are used.

The BP-tree (Ball-and-Plane tree) [1] is constructed by recursively dividing the data set into compact, low-overlap clusters. It is static and was designed to deal with high dimensional data, where a data distribution analysis is used to search in clusters.

Several dynamic, disk-based MAMs were proposed in the literature[13][4]. A disk-based MAM requires the structure to hold many elements per node, in order to decrease the number of disk accesses. They employ a bottom up strategy to construct the trees, assuring the creation of balanced trees. The M-tree [4] was the first of such trees proposed, followed by the Slim-tree [13], which includes the Slim-down algorithm to reduce node overlaps. The OMNI concept [11] increases the pruning power of search operations using a few elements strategically positioned as pivots, the foci set. These methods store distances among the elements and the pivots, so the triangular inequality property can be used to prune nodes and reduce the number of sub-tree accesses.

Most of the indexing structures presented above are based on ball partitioning or pivot-based structures. Hashing, the "key to address" mapping, is the basis for D-Index [5] and SH [7]. The LAESA algorithm [9] is a pivot table that uses a matrix of distances between all pairs of pivots selected from the dataset. When processing queries, it sequentially process the entire distance matrix (or parts of it in multiple passes), pruning by using the triangle inequality property. But, the internal cost of LAESA can be so high that it can be equivalent to perform a sequential search when indexing high dimension datasets at low cost metrics [12].

Both ball and hash based methods have particular advantages that can be combined to achieve better metric structures. The way Ball-based MAM partitions the metric space leads to a better organization of the data structure, so every resulting partition of the metric space groups similar elements. However, the best dynamic approaches produce regions that overlap, imposing to the search algorithm to visit many regions. The pivot partitioning methods are affected by the pivots selection policy and how they are combined to prune regions. Hash-based methods usually partition the data into subsets that are addressed later to answer the queries.

Our proposal is innovative as it exploits the best properties from ball-based and from pivot-based methods. Specifically, our method Bucket-Slim-Tree (BST) uses the Slim-tree as a hash function to search within a bucket file to improve performance. BST merges the usage of buckets of elements with the Slim-tree, enabling to explore properties from both structures that results in an overall reduced consumption of computational resources.

3   The Bucket-Slim-Tree

The Bucket-Slim-Tree (BST) is composed of a Slim-tree and a set of buckets pointed by the Slim-tree leaf nodes. The Slim-tree acts as a hash function that during query answering determines which bucket should be visited next. The basic structure of a BST is shown in Figure 1. Each element in a leaf node has a pointer to its respective bucket. Although each Slim-tree node have a limited capacity, each bucket is (theoretically) limitless.

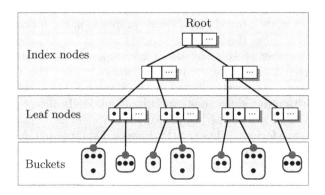

Fig. 1. Structure of a Bucket-Slim-Tree.

The elements indexed in the index and leaf nodes are considered search keys, and act like pivots in a hash structure. Inserting new elements or answering queries require to traverse the tree structure using the search keys to determine which leaf nodes will have their buckets accessed for inspection. The query result will be composed of keys from the Slim-tree and also of elements stored in the corresponding buckets that match the search criteria.

3.1   The Structure of the Buckets

A BST is constructed for a specific bucket radius η, given beforehand. A bucket $B^\eta(bc_i)$ represents a ball of radius η in the metric space, whose center is the

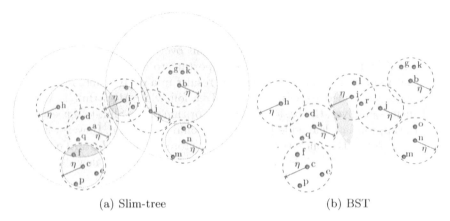

(a) Slim-tree   (b) BST

Fig. 2. Ball partitioning comparison between regular Slim-tree and BST.

element bc_i stored in a Slim-tree leaf node pointing to that bucket. Thus, a bucket stores elements s_i such that $\forall s_i \in B^\eta(bc_i) : d(s_i, bc_i) \leq \eta$. Each element s_i that belongs to a bucket does not belong to other buckets.

For example, consider Figure 2. In Figure 2(a) it shows a set of points in $\mathbb{R}^2$ indexed in a regular Slim-tree with four levels. But, in Figure 2(b) it shows the same set of points with buckets of a fixed radius η centered at elements $\{a, b, c, h, i, j, n\}$, indexed on a BST. As it can be seen, there can be empty buckets, such as $B^\eta(h)$, and elements that are covered by more than one bucket are stored in only one bucket, such as element r. Note that, comparing both figures, using buckets reduces the overall covering radius of the index elements, reducing the overlap in the structure.

The bucket radius plays an important role in the performance of the Bucket-Slim-Tree. Setting too large η values will result in large buckets, thus creating long subsets of elements to be analyzed during queries. This degenerates into sequential scans in the buckets and few key elements to filter the buckets in the tree. Furthermore, the larger is the region covered by a bucket the more the overlap between sibling buckets, which leads to potentially more unnecessary accesses. On the other hand, choosing too small values for η will produce many small or empty buckets, leading the search cost to occur mostly in the Slim-tree and an added internal cost to manage the buckets. Choosing the bucket radius $\eta = 0$ the result is the Slim-tree itself, thus the Bucket-Slim-Tree can be seen as a generalization of the Slim-tree.

Next we will discuss how to build the Bucket-Slim-Tree, i.e., how to choose the keys and how to create the buckets.

3.2   Building the Bucket-Slim-Tree

The Bucket-Slim-Tree is a dynamic MAM able to be constructed either using bulk-loading or adding elements one at a time. The BST is designed to group similar elements into buckets centered at the key elements stored in the Slim-tree. As elements are added, some of them are stored in the Slim-tree leaf nodes, thus becoming keys to the buckets, and others are stored in the buckets. The way that those keys are organized in the Slim-tree affects how many buckets are necessary to answer each query. The more bucket regions overlap the more buckets need to be visited in a query. Therefore, it is important to choose an index creation policy that reduces such overlap, even if it results in a deeper tree.

Elements are added to a Bucket-Slim-Tree following Algorithm 1. When a new element s_n arrives, the basic insertion algorithm of the Slim-tree is executed to find the appropriate leaf node L_m where it would be inserted. However it is not inserted yet. Next, the buckets from the keys stored in node L_m are evaluated looking for the keys $bc_i \in L_m$ such that $d(bc_i, s_n) \leq \eta$. The new element s_n is stored in the qualifying bucket whose center is the closest to s_n, along with the distance from the bucket center. If no bucket qualifies, s_n is stored in L_m splitting the node if required, as in the regular Slim-tree insertion algorithm, and s_n becomes a new bucket center. However, the corresponding bucket is not created now – it will be created only when another element is stored in it. Once the structure is constructed, similarity queries can be performed considering η as an additional pruning radius, as is explained following.

Algorithm 1. BST:ADD(s_n)

Input: new element s_n
var candidate : bucket center that covers s_n
var leaf : leaf node of Slim-tree
Set *chooseSubTree* policy of Slim-tree to 'MINDIST'
Set leaf to the proper leaf node that covers s_n
foreach *element bc_i in leaf* **do**
| **if** $d(bc_i, s_n) \leq \eta$ **then**
| | Add bc_i as a candidate
if *there are candidates* **then**
| Choose the first center bc_i where $d(bc_i, s_n)$ is minimum
| Insert s_n in the bucket of bc_i and store $d(bc_i, s_n)$
else
| Add s_n to leaf
| Split leaf if necessary
End

3.3   Querying the Bucket-Slim-Tree

The BST structure allows performing both range (Rng) and k-nearest neighbor (k-NN) similarity queries. The algorithms to answer those queries visit both the nodes in the tree and the buckets. For both query types, radius η must be taken into account to correctly prune subtrees at each index level. As radius η is a fixed value defined beforehand of the BST construction, it is possible to avoid the need to adjust each region formed in the Slim-tree during construction by adding η to every query radius.

An example of a range query $Rng(s_q, \xi)$ is shown in Figure 3, considering a two-dimensional set of points using the Euclidean distance. The element ds_1 shown in the figure is a representative in an index node, so it is also stored in a leaf node L_m. Elements bc_1 to bc_4 are elements stored in leaf node L_m of the Slim-tree. Thus, node L_m has five elements stored: $\{ds_1, bc_1, bc_2, bc_3, bc_4\}$, and each one is the center of a bucket.

Each bucket is shown as a dashed ball in Figure 3, which represents the space region whose corresponding elements (e.g. $s_1, s_2 \ldots$) will be stored. Thus, to avoid pruning valid buckets (like the one centered at bc_1 in the subtree centered at ds_1) the query radius ξ must be adjusted to $\xi + \eta$. In Figure 3, this corresponds to change the query ball drawn in solid line centered at s_q to the one drawn in dotted line. In this example, only the bucket centered at element bc_1 must be evaluated, adding element s_1 to the result. The same idea applies to the k-nearest neighbor query, which requires to enlarge the dynamic radius by η.

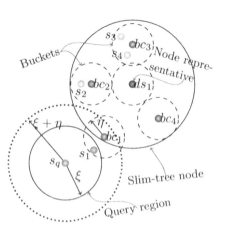

Fig. 3. Querying a Bucket-Slim-Tree.

The similarity query algorithms use the triangular inequality to prune subtrees as in the original Slim-tree and also inside each bucket. In Figure 3 example, instead of calculating every distance $d(s_q, bc_i), bc_i \in L_m$, just evaluate the lower bound of the required distance using the triangular inequality, avoiding a calculation whenever $d(s_q, bc_i) \geq |d(s_q, ds_1) - d(ds_1, bc_i)|$. Notice that the values $d(s_q, ds_i)$ are already stored in BST, and that $d(s_q, ds_i)$ is evaluated only once for each leaf node. Thus, assuming the pruning radius $r_p = \eta + \xi$, whenever $|d(s_q, ds_1) - d(ds_1, bc_i)| > r_p + \eta$ then bucket bc_i can be safely pruned without evaluating $d(s_q, bc_i)$.

The steps to evaluate a range query $Rng(s_q, \xi)$ is shown in Algorithm 2, where s_q is the query center and ξ is the query radius. To traverse the tree, the algorithm evaluates the index nodes using both η and ξ to qualify the subtrees that must be visited. To process the leaf nodes, only the radius η is used.

The procedure of a k-nearest neighbors query $k\text{-}NN(s_q, k)$ in the BST is analogous to the range query, but now we update the result list maintaning k elements and updating the active radius. the technique of a shrinking active radius starts with a value larger than the dataset diameter (or infinity), and reduces when the ongoing result list achieves k elements and updates.

4 Experiments

In this section we show experiments to evaluate the proposed index structure, the Bucket-Slim-Tree. We compare it with the original Slim-Tree, using different values for bucket radius (η). The experiments show that using the bucket-based

---

Algorithm 2. $RangeQuery(s_q, \xi, root)$

Input: Query center s_q, Query radius ξ, Slim-tree $root$

if $root$ is $index$ $node$ **then**

   foreach $ds_i \in root$ **do**

      //Evaluate if the triangular inequality allows pruning;

      if $|d(ds_i, root) - d(root, s_q)| > \eta + \xi + ds_i.Radius$ **then**

      | Prune subtree of ds_i;

      foreach $element$ ds_i not $pruned$ **do**

         if $d(ds_i, s_q) \leq \eta + \xi + ds_i.Radius$ **then**

         | RangeQuery(s_q, ξ, ds_i.Subtree);

if $root$ is a $leaf$ $node$ **then**

   foreach $bc_i \in root$ **do**

      if $d(bc_i, s_q) \leq \xi$ **then**

      | add bc_i to result;

      //Evaluate if the bucket can be pruned

      if $d(bc_i, s_q) \leq \eta + \xi$ **then**

         foreach $s_i \in B(bc_i)$ **do**

            //Evaluate if the triangular inequality allows pruning;

            if $|d(s_i, root) - d(root, s_q)| \leq \xi$ **then**

               if $d(s_i, s_q) \leq \xi$ **then**

               | add s_i to result;

---

approach increases the query answering performance being up to twice faster than slim-tree. They also show how the query performance is affected when different bucket sizes are employed.

We used three datasets for the experiments. The `Corel` Dataset consists of 10 thousand color histograms in a 32 dimension space extracted from an image set, using the L_1 distance. The `USCities` Dataset consists of the latitude and longitude coordinates of 25,376 cities in the USA, using the great-circle distance modified to return distances in kilometers. The `HCimages` Dataset was obtained from a collection of 500,000 DICOM images from the Ribeirão Preto Medical School Clinics Hospital of the University of São Paulo (HCFMRP-USP). From each image, we extracted a 256-bin grayscale normalized histogram. All the experiments were performed in a machine with a Intel Core i7 920 processor with 8 Gb RAM of memory.

The first experiment measured how the buckets are filled with elements according its radius η. The plots in Figure 4 show the percentage of element distributed among the slim-tree and the buckets, with bucket radius η varying from 0.10 to 0.15 for Corel (Figure 4(a)), from 10 Km to 60 Km for USCities (Figure 4(b)) and from 10 to 60 for HCimages (Figure 4(c)). The plots for Corel show that, as the bucket radius increases, the percentile of elements stored in the buckets increases from 36% to 75%. Similar behavior occur in the plot for HCimages, but in this case the number of elements in the buckets increase slower. Exemplifying the case where a high value for η produces dense buckets, the plot for USCities shows that as the radius η we increases from 10 to 60 Km for the `USCities` dataset, the number of elements stored in the buckets reaches almost 95%, meaning that almost all elements are

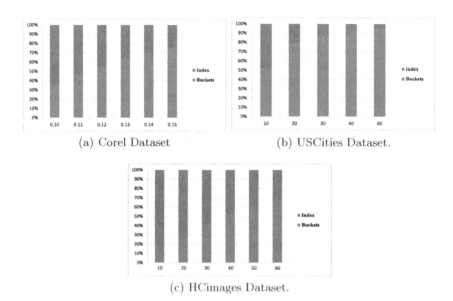

(a) Corel Dataset

(b) USCities Dataset.

(c) HCimages Dataset.

Fig. 4. Distribution of elements in the Bucket-Slim-Tree components varying the bucket radius η.

stored in the buckets, probably degenerating the structure, where queries would sequentially scan dense buckets.

The performance of a BST depends on the chosen value for η. This value changes for different datasets and should be set close to the frequently used radius on range queries in order to achieve good results. An initial value can be given by a percentage of the value of the dataset maximum radius, or estimating the mean distance from all elements in the dataset. All experiments were performed using different values of η for both range and k-nearest neighbor queries. As previously noted, the BST uses a modified Slim-tree with the *mindist* policy for the *ChooseSubTree* algorithm. For comparison purposes, we also evaluated the results if it is employed a Slim-tree with the usual *minoccup* policy. Every query was performed 500 times with the same radius ξ or k but different centers, in order to evaluate the average of the number of performed distance calculations and the total time spent.

The plots in Figure 5 show the results of measuring the performance for both types of queries using the Corel Dataset. They show that in the beginning, as the value of η increases, the query performance increases. However, if η becomes too high, the performance is decreased, as shown in Figure 5(c) when $\eta > 0.12$. This is because buckets become larger and the sequential scans inside each bucket spend more time.

The plots in Figures 6 show the performance measurements for both types of queries using the USCities dataset, obtained using η set to 10, 20 and 40 Km. As it can be noticed, all configurations lead to BST with a performance better than that of a slim-tree for all queries, where the value of 20 Km produced the best one. It is important to notice that for $\eta = 40$km, the performance was worse than for $\eta = 20$km. This is because for radius larger than $\eta = 20$km, the number of elements in the buckets tends to increase too much, as shown in Figure 4(b).

The plots in Figures 7 show the performance results for queries using the HCimages dataset, using η with the values 10, 20 and 30. As this dataset has a high dimensionality, any variation of the radius will strongly change the covering of elements, as expected of the curse of the high dimensionality. From the results we can note that our technique still enhances the performance of queries when choosing η next to the query values. This is because any decrease in the index level covering radius greatly reduces the overlap on nodes.

5   Conclusion

In this paper we proposed the Bucket-Slim-tree (BST), a MAM based on hash and ball partitioning that aims at reducing the number of distance calculations required to answer similarity queries. The BST is composed of a slim-tree and a set of buckets assigned to each element in the Slim-tree leaf nodes. The slim-tree acts as a hash function which maps the stored elements to the buckets that will be visited during search. The leaf nodes contain all key elements associated with a bucket of radius η, and all of them must be compared to the query element during query executions.

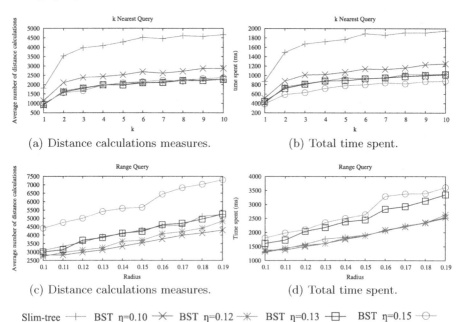

(a) Distance calculations measures.

(b) Total time spent.

(c) Distance calculations measures.

(d) Total time spent.

Slim-tree ──+── BST η=0.10 ──✕── BST η=0.12 ──✳── BST η=0.13 ──⊟── BST η=0.15 ──⊖──

Fig. 5. Results using the Corel dataset indexed in a slim-tree and BSTs with $\eta = 0.10$, 0.12, 0.13 and 0.15. (a) Number of distance calculations for k-NN queries; (b) Time spent for k-NN queries; (c) Number of distance calculations for Rq queries; (d) Time spent for Rq queries;

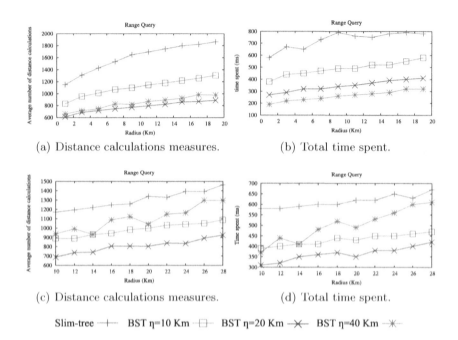

(a) Distance calculations measures.   (b) Total time spent.

(c) Distance calculations measures.   (d) Total time spent.

Slim-tree ⊢   BST η=10 Km ⊟   BST η=20 Km ⨯   BST η=40 Km ✳

Fig. 6. Results using the USCities dataset. (a) and (b): Nearest Neighbor query evaluation; (c) and (d): Range query evaluation.

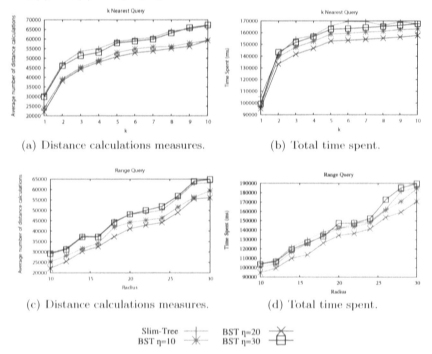

(a) Distance calculations measures.   (b) Total time spent.

(c) Distance calculations measures.   (d) Total time spent.

Slim-Tree ⊢   BST η=20 ⨯
BST η=10 ✳   BST η=30 ⊟

Fig. 7. Results using the HCimages dataset. (a) and (b): Nearest Neighbor query evaluation; (c) and (d): Range query evaluation.

Experiments performed over real data sets show that the proposed MAM was able to reduce up to half the execution time of both range and k-nearest queries, reducing also the number of distances calculations under different values of η.

Acknowledgments. This research has been partially suported by CAPES, CNPq and by FAPESP.

References

1. Almeida, J., Torres, R.d.S., Leite, N.J.: Bp-tree: an efficient index for similarity search in high-dimensional metric spaces. In: Proceedings of the 19th ACM International Conference on Information and Knowledge Management, CIKM 2010, pp. 1365–1368. ACM, New York (2010)
2. Bozkaya, T., Özsoyoglu, Z.M.: Distance-based indexing for high-dimensional metric spaces. In: ACM SIGMOD International Conference on Management of Data, Tucson, AZ, pp. 357–368. ACM Press (1997)
3. Brin, S.: Near neighbor search in large metric spaces. In: Dayal, U., Gray, P.M.D., Nishio, S. (eds.) International Conference on Very Large Databases (VLDB), break pp. 574–584. Morgan Kaufmann, Zurich (1995)
4. Ciaccia, P, Patella, M., Rabitti, F., Zezula, P.: Indexing metric spaces with m-tree. In: Atti del Quinto Convegno Nazionale SEBD, Verona, Italy, pp. 67–86 (1997)
5. Dohnal, V., Gennaro, C., Savino, P., Zezula, P.: D-index: Distance searching index for metric data sets. Multimedia Tools and Applications Journal (MTAJ) **21**(1), 9–33 (2003)
6. Faloutsos, C.: Indexing of multimedia data. In: Multimedia Databases in Perspective, pp. 219–245. Springer Verlag (1997)
7. Gennaro, C., Savino, P., Zezula, P.: Similarity search in metric databases through hashing. In: 3rd International Workshop on Multimedia Information Retrieval, Ottawa, Canada, pp. 1–5 (2001)
8. Kelley, J.L.: General Topology. Springer (1955)
9. Micó, L., Oncina, J., Vidal, E.: A new version of the nearest-neighbor approximating and eliminating search (aesa) with linear processing-time and memory requirements. Pattern Recognition Letters **15**, 9–17 (1994)
10. Navarro, G., Uribe-Paredes, R.: Fully dynamic metric access methods based on hyperplane partitioning. Inf. Syst. **36**, 734–747 (2011)
11. Santos Filho, R.F., Traina, A.J.M., Traina Jr., C., Faloutsos, C.: Similarity search without tears: the omni family of all-purpose access methods. In: IEEE International Conference on Data Engineering (ICDE), Heidelberg, Germany, pp. 623–630. IEEE Computer Society (2001)
12. Skopal, T.: Where are you heading, metric access methods?: a provocative survey. In: Proceedings of the Third International Conference on SImilarity Search and APplications, SISAP 2010, pp. 13–21. ACM, New York (2010)
13. Traina Jr, C., Traina, A.J.M., Faloutsos, C., Seeger, B.: Fast indexing and visualization of metric datasets using slim-trees. IEEE Transactions on Knowledge and Data Engineering (TKDE) **14**(2), 244–260 (2002)
14. Yianilos, P.N.: Data structures and algorithms for nearest neighbor search in general metric spaces. In: Fourth Annual ACM/SIGACT-SIAM Symposium on Discrete Algorithms (SODA), Austin, TX, pp. 311–321 (1993)

Faster Dual-Tree Traversal
for Nearest Neighbor Search

Ryan R. Curtin[✉]

Georgia Institute of Technology, Atlanta, GA 30332, USA
ryan@ratml.org

Abstract. Nearest neighbor search is a nearly ubiquitous problem in computer science. When nearest neighbors are desired for a query set instead of a single query point, dual-tree algorithms often provide the fastest solution, especially in low-to-medium dimensions (i.e. up to a hundred or so), and can give exact results or absolute approximation guarantees, unlike hashing techniques. Using a recent decomposition of dual-tree algorithms into modular pieces, we propose a new piece: an improved traversal strategy; it is applicable to any dual-tree algorithm. Applied to nearest neighbor search using both kd-trees and ball trees, the new strategy demonstrably outperforms the previous fastest approaches. Other problems the traversal may easily be applied to include kernel density estimation and max-kernel search.

1 Introduction

The task of nearest neighbor search arises continually in machine learning, data mining, and related domains. For instance, many computer vision algorithms require forms of similarity search [1]; recommendation systems may use k-nearest-neighbor search internally: BellKor's Netflix prize solution does this [2]. Nearest neighbors are also often used in machine learning applications as simple classifiers [3]; more advanced machine learning techniques may also depend on the calculation of nearest neighbors [4].

To formally describe the problem, take S_r to be the reference set. The nearest neighbor search task is, for a given query point p_q, find $\mathrm{argmin}_{p_r \in S_r} d(p_q, p_r)$ for some metric $d(\cdot, \cdot)$.[1] The most straightforward technique for solving this problem is a linear scan over all points in S_r, but for large S_r—or for situations where answers are desired not just for one query point p_q but instead an entire query set S_q—this approach is computationally infeasible. Given $|S_r| = N$, a result for a single query point p_q takes $O(N)$ time.

Owing to both this computational difficulty and the wide applicability of nearest neighbor search, much ink has been spilled describing fast algorithms to solve the nearest neighbor search problem. The first fast algorithms for

---

The rights of this work are transferred to the extent transferable according to title 17 U.S.C. 105.

[1] Extending this to the k-nearest neighbor search task is straightforward: replace argmin with k argmin.

© Springer International Publishing Switzerland 2015
G. Amato et al. (Eds.): SISAP 2015, LNCS 9371, pp. 77–89, 2015.
DOI: 10.1007/978-3-319-25087-8_7

nearest-neighbor search were based on tree structures [5] [6], where some type of tree structure is built on the reference set S_r and then, to find the nearest neighbor of a query point p_q, a branch-and-bound algorithm is used. Other popular approaches include the use of nets [7] and also locality-sensitive hashing [8] [9] [10]. In general, nets and hashing give approximate solutions, whereas tree-based approaches can give both approximate and exact solutions.

In the situation where there is a query set S_q and not just a single query point p_q, it often makes sense to build a tree on *both* the reference set S_r and the query set S_q, and simultaneously traverse both the query and reference trees. This type of approach is known as a *dual-tree algorithm* [11] [12], and is generally the fastest known way to perform nearest-neighbor search, for sufficiently large query sets in low-to-medium dimensions (i.e. up to a hundred or so, depending on the type of tree and the properties of the dataset). Further, when cover trees are used, and $S_q \sim O(N)$, search time for *all* points in S_q is worst-case $O(N)$ [13] [14]; though, this bound depends on dataset-dependent quantities.

Dual-tree algorithms exist for problems other than nearest neighbor search; some examples include range search [12], kernel density estimation [15], minimum spanning tree calculation [16], mean shift clustering [17], kernel summations [18], max-kernel search [19], and other problems [20] [21] [22]. Thus, results for any dual-tree algorithm are often readily applied to other dual-tree algorithms.

Curtin et al. recently proposed a generalizing abstraction for all dual-tree algorithms, which allows dual-tree algorithms to be understood as four separate components: a type of tree, a dual-tree traversal, a problem-specific pruning rule, and a problem-specific base case [12]. This convenient, modular abstraction lets us focus on only one component at a time, independent of the other three pieces.

For tree-based nearest neighbor search, whether single-tree or dual-tree, the order that tree nodes are visited makes a noticeable difference in both the quality of the results (for approximate search) and the speed of the results. This is why single-tree algorithms such as the original kd-tree nearest neighbor search algorithm [5] first recurse into the nearest node to a query point.

In this paper, we exploit the tree-independent dual-tree algorithm abstraction in order to develop an improved general depth-first dual-tree traversal. By applying this traversal to the problem of nearest-neighbor search, we obtain significant speedup over previous dual-tree traversal strategies, and outperform competing strategies, such as single-tree search and LSH, in both the approximate and exact nearest neighbor search tasks. Because of the traversal's generality, it can be applied to problems other than just nearest neighbor search.

2   Trees

First, we must introduce the concepts underlying dual-tree algorithms more formally, and we must also introduce notation. As in [12] and more recent contributions [14] [23], we will use the tree-independent dual-tree algorithm framework. This means that given some dual-tree algorithm that works on a set of query points S_q and a set of reference points S_r, we may understand this algorithm as the combination of four distinct parts:

- A type of *space tree*.
- A *pruning dual-tree traversal*, which visits combinations of nodes from the query tree and reference tree, and is parameterized by a `BaseCase()` and `Score()` function.
- A `Score()` function, which determines if a combination of two nodes can be pruned.
- A `BaseCase()` function, which defines the action to take on a combination of query point and reference point.

If we have each of these four pieces, then, we may assemble a dual-tree algorithm: using a pruning dual-tree traversal with the given `BaseCase()` and `Score()` functions on two space trees that are built on the query and reference sets will yield a working dual-tree algorithm. A formal definition of each of these components is necessary for complete understanding. These definitions are taken from the original introduction of Curtin et. al. [12].

Definition 1. *A* space tree *on a dataset $S \in R^{n \times d}$ is a rooted, undirected, connected, acyclic simple graph satisfying the following properties:*

- *Each node (or vertex) holds a number of points (possibly zero) and is connected to one parent node and a number of child nodes (possibly zero).*
- *There is one node with no parent; this is the root of the tree.*
- *Each point in S is contained in at least one node of the tree.*
- *Each node $\mathscr{N}$ of the tree corresponds to a convex subset of R^d that contains each of the points in the node as well as each of the convex subsets corresponding to each child of the node.*

Most tree structures in the literature fall under the umbrella definition of a space tree: *kd*-trees [5], PCA trees, metric trees, cover trees, R trees and variants, and even spill trees [24], where the subsets of child nodes are allowed to overlap. In this document formal script letters will be used to notate trees and corresponding quantities; this is the same notation used in [12]. In specific,

- A node in a tree will be denoted with the letter $\mathscr{N}$.
- For some node $\mathscr{N}_i$, the set of children of $\mathscr{N}_i$ will be denoted $\mathscr{C}_i$.
- For some node $\mathscr{N}_i$, the set of points contained in $\mathscr{N}_i$ will be denoted $\mathscr{P}_i$.
- The convex subset of R^d corresponding to the node $\mathscr{N}_i$ will be denoted $\mathscr{S}_i$.
- For some node $\mathscr{N}_i$, the set of descendant nodes of $\mathscr{N}_i$ will be denoted $\mathscr{D}_i^n$. This set is defined as $\mathscr{C}(\mathscr{N}_i) \cup \mathscr{C}(\mathscr{C}(\mathscr{N}_i)) \cup \ldots$.
- For some node $\mathscr{N}_i$, the set of descendant points of $\mathscr{N}_i$ will be denoted $\mathscr{D}_i^p$. This set is defined as $\mathscr{P}_i \cup \mathscr{P}(\mathscr{D}_i^p)$.

The utility of trees stems from the ability to quickly place bounds on various distance-related quantities for a single node. Consider two space tree nodes $\mathscr{N}_i$ and $\mathscr{N}_j$, and suppose our task is to find the minimum distance between any two descendant points in the nodes:

$$d_{\min}(\mathscr{N}_i, \mathscr{N}_j) = \min_{p_i \in \mathscr{D}_i^p, p_j \in \mathscr{D}_j^p} d(p_i, p_j). \tag{1}$$

Now, suppose $\mathscr{S}_i$ is a ball centered at some point $\mu_i \in \mathcal{R}^d$ with radius λ_i, and $\mathscr{S}_j$ is a ball centered at some point $\mu_j \in \mathcal{R}^d$ with radius λ_j. Then, we may easily place a lower bound: $d_{\min}(\mathscr{N}_i, \mathscr{N}_j) \geq \max(d(\mu_i, \mu_j) - \lambda_i - \lambda_j, 0)$. Note that the bound truncates to 0 when the balls are overlapping. This bound may be calculated with just one distance calculation, instead of $|\mathscr{D}_i^p||\mathscr{D}_j^p|$ distance calculations. During the traversal, bounds like the one on $d_{\min}(\mathscr{N}_i, \mathscr{N}_j)$ are often used to prune away large amounts of work.

3   Traversals

Next, we formally introduce the notion of a dual-tree traversal, again from [12].

Definition 2. *A pruning dual-tree traversal is a process that, given two space trees $\mathscr{T}_q$ (the query tree, built on the query set S_q) and $\mathscr{T}_r$ (the reference tree, built on the reference set S_r), will visit combinations of nodes $(\mathscr{N}_q, \mathscr{N}_r)$ such that $\mathscr{N}_q \in \mathscr{T}_q$ and $\mathscr{N}_r \in \mathscr{T}_r$ no more than once, and call a function $\texttt{Score}(\mathscr{N}_q, \mathscr{N}_r)$ to assign a score to that node. If the score is ∞ (or above some bound), the combination is pruned and no combinations $(\mathscr{N}_{qc}, \mathscr{N}_{rc})$ such that $\mathscr{N}_{qc} \in \mathscr{D}_q^n$ and $\mathscr{N}_{rc} \in \mathscr{D}_r^n$ are visited. Otherwise, for every combination of points (p_q, p_r) such that $p_q \in \mathscr{P}_q$ and $p_r \in \mathscr{P}_r$, a function $\texttt{BaseCase}(p_q, p_r)$ is called. If no node combinations are pruned during the entire traversal, $\texttt{BaseCase}(p_q, p_r)$ is called at least once on each combination of $p_q \in S_q$ and $p_r \in S_r$.*

Although the definition is quite complex, real-world dual-tree traversals tend to be straightforward. The standard depth-first dual-tree traversal is shown in Algorithm 1; this is the same traversal used in most dual-tree algorithms that use the kd-tree [11] [16] [18][2] and is often used in practice [25]. Generally, a depth-first traversal is preferred because many space trees in practice only hold points in the leaves; breadth-first traversals do not perform well in these situations.

The traversal is originally called with the root of the query and reference trees: $(\mathscr{T}_q, \mathscr{T}_r)$. First, $\texttt{BaseCase()}$ is called with every pair of query and reference points (lines 4–6). Then, for recursion, we collect a list of combinations to recurse into, sorted by their score. Any combinations with score ∞ are not recursed into. If both nodes have children, then we recurse into combinations of query children and reference children. If only the reference node has children, we recurse into combinations of the query node and the reference children. If only the query node has children, we recurse into combinations of the query children and the reference node. If neither node has children, there is no need to recurse.

The algorithm first recurses into those node combinations with lowest score. Depending on the task being solved (that is, which $\texttt{Score()}$ and $\texttt{BaseCase()}$ functions are being used), this prioritized approach to recursion can provide significant speedup over unprioritized recursion. For nearest neighbor search, a prioritized recursion gives significant speedup; Section 6 will demonstrate this.

---

[2] The algorithms in each of the referenced papers tend to look very different because they are not derived in a tree-independent form, but using the kd-tree with the traversal in Algorithm 1 and simplifying will yield the same algorithm.

Algorithm 1 DualDepthFirstTraversal($\mathcal{N}_q$, $\mathcal{N}_r$).

1: **Input:** query node $\mathcal{N}_q$, reference node $\mathcal{N}_r$
2: **Output:** none

3: {Perform base cases for points in node combination.}
4: **for all** $p_q \in \mathscr{P}_q$ **do**
5:     **for all** $p_r \in \mathscr{P}_r$ **do**
6:         BaseCase(p_q, p_r)

7: {Assemble list of combinations to recurse into.}
8: $q \leftarrow$ empty priority queue
9: **if** $\mathcal{N}_q$ and $\mathcal{N}_r$ both have children **then**
10:     **for all** $\mathcal{N}_{qc} \in \mathscr{C}_q$ **do**
11:         **for all** $\mathcal{N}_{rc} \in \mathscr{C}_r$ **do**
12:             $s_i \leftarrow$ Score($\mathcal{N}_{qc}, \mathcal{N}_{rc}$)
13:             **if** $s_i \neq \infty$ **then** push $(\mathcal{N}_{qc}, \mathcal{N}_{rc})$ into q with priority $1/s_i$
14: **else if** $\mathcal{N}_q$ has children but $\mathcal{N}_r$ does not **then**
15:     **for all** $\mathcal{N}_{qc} \in \mathscr{C}_q$ **do**
16:         $s_i \leftarrow$ Score($\mathcal{N}_{qc}, \mathcal{N}_r$)
17:         **if** $s_i \neq \infty$ **then** push $(\mathcal{N}_{qc}, \mathcal{N}_r)$ into q with priority $1/s_i$
18: **else if** $\mathcal{N}_q$ does not have children but $\mathcal{N}_r$ does **then**
19:     **for all** $\mathcal{N}_{rc} \in \mathscr{C}_r$ **do**
20:         $s_i \leftarrow$ Score($\mathcal{N}_q, \mathcal{N}_{rc}$)
21:         **if** $s_i \neq \infty$ **then** push $(\mathcal{N}_q, \mathcal{N}_{rc})$ into q with priority $1/s_i$

22: {Recurse into combinations with highest priority first.}
23: **for all** $(\mathcal{N}_{qi}, \mathcal{N}_{ri}) \in q$, highest priority first **do**
24:     DualDepthFirstTraversal($\mathcal{N}_{qi}$, $\mathcal{N}_{ri}$)

---

4   Nearest Neighbor Search

With the notions of space tree and dual-tree traversal established, we may now introduce the problem-specific BaseCase() and Score() functions used to perform dual-tree nearest neighbor search. These are the same rules introduced by Curtin et. al. [12] and used in **mlpack** [25]. The rules depend on auxiliary arrays N and D; during the traversal, $N[p_q]$ holds the current nearest neighbor candidate for query point p_q, and $D[p_q]$ holds $d(p_q, N[p_q])$. At the beginning of the traversal, each element in D is initialized to ∞. At the end of the traversal, $N[p_q]$ will hold the nearest neighbor of p_q, and $D[p_q]$ will hold the distance between p_q and its nearest neighbor.

The BaseCase() function (Algorithm 2) receives a query point p_q and a reference point p_r as input. The distance between the points is calculated, and if this is better than the current best candidate distance for p_q, $d(p_q, p_r)$ is taken as the new best candidate distance and p_r as the new nearest neighbor candidate.

Algorithm 2. BaseCase(p_q, p_r) for nearest neighbor search.

1: **Input:** query point p_q, reference point p_r, candidate point $N[p_q]$, candidate distance $D[p_q]$

2: **Output:** none

3: **if** $d(p_q, p_r) < D[p_q]$ **then**
4:     $N[p_q] \leftarrow p_r$
5:     $D[p_q] \leftarrow d(p_q, p_r)$

Algorithm 3. Score($\mathcal{N}_q$, $\mathcal{N}_r$) for nearest neighbor search.

1: **Input:** query node $\mathcal{N}_q$, reference node $\mathcal{N}_r$
2: **Output:** a score for the node combination $(\mathcal{N}_q, \mathcal{N}_r)$, or ∞ if it should be pruned

3: **if** $d_{\min}(\mathcal{N}_q, \mathcal{N}_r) > B_{df}(\mathcal{N}_q)$ **then**
4:     **return** ∞
5: **return** $d_{\min}(\mathcal{N}_q, \mathcal{N}_r)$

The Score() function is significantly more complex due to the bound function $B_{df}(\mathcal{N}_q)$[3]. Given a query node $\mathcal{N}_q$ and a reference node $\mathcal{N}_r$, we can prune if we can determine that no descendant point of $\mathcal{N}_r$ can possibly be the nearest neighbor of any descendant point of $\mathcal{N}_q$. If we had perfect knowledge, this condition is easily expressed; we would prune if

$$d_{\min}(\mathcal{N}_q, \mathcal{N}_r) > \max_{p_q \in \mathscr{D}_q^p} D[p_q]. \tag{2}$$

But of course, because this requires looping over every descendant point in $\mathcal{N}_q$, we cannot calculate this every time Score() is called. Instead, we can use caching. Define the depth-first traversal bound function, $B_{df}(\mathcal{N}_q)$, recursively:

$$B_{df}(\mathcal{N}_q) = \max \left\{ \max_{p_q \in \mathscr{P}_q} D[p_q], \max_{\mathcal{N}_{qc} \in \mathscr{C}_q} B_{df}(\mathcal{N}_{qc}) \right\}. \tag{3}$$

When we visit a node combination $(\mathcal{N}_q, \mathcal{N}_r)$, we may cache the result of the calculation $B_{df}(\mathcal{N}_q)$, for use by subsequent calls to Score(). Then, a call to Score() takes just one distance calculation $(d_{\min}(\mathcal{N}_q, \mathcal{N}_r))$ and $|\mathscr{P}_q| + |\mathscr{C}_q|$ accesses. Proving the correctness of this algorithm is straightforward.

We may use this to construct a generalized dual-tree algorithm for nearest neighbor search. Any type of space tree can be paired with any type of pruning dual-tree traversal that uses the BaseCase() and Score() above, and correct nearest-neighbor search results will be obtained. With this algorithm established, we will now turn towards improving the depth-first dual-tree recursion strategy.

---

[3] Our formulation here is specialized for depth-first traversals, unlike some more general formulations [12]. We are only considering depth-first traversals in this work, though, so there is no need to introduce a more complicated bound function.

5   Delaying Reference Recursion

Algorithm 1 is the standard depth-first dual-tree traversal that is used in practice, and it prioritizes recursions: node combinations with lower scores (from Score()) are recursed into first. So, for instance, consider nearest neighbor search, where the result of Score(), if the node combination is not pruned, is $d_{\min}(\mathscr{N}_q, \mathscr{N}_r)$. In the situation depicted in Figure 1(a), combination $(\mathscr{N}_q, \mathscr{N}_{r1})$ should be visited before combination $(\mathscr{N}_q, \mathscr{N}_{r2})$. It is clear that this is the right choice, because a depth-first traversal of $(\mathscr{N}_q, \mathscr{N}_{r1})$ is more likely to tighten the bound $B_{df}(\mathscr{N}_q)$ such that $(\mathscr{N}_q, \mathscr{N}_{r2})$ can be pruned when it is recursed into.

But, consider a more tricky case, depicted in Figure 1(b). Here, $d_{\min}(\mathscr{N}_q, \mathscr{N}_{r1}) = d_{\min}(\mathscr{N}_q, \mathscr{N}_{r2}) = 0$, so we are unable to tell whether it is better to recurse into $(\mathscr{N}_q, \mathscr{N}_{r1})$ first or into $(\mathscr{N}_q, \mathscr{N}_{r2})$ first. Indeed Algorithm 1 will select arbitrarily. This situation may occur in Algorithm 1 from lines 11 to 13 if, for a given child query node $\mathscr{N}_{qc}$, two or more reference children $\mathscr{N}_{rc}$ have the same score s_i.

We can do better than arbitrary selection. Consider some child $\mathscr{N}_{qc}$ of $\mathscr{N}_q$. Figure 1(c) shows an example $\mathscr{N}_{qc}$. In this example, the choice is now clear: the combination $(\mathscr{N}_{qc}, \mathscr{N}_{r1})$ should be recursed into before $(\mathscr{N}_{qc}, \mathscr{N}_{r2})$. Thus, the correct answer to the question "should we recurse into $(\mathscr{N}_q, \mathscr{N}_{r1})$ or $(\mathscr{N}_q, \mathscr{N}_{r2})$ first?" is to sidestep the question entirely: we should not recurse in the reference node, but instead in the query node. Then, at the level of the query child, the decision may be clearer.

In essence, the strategy is to delay recursion in the reference nodes until it is clear which reference node should be recursed into first. This improvement, once generalized, is encapsulated in Algorithm 4. Lines 15–20 check if reference recursion should be delayed because the scores of all reference children are identical. If so, the recursion will proceed by recursing only in the queries. If necessary, this reference recursion delay will continue until no longer possible. This delay is not possible when the query node does not have any children. This improved strategy can make a huge difference in the performance of the algorithm; recursing into a suboptimal reference child first can cause the bound $B_{df}(\cdot)$ to be unnecessarily loose, whereas first recursing into the best reference child will tighten $B_{df}(\cdot)$ more quickly and possibly allow other reference children to be pruned entirely.

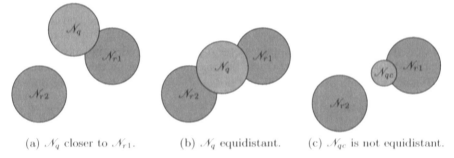

(a) $\mathscr{N}_q$ closer to $\mathscr{N}_{r1}$.　　(b) $\mathscr{N}_q$ equidistant.　　(c) $\mathscr{N}_{qc}$ is not equidistant.

Fig. 1. Different situations for recursion.

Algorithm 4. ImprovedDualDepthFirstTraversal($\mathcal{N}_q$, $\mathcal{N}_r$).

1: **Input:** query node $\mathcal{N}_q$, reference node $\mathcal{N}_r$
2: **Output:** none

3: {Perform base cases for points in node combination.}
4: **for all** $p_q \in \mathcal{P}_q$ **do**
5:    **for all** $p_r \in \mathcal{P}_r$ **do**
6:       BaseCase(p_q, p_r)

7: {Assemble list of combinations to recurse into.}
8: $q \leftarrow$ empty priority queue
9: **if** $\mathcal{N}_q$ and $\mathcal{N}_r$ both have children **then**
10:    **for all** $\mathcal{N}_{qc} \in \mathcal{C}_q$ **do**
11:       $q_{qc} \leftarrow \{\}$
12:       **for all** $\mathcal{N}_{rc} \in \mathcal{C}_r$ **do**
13:          $s_i \leftarrow$ Score($\mathcal{N}_{qc}, \mathcal{N}_{rc}$)
14:          **if** $s_i \neq \infty$ **then** push $(\mathcal{N}_{qc}, \mathcal{N}_{rc}, s_i)$ into q_{qc}
15:       **if** all elements of q_{qc} have identical score **then**
16:          $s_i \leftarrow$ Score($\mathcal{N}_{qc}, \mathcal{N}_r$)
17:          push $(\mathcal{N}_{qc}, \mathcal{N}_r)$ into q with priority $1/s_i$
18:       **else**
19:          **for all** $(\mathcal{N}_{qi}, \mathcal{N}_{ri}, s_i) \in q_{qc}$ **do**
20:             push $(\mathcal{N}_{qi}, \mathcal{N}_{ri})$ into q with priority $1/s_i$
21: **else if** $\mathcal{N}_q$ has children but $\mathcal{N}_r$ does not **then**
22:    **for all** $\mathcal{N}_{qc} \in \mathcal{C}_q$ **do**
23:       $s_i \leftarrow$ Score($\mathcal{N}_{qc}, \mathcal{N}_r$)
24:       **if** $s_i \neq \infty$ **then** push $(\mathcal{N}_{qc}, \mathcal{N}_r)$ into q with priority $1/s_i$
25: **else if** $\mathcal{N}_q$ does not have children but $\mathcal{N}_r$ does **then**
26:    **for all** $\mathcal{N}_{rc} \in \mathcal{C}_r$ **do**
27:       $s_i \leftarrow$ Score($\mathcal{N}_q, \mathcal{N}_{rc}$)
28:       **if** $s_i \neq \infty$ **then** push $(\mathcal{N}_q, \mathcal{N}_{rc})$ into q with priority $1/s_i$

29: {Recurse into combinations with highest priority first.}
30: **for all** $(\mathcal{N}_{qi}, \mathcal{N}_{ri}) \in q$, highest priority first **do**
31:    ImprovedDualDepthFirstTraversal($\mathcal{N}_{qi}$, $\mathcal{N}_{ri}$)

For trees such as the kd-tree where each node has two children only, the extra implementation overhead for this strategy is trivial and simplifies to the addition of a single **if** statement. However, note that there are some situations where the modified traversal will not outperform the original prioritized traversal. For instance, for nearest neighbor search, if the query tree is identical to the reference tree and nodes in the tree cannot overlap, then it is very unlikely that the situation described in Figure 1(a) will be encountered: during the recursion, the query node will only overlap itself and possibly be adjacent to a sibling node.

6   Experiments

To test the efficiency of this strategy, we will observe the performance of our recursion strategy on the tasks of exact and approximate nearest neighbor search, with multiple types of trees, and with many different datasets. For approximate search, we compare with LSH (locality-sensitive hashing). The datasets utilized in these experiments are described in Table 1. Each dataset is from the UCI dataset repository [26], with the exception of the birch3 dataset [27], LCDM dataset [28], and SDSS-DR6 dataset [29].

The first test focuses on exact nearest neighbor search: Algorithms 2 and 3 paired with a type of tree and traversal. Using the flexible **mlpack** library [25], we test with the kd-tree and the ball tree, using three dual-tree traversal strategies: a depth-first unordered recursion (equivalent to Algorithm 1 where the recursion priority is ignored); the standard depth-first prioritized recursion (Algorithm 1); and our improved recursion (Algorithm 4). In addition, a single-tree algorithm is used; this is the canonical tree-based nearest neighbor search algorithm [5] with a prioritized recursion, run once for each query point. The dataset is randomly split into 60% reference set and 40% query set, and the algorithm is run ten times. The number of distance evaluations and the total runtime are collected. Table 2 shows the average number of distance calculations and the average runtime for each algorithm. Preprocessing time (tree building) is not included, but generally was a minor fraction of search time.

We can see from the results that our improvement is, in many cases, significant. In the best case, it gives more than 2x speedup over the next fastest strategy. This effect is especially pronounced on larger datasets, which will have deeper trees: a bad recursion decision early on can significantly affect the ability to prune during the algorithm. Ball trees exhibit less pronounced effects. This is because the bounding structure is a ball of fixed radius, whereas the kd-tree is adaptive in all dimensions. Therefore, two child nodes of a ball tree node may overlap,

Table 1. Datasets.

Dataset	n	d
cloud	2048	10
winequality	6497	11
birch3	100000	2
miniboone	130064	50
covertype	581012	55
power	2075259	7
lcdm	16777216	3
sdss-dr6	39761242	4

Table 2. Runtime (distance evaluations) for exact nearest neighbor search.

algorithm	cloud	winequality	birch3	miniboone
kd-tree, unordered	0.036s (270k)	0.288s (2.15M)	7.310s (62.2M)	62.481s (214M)
kd-tree, prioritized	0.005s (34.2k)	0.039s (222k)	0.419s (2.90M)	25.081s (78.8M)
kd-tree, improved	0.005s (**27.7k**)	0.021s (**104k**)	0.201s (**1.10M**)	12.643s (34.5M)
single kd-tree	0.005s (32.9k)	**0.017s** (112k)	0.262s (1.65M)	**6.637s (19.2M)**
ball tree, unordered	0.011s (356k)	0.104s (3.08M)	1.817s (71.6M)	32.947s (616M)
ball tree, prioritized	0.003s (104k)	0.023s (666k)	0.285s (10.9M)	27.934s (514M)
ball tree, improved	0.003s (86.8k)	0.017s (455k)	**0.160s** (5.65M)	**22.332s** (351M)
single ball tree	**0.002s** (69.6k)	**0.012s** (315k)	0.165s (**5.38M**)	26.357s (**254M**)

algorithm	covertype	power	lcdm	sdss-dr6
kd-tree, unordered	302.8s (1.09B)	1163.0s (18.7B)	5628.7s (41.5B)	24717s (156B)
kd-tree, prioritized	15.823s (52.5M)	30.072s (302M)	319.871s (1.87B)	9069s (50.3B)
kd-tree, improved	**4.469s (12.8M)**	**12.714s (200M)**	**71.587s (350M)**	**428.9s (2.14B)**
single kd-tree	6.207s (16.3M)	19.684s (232M)	120.6s (476M)	471.4s (2.24B)
ball tree, unordered	163.027s (2.90B)	771.975s (25.3B)	1861.9s (71.1B)	9444s (363B)
ball tree, prioritized	52.487s (902M)	113.437s (3.90B)	386.74s (14.4B)	5202s (192B)
ball tree, improved	**27.251s** (392M)	**83.744s** (2.58B)	**195.175s** (6.46B)	**5150s** (136B)
single ball tree	29.948s (**228M**)	138.422s (**2.49B**)	402.6s (**5.93B**)	7226s (**101B**)

Table 3. Runtime (distance calculations) [ε or M/W] for approximate NN search.

algorithm	cloud	winequality	birch3	miniboone
kd-tree, unordered	0.005s (34.5k) [1.5]	0.025s (148k) [1.44]	0.267s (2.14M) [1.44]	6.831s (22.6M) [1.38]
kd-tree, prioritized	0.003s (17.4k) [1.5]	0.012s (74.5k) [1.5]	0.140s (1.16M) [1.5]	4.863s (15.5M) [1.38]
kd-tree, improved	**0.002s (13.7k)** [1.7]	**0.010s (51.2k)** [1.63]	**0.107s (654k)** [1.63]	3.360s (9.28M) [1.38]
single kd-tree	0.003s (23.2k) [2.45]	0.013s (78.0k) [2.33]	0.198s (1.47M) [2.33]	**1.845s (5.75M)** [1.5]
ball tree, unordered	**0.002s** (50.8k) [27.6]	0.007s (186k) [32.3]	0.079s (2.72M) [11.5]	**2.942s** (50.4M) [285]
ball tree, prioritized	**0.002s** (49.2k) [27.6]	**0.006s** (167k) [32.3]	**0.072s** (2.46M) [11.5]	3.266s (54.2M) [249]
ball tree, improved	**0.002s** (45.1k) [27.6]	**0.006s (161k)** [32.3]	**0.072s (2.25M)** [11.5]	3.494s (50.3M) [99]
single ball tree	**0.002s** (43.2k) [999]	**0.006s** (176k) [36.0]	0.111s (3.56M) [10.1]	3.812s (**36.1M**) [99]
multiprobe LSH	0.031s (19.3k) [20/122]	0.011s (472k) [37/33]	1.614s (8.85M) [8/16k]	175.995s (1.77M) [13/328]

algorithm	covertype	power	lcdm	sdss-dr6
kd-tree, unordered	7.796s (27.4M) [1.5]	419.725s (13.0B) [1.27]	75.432s (508M) [1.33]	512.829s (2.89B) [1.27]
kd-tree, prioritized	2.954s (10.6M) [1.5]	**8.392s (189M)** [1.44]	44.187s (306M) [1.38]	380.047s (2.17B) [1.27]
kd-tree, improved	**2.045s (6.25M)** [1.5]	11.044s (191M) [1.56]	**29.069s (160M)** [1.44]	242.624s (1.11B) [1.27]
single kd-tree	3.869s (11.2M) [1.86]	16.674s (226M) [2.33]	85.821s (397M) [1.78]	329.663s (1.58B) [1.27]
ball tree, unordered	2.187s (33.0M) [99]	415.964s (13.0B) [11.5]	**19.776s** (668M) [19]	**73.638s** (239M) [49]
ball tree, prioritized	**2.183s (32.3M)** [75.9]	**6.753s (233M)** [13.3]	20.158s (**660M**) [19]	75.687s (**237M**) [49]
ball tree, improved	2.539s (33.8M) [49]	8.269s (248M) [15.7]	25.749s (702M) [21.2]	299.8s (451M) [49]
single ball tree	5.496s (40.3M) [27.6]	19.097s (431M) [15.7]	113.299s (1.46B) [21.2]	2054.8s (3.06B) [19]
multiprobe LSH	130.699s (963M) [0.51]	1181.32s (14.0B) [63/9.6]	*timeout* [14/0.968]	*timeout* [7/0.29]

causing the improved strategy of delaying reference recursions to not pay off at lower levels. Nonetheless, especially for large datasets, where the dual-tree strategy is faster than the single-tree strategy, the improved traversal is a clear best choice.

The second task is approximate nearest neighbor search, and in this situation we will also be able to compare with locality-sensitive hashing. Relative-value approximation means that for an approximation parameter ϵ, we are guaranteed for a query point p_q with true nearest neighbor p_r^*, the algorithm will return an approximate nearest neighbor $\hat{p}_r$ such that $d(p_q, \hat{p}_r) \leq (1 + \epsilon)d(p_q, p_r^*)$. It is easy to modify the given Score() function to enforce this condition; replace the equation in line 3 with $d_{\min}(\mathcal{N}_q, \mathcal{N}_r) > (1/(1 + \epsilon))B(\mathcal{N}_q)$.

After applying this change, testing is performed in the same way as for exact nearest neighbor search. ϵ for each tree-based approach is selected to give an average per-point relative error of 0.1 (± 0.01) for each dataset. Because our scheme does not allow the error for an individual point to exceed ϵ, the actual relative error for an individual query point is often much lower. Thus, it is often necessary to set ϵ far higher than the target average error of 0.1. For LSH, the LSHKIT package is used, which implements multi-probe LSH and autotunes the hashing parameters [30]. We use the suggested number of hash tables ($L = 10$) and probes ($T = 20$), and then autotune to select the number of hash functions (M) and bin width (W). Autotuning failed for the larger power, lcdm, and sdss-dr6 datasets; in these cases suggestions of the LSHKIT authors are used [31].

The results are given in Table 3. With approximation, the improved dual-tree traversal performs fewer distance calculations on smaller datasets, and is still dominant for the larger datasets with kd-trees. But with ball trees, the bounds are looser and thus nodes are more likely to be overlapping. Because only an approximate nearest neighbor is required, finding the absolute best reference child to recurse into is of less importance, and the added overhead of delaying query recursions may not necessarily be helpful. Thus, the benefit of the improved traversal may be related to the type of tree being used and the problem being solved. LSH is not competitive on the larger datasets, and on the largest datasets LSH did not complete within 3 days, but it should be noted that the low-dimensional setting is where trees are most effective.

Overall, for large datasets in low-to-medium dimensions, dual-tree search is faster, and the improved traversal we have proposed is the fastest. These experiments, as well as further investigations (not shown here due to space constraints) seem to show for smaller datasets, single-tree search may be fastest; for sufficiently high dimensions, LSH is faster. This corroborates existing results [32]; as the dimension of data gets higher, pruning rules become less effective. Regardless, in low-to-medium dimensions, the improved dual-tree traversal is dominant.

7   Conclusion

Using the recent abstraction of tree-independent dual-tree algorithms, we have proposed a novel depth-first dual-tree traversal which compares favorably against other techniques for exact and approximate nearest neighbor search. Additionally, because of the generic nature of the traversal, it may be applied to many problems: the traversal simply needs to be paired with a type of space tree and `Score()` and `BaseCase()` functions. Examples of problems with existing `Score()` and `BaseCase()` functions include kernel density estimation [12] and max-kernel search [23]. These problems, as well as nearest neighbor search, all stand to benefit from the improved traversal strategy we have proposed.

Acknowledgements. Thanks to Rich W. Vuduc and Chad D. Kersey for helpful discussions and comments during the preparation of this work. This material is based on work supported by the U.S. National Science Foundation (NSF) Award Number 1339745. Any opinions, findings and conclusions or recommendations expressed in this material are those of the author and do not necessarily reflect those of NSF.

References

1. Kumar, N., Zhang, L., Nayar, S.K.: What is a good nearest neighbors algorithm for finding similar patches in images? In: Forsyth, D., Torr, P., Zisserman, A. (eds.) ECCV 2008, Part II. LNCS, vol. 5303, pp. 364–378. Springer, Heidelberg (2008)
2. Koren, Y.: The BellKor solution to the Netflix Grand Prize (2009)
3. Cunningham, P., Delany, S.J.: k-nearest neighbour classifiers. Technical Report UCD-CSI-2007-4, University College Dublin (2007)
4. Weinberger, K.Q., Blitzer, J., Saul, L.K.: Distance metric learning for large margin nearest neighbor classification. In: Advances in Neural Information Processing Systems 18 (NIPS 2005), pp. 1473–1480 (2005)
5. Bentley, J.L.: Multidimensional binary search trees used for associative searching. Communications of the ACM **18**(9), 509–517 (1975)
6. Fukunaga, K., Narendra, P.M.: A branch and bound algorithm for computing k-nearest neighbors. IEEE Transactions on Computers **100**(7), 750–753 (1975)

7. Clarkson, K.L.: Nearest neighbor queries in metric spaces. Discrete & Computational Geometry **22**(1), 63–93 (1999)
8. Andoni, A., Indyk, P.: Near-optimal hashing algorithms for approximate nearest neighbor in high dimensions. In: Forty-Seventh Annual IEEE Symposium of Foundations of Computer Science (FOCS 2006), pp. 459–468. IEEE (2006)
9. Indyk, P., Motwani, R.: Approximate nearest neighbors: towards removing the curse of dimensionality. In: Proceedings of the Thirtieth Annual ACM Symposium on Theory of Computing (STOC 1998), pp. 604–613. ACM (1998)
10. Datar, M., Immorlica, N., Indyk, P., Mirrokni, V.S.: Locality-sensitive hashing scheme based on p-stable distributions. In: Proceedings of the Twentieth Annual Symposium on Computational Geometry (SoCG 2004), pp. 253–262. ACM (2004)
11. Gray, A.G., Moore, A.W.: 'N-body' problems in statistical learning. In: Advances in Neural Information Processing Systems, vol. 14, no. 4, pp. 521–527 (2001)
12. Curtin, R.R., March, W.B., Ram, P., Anderson, D.V., Gray, A.G., Isbell Jr., C.L.: Tree-independent dual-tree algorithms. In: Proceedings of the 30th International Conference on Machine Learning (ICML 2013) (2013)
13. Ram, P., Lee, D., March, W.B., Gray, A.G.: Linear-time algorithms for pairwise statistical problems. In: Advances in Neural Information Processing Systems, vol. 22 (2009)
14. Curtin, R.R., Lee, D., March, W.B., Ram, P.: Plug-and-play runtime analysis for dual-tree algorithms. The Journal of Machine Learning Research (2015)
15. Gray, A.G., Moore, A.W.: Nonparametric density estimation: toward computational tractability. In: Proceedings of the 3rd SIAM International Conference on Data Mining (SDM 2003), San Francisco, pp. 203–211 (2003)
16. March, W.B., Ram, P., Gray, A.G.: Fast Euclidean minimum spanning tree: algorithm, analysis, and applications. In: Proceedings of the 16th ACM SIGKDD International Conference on Knowledge Discovery and Data Mining (KDD 2010), Washington, D.C., pp. 603–612 (2010)
17. Wang, P., Lee, D., Gray, A.G., Rehg, J.M.: Fast mean shift with accurate and stable convergence. In: Proceedings of the Eleventh International Conference on Artificial Intelligence and Statistics (AISTATS 2007), pp. 604–611 (2007)
18. Lee, D., Gray, A.G.: Faster gaussian summation: theory and experiment. In: Proceedings of the 22nd Conference on Uncertainty in Artificial Intelligence (2006)
19. Curtin, R.R., Ram, P., Gray, A.G.: Fast exact max-kernel search. In: SIAM International Conference on Data Mining (SDM 2013), pp. 1–9 (2013)
20. Klaas, M., Briers, M., De Freitas, N., Doucet, A., Maskell, S., Lang, D.: Fast particle smoothing: if I had a million particles. In: Proceedings of the 23rd International Conference on Machine Learning (ICML 2006), pp. 25–29 (2006)
21. Van Der Maaten, L.: Accelerating t-sne using tree-based algorithms. The Journal of Machine Learning Research **15**(1), 3221–3245 (2014)
22. Moore, D.A., Russell, S.J.: Fast Gaussian process posteriors with product trees. In: Proceedings of the Thirtieth Conference on Uncertainty in Artificial Intelligence (UAI 2014), Quebec City, July 2014
23. Curtin, R.R., Ram, P.: Dual-tree fast exact max-kernel search. Statistical Analysis and Data Mining **7**(4), 229–253 (2014)
24. Liu, T., Moore, A.W., Yang, K., Gray, A.G.: An investigation of practical approximate nearest neighbor algorithms. In: Advances in Neural Information Processing Systems 17 (NIPS 2004), pp. 825–832 (2004)
25. Curtin, R.R., Cline, J.R., Slagle, N.P., March, W.B., Ram, P., Mehta, N.A., Gray, A.G.: mlpack: A scalable C++ machine learning library. Journal of Machine Learning Research **14**, 801–805 (2013)

26. Lichman, M.: UCI machine learning repository (2013)
27. Zhang, T., Ramakrishnan, R., Livny, M.: BIRCH: A new data clustering algorithm and its applications. Data Mining and Knowledge Discovery **1**(2), 141–182 (1997)
28. Lupton, R., Gunn, J.E., Ivezic, Z., Knapp, G.R., Kent, S.: The SDSS imaging pipelines. In: Astronomical Data Analysis Software and Systems X, vol. 238, p. 269 (2001)
29. Adelman-McCarthy, J.K., Agüeros, M.A., Allam, S.S., Prieto, C.A., Anderson, K.S.J., et al.: The sixth data release of the Sloan Digital Sky Survey. The Astrophysical Journal Supplement Series **175**(2), 297 (2008)
30. Dong, W., Wang, Z., Josephson, W.K., Charikar, M., Li, K.: Modeling LSH for performance tuning. In: Proceedings of the 17th ACM Conference on Information and Knowledge Management (CIKM 2008), pp. 669–678. ACM (2008)
31. Dong, W.: Personal communication (2015)
32. Moore, A.W.: The anchors hierarchy: using the triangle inequality to survive high dimensional data. In: Proceedings of the Sixteenth Conference on Uncertainty in Artificial Intelligence (UAI 2000), pp. 397–405 (2000)

Optimizing the Distance Computation Order of Multi-Feature Similarity Search Indexing

Marcel Zierenberg[(✉)] and Ingo Schmitt

Institute of Computer Science, Information and Media Technology, Chair of Database and Information Systems, Brandenburg University of Technology Cottbus - Senftenberg, P.O. Box 10 13 44, 03013 Cottbus, Germany
{zieremar,schmitt}@b-tu.de

Abstract. Multi-feature search is an effective approach to similarity search. Unfortunately, the search efficiency decreases with the number of features. Several indexing approaches aim to achieve efficiency by incrementally reducing the approximation error of aggregated distance bounds. They apply heuristics to determine the distance computations order and update the object's aggregated bounds after each computation. However, the existing indexing approaches suffer from several drawbacks. They use the same computation order for all objects, do not support important types of aggregation functions and do not take the varying CPU and I/O costs of different distance computations into account. To resolve these problems, we introduce a new heuristic to determine an efficient distance computation order for each individual object. Our heuristic supports various important aggregation functions and calculates cost-benefit-ratios to incorporate the varying computation costs of different distance functions. The experimental evaluation reveals that our heuristic outperforms state-of-the-art approaches in terms of the number of distance computations as well as search time.

1 Introduction

Similarity search means to find objects that are similar to the given query object. Object (dis-)similarity is computed by a *distance function* δ based on feature data extracted from a query object q and each database object o^i. The effectiveness of the search is frequently improved by combining *partial distances* $d^i_j = \delta_j(q, o^i)$ of m different features into a *aggregated distance* d^i_{agg} by *aggregation function* agg $: \mathbb{R}^m_{\geq 0} \mapsto \mathbb{R}_{\geq 0}$. However, the benefits of *multi-feature search* come at the expense of efficiency, since the number of required distance computations increases with each added feature.

Indexing approaches to single-feature similarity search aim to reduce the number of required distance computations by estimating lower and/or upper bounds of a *exact distance* d^i for each object o^i with low computational costs [1]. Objects with a lower distance bound lb^i greater than current pruning threshold t_{max} cannot belong to the query result. They are excluded from the search without computing their exact distance. Indexing approaches for multi-feature

G. Amato et al. (Eds.): SISAP 2015, LNCS 9371, pp. 90–96, 2015.
DOI: 10.1007/978-3-319-25087-8_8

similarity search compute *partial bounds* lb_j^i and ub_j^i for each partial distance d_j^i separately [2–6]. They dynamically combine these partial bounds into *aggregated bounds* lb_{agg}^i and ub_{agg}^i of the exact aggregated distance d_{agg}^i.

Unfortunately, the approximation error ϵ_{agg}^i increases with the number of features. Thereby, the efficiency of indexing approaches decreases significantly. Several techniques aim to solve this problem by incrementally reducing the approximation error ϵ_{agg}^i of objects [4–6]. They compute each partial distance d_j^i of object o^i separately and recalculate the aggregated distance bounds lb_{agg}^i and ub_{agg}^i after each computation. This allows them to reduce the approximation error ϵ_{agg}^i in a step-by-step manner and thus, to exclude many objects, before all of their exact partial distances have been computed.

1.1   Contribution

Because the indexability of features can vary greatly, the existing indexing approaches [4–6] use different heuristics to determine the *computation order* of partial distances. However, we argue that these heuristics have several drawbacks in terms of their applicability and efficiency (see Section 2).

The main contribution of this paper is the introduction of a new heuristic to determine the order of partial distance computations for multi-feature queries (see Section 3). Our proposed heuristic has several advantages over state-of-the-art approaches. In contrast to *ASAP* [5], our heuristic is not restricted to vector spaces and individually adapts the computation order to each object. Additionally, it efficiently supports a greater variety of important aggregation functions than *Quick-Combine* [4] and *Partial Refinement* [6]. Furthermore, our heuristic is the only approach that takes the computation costs of distance functions (CPU and I/O costs) into account. This enables us to further decrease the search time in cases where these costs vary greatly between distance functions.

We extended our recently published *Partial Refinement* approach [6] with the heuristic and experimentally compared it against state-of-the-art approaches. Even though our evaluation (see Section 4) is rather preliminary, it already suggests that the new heuristic frequently outperforms other approaches (linear scan, *Onion-tree* [7], *FlexiDex* [3], *Partial Refinement* [6]) in terms of distance computations and search time.

2   Related Work

The following section describes several indexing approaches for multi-feature search briefly and discusses their drawbacks. We focus on approaches for *k-Nearest-Neighbor* (*k*NN) queries [1] and their heuristics for determining the distance computation order.

ASAP [5] stores compact bit signatures of multi-feature objects in a B+-tree. The computation order of partial distances is based on the relative energy of each feature. *ASAP* is restricted to vector spaces and uses the same computation order for all objects, which is usually less efficient than individually adapting the order to each object.

Quick-Combine [4] merges the results of incremental subqueries for each separate feature into an aggregated result. The computation order of these subqueries relies on the partial derivatives of the aggregation function. It prioritizes partial distance computations with the most impact on result of the aggregation function in the last iterations. However, it is not applicable to several important types of aggregation functions that are non-differentiable (e.g., the minimum/maximum function used in logic-based search [3]).

Partial Refinement [6] is divided into a filtering and a refinement phase. It is based on precomputed distances to pivot objects. Candidate objects are managed with a priority queue ordered according to the aggregated lower bounds. In each iteration, the refinement phase computes one exact partial distance for a candidate object, updates its aggregated bounds and subsequently reinserts it into the candidate list.

To determine the distance computation order, *Partial Refinement* assumes that computing the partial distance with the highest individual approximation error ϵ_j^i reduces the approximation error ϵ_{agg}^i the most. The approach is applicable to various important types of aggregation functions (including minimum and maximum function) and performs reasonably well for many of them, especially for linear functions (e.g., arithmetic mean or sum). However, it is inefficient for non-linear aggregation functions (e.g., harmonic or quadratic mean). Even though the approximation error ϵ_j^i of a partial bound may be small, computing it's corresponding exact partial distance d_j^i can cause a large reduction of the approximation error ϵ_{agg}^i for non-linear aggregation functions.

Finally, none of the existing approaches takes the varying computation costs of different distance functions (CPU and I/O cost) into consideration when determining the computation order. However, in cases where these costs vary significantly (e.g., euclidean vs. quadratic form distance function), it is frequently possible to reduce the overall search time at the expense of additional (but faster) partial distance computations.

3 Optimizing the Distance Computation Order

This sections describes our main contribution, a new heuristic that optimizes the order of partial distance computations for multi-feature queries. After a brief overview of some preliminaries, we explain how the computation order of each object is determined based on the *expected approximation error*. Afterward, we show how to incorporate the computation costs of partial distance functions into the heuristic, in order to further decrease the search time.

3.1 Partial and Aggregated Bounds

For reasons of simplicity we only consider *metric* distance functions [1] and *globally monotone increasing* aggregation functions [3] in the following. However, our approach is also applicable to other types of distance and aggregation functions (e.g., locally monotone aggregation functions [3]).

Lower lb_j^i and upper bounds ub_j^i (*partial bounds*) for the partial distance $d_j^i = \delta_j(q, o^i)$ of a metric distance function δ_j can be derived from the triangle inequality and precomputed distances to a reference object p (pivot object) as follows [1]:

$$lb_j^i = |\delta_j(q, p) - \delta_j(p, o^i)| \le d_j^i \le \delta_j(q, p) + \delta_j(p, o^i) = ub_j^i . \tag{1}$$

The lower lb_{agg}^i and upper bounds ub_{agg}^i (*aggregated bounds*) of the aggregated distance $d_{agg}^i = agg(d_1^i, \ldots, d_m^i)$ are then given by [3]:

$$lb_{agg}^i = agg\left(lb_1^i, \ldots, lb_m^i\right) \le d_{agg}^i \le agg\left(ub_1^i, \ldots, ub_m^i\right) = ub_{agg}^i . \tag{2}$$

3.2 Expected Approximation Error

Our goal is to determine a computation order of partial distances that minimizes the approximation error $\epsilon_{agg}^i = ub_{agg}^i - lb_{agg}^i$ with each consecutive partial distance computation, in order to maximize the chances of excluding each object. Therefore, we have to estimate the potential impact of computing a partial distance d_j^i on the approximation error ϵ_{agg}^i for each feature.

As described in Section 2, computation orders based on the individual approximation error $\epsilon_j^i = ub_j^i - lb_j^i$ of partial distances tend to be inefficient for non-linear aggregation functions (*Partial Refinement* [6]). Instead, our new heuristic estimates the potential impact of each partial distance computation with the help of inexpensively computed *expected partial distances* $\widetilde{d}_j^i$ and the according *expected approximation error* $\widetilde{\epsilon}_j^i$ as follows:

$$\widetilde{d}_j^i = \frac{1}{2}\left(ub_j^i - lb_j^i\right) , \tag{3}$$

$$\widetilde{\epsilon}_j^i = agg\left(ub_1^i, \ldots, ub_{j-1}^i, \widetilde{d}_j^i, ub_{j+1}^i, \ldots, ub_m^i\right) - agg\left(lb_1^i, \ldots, lb_{j-1}^i, \widetilde{d}_j^i, lb_{j+1}^i, \ldots, lb_m^i\right) . \tag{4}$$

From the monotonicity of the aggregation function follows that $\widetilde{\epsilon}_j^i \le \epsilon_{agg}^i$. Consequently, we can derive an individual computation order of partial distances for object o^i that minimizes the expected approximation error $\widetilde{\epsilon}_j^i$ with each partial distance computation (i.e., sorted from lowest to highest expected approximation error $\widetilde{\epsilon}_j^i$).

In contrast to *Quick-Combine* [4], our heuristic relies only on the aggregation function itself, instead of its partial derivatives, and is therefore also applicable to non-differentiable aggregation functions (e.g., minimum function).

Example 1. Consider a multi-feature search with three features ($m = 3$) and the harmonic mean as the aggregation function. Let exemplary partial bounds between query object q and an object o^i be given by: $(lb_1^i, ub_1^i) = (2.0, 5.0)$, $(lb_2^i, ub_2^i) = (2.5, 4.5)$ and $(lb_3^i, ub_3^i) = (1.5, 4.0)$. The resulting aggregated bounds are $(lb^i_{agg}, ub^i_{agg}) = (1.91, 4.46)$ and the approximation error is $\epsilon^i_{agg} = 2.55$. Now, the expected partial distances are $\widetilde{d^i_1} = 3.5$, $\widetilde{d^i_2} = 3.5$, and $\widetilde{d^i_3} = 2.75$. The corresponding expected approximation errors are $\widetilde{\epsilon^i_1} = 1.74$, $\widetilde{\epsilon^i_2} = 2.01$ and $\widetilde{\epsilon^i_3} = 1.44$. Consequently, partial distance $d^i_3 = \delta_3(q, o^i)$ will be computed first, since it minimizes the expected approximation error for object o^i.

3.3 Computation Costs of Distance Functions

The computation costs (CPU and I/O costs) can vary significantly between different distance functions (e.g., euclidean vs. quadratic form distance function). Therefore, it can be beneficial to prioritize a partial distance computation that has a smaller impact on the the aggregated bounds, if the respective computation costs are lower than the costs of the other partial distance computations.

To incorporate the computation costs, we calculate a *cost-benefit-ratio* C^i_j between the expected change of the approximation error of object o^i for each feature j and the average computation costs T_j of the corresponding partial distance function δ_j:

$$C^i_j = \frac{1}{T_j} \left(\epsilon^i_{agg} - \widetilde{\epsilon^i_j} \right) . \tag{5}$$

The according computation order of object o^i is determined by sorting the cost-benefit-ratios C^i_j from highest to lowest.

Example 2. Given the same situation as in Example 1, we now assume the exemplary average computation costs of $T_1 = 1$, $T_2 = 2$ and $T_3 = 5$. The resulting cost-benefit-ratios are $C^i_1 = 0.81$, $C^i_2 = 0.27$ and $C^i_3 = 0.22$. Accordingly, partial distance d^i_1 will be computed first, because it allows the fastest improvement of the aggregated bounds.

4   Experimental Evaluation

This section describes our experimental setup and explains the results of our evaluation.

We extended our recently published *Partial Refinement* approach [6] with our new heuristic and utilized an image-retrieval system with the *Caltech-256 Object Category Dataset* [8] image collection (30,607 images) for the experimental evaluation.

The efficiency was measured by determining the average number of required distance computations and the average search time (wall-clock time) of 100 randomly selected 10-Nearest-Neighbor queries. The queries were based on several aggregation functions (quadratic mean, median, maximum, minimum) for five features ($m = 5$) and their respective distance functions: δ_1 – CEDD with Minkowski distance function L_2 ($T_1 = 8.7\,\mu s$), δ_2 – FCTH with L_1 ($T_2 = 5.8\,\mu s$), δ_3 – EdgeHistogram with weighted L_1 ($T_3 = 6.9\,\mu s$), δ_4 – DominantColor with Earth Mover's distance function ($T_4 = 295.5\,\mu s$), and δ_5 – ColorHistogram with quadratic form distance function ($T_5 = 686.6\,\mu s$).

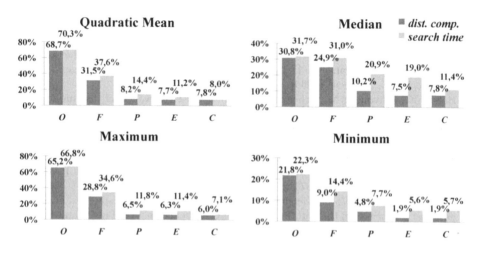

Fig. 1. Efficiency of 10-Nearest-Neighbor queries (relative to linear scan)

For indexing we used an *Onion-tree* [7] (*O*), *FlexiDex* [3] (*F*), *Partial Refinement* [6] (*P*) and both variants of our new heuristic (*E*, see Section 3.2) and (*C*, see Section 3.3). The pivot-based indexes were build with the same 16 pivot objects and we kept all index and feature data in main memory.

Figure 1 shows the results of our experiments relative to the performance of the linear scan of all objects, which required 153,035 distance computations and an average search time of 31,604 ms. The first variant of our heuristic (*E*) significantly decreased the required number of distance computations and search time for all aggregation functions in comparison to original *Partial Refinement* [6] (*P*) and the other state-of-the-art approaches. This means that our heuristic is inexpensive to compute and efficiently applicable to several important aggregation functions.

The second variant (*C*) took the computation costs of distance functions into account. It required more distance computations than the first variant in most cases. However, these additional distance computations had the benefit of a lower search time for quadratic mean and the median function. The search time increased only slightly in case of the minimum function. We conclude that this variant of our heuristic can further improve the efficiency, if the computation costs T_j vary significantly between distance functions.

5   Summary and Outlook

This paper introduced a new heuristic to determine the computation order of partial distance computations for multi-feature similarity search indexing. Even though our experimental evaluation is rather small and therefore preliminary, the new heuristic shows promising results and is widely applicable. Future research will focus on an extended evaluation and the introduction of an analytical cost formula for the heuristic.

References

1. Zezula, P., Amato, G., Dohnal, V., Batko, M.: Similarity Search: The Metric Space Approach. Advances in Database Systems, vol. 32, pp. 1–191. Springer-Verlag New York Inc., Secaucus (2006)
2. Böhm, K., Mlivoncic, M., Schek, H.-J., Weber, R.: Fast evaluation techniques for complex similarity queries. In: Proc. of the 27th International Conference on Very Large Data Bases, VLDB 2001, pp. 211–220. Morgan Kaufmann Publishers Inc., San Francisco (2001)
3. Zierenberg, M., Bertram, M.: FlexiDex: flexible indexing for similarity search with logic-based query models. In: Catania, B., Guerrini, G., Pokorný, J. (eds.) ADBIS 2013. LNCS, vol. 8133, pp. 274–287. Springer, Heidelberg (2013)
4. Güntzer, U., Balke, W.-T., Kießling, W.: Optimizing multi-feature queries for image databases. In: Proc. of the 26th International Conference on Very Large Data Bases, VLDB 2000, pp. 419–428. Morgan Kaufmann Publishers Inc., San Francisco (2000)
5. Jagadish, H.V., Ooi, B.C., Shen, H.T., Tan, K.-L.: Toward Efficient Multifeature Query Processing. IEEE Trans. on Knowl. and Data Eng. **18**, 350–362 (2006)
6. Zierenberg, M.: Partial refinement for similarity search with multiple features. In: Traina, A.J.M., Traina Jr, C., Cordeiro, R.L.F. (eds.) SISAP 2014. LNCS, vol. 8821, pp. 13–24. Springer, Heidelberg (2014)
7. Carélo, C.C.M., Pola, I.R.V., Ciferri, R.R., Traina, A.J.M., Traina Jr, C., de Aguiar Ciferri, C.D.: Slicing the Metric Space to Provide Quick Indexing of Complex Data in the Main Memory. Inf. Syst. **36**(1), 79–98 (2011)
8. Griffin, G., Holub, A., Perona, P.: Caltech-256 Object Category Dataset. Tech. rep. 7694. California Institute of Technology (2007)

Dynamic Permutation Based Index
for Proximity Searching

Karina Figueroa[1]([⊠]) and Rodrigo Paredes[2]([⊠])

[1] Facultad de Ciencias Físico-Matemáticas, Universidad Michoacana, Morelia, Mexico
karina@fismat.umich.mx
[2] Departamento de Ciencias de la Computación, Universidad de Talca, Curicó, Chile
raparede@utalca.cl

Abstract. Proximity searching consists in retrieving objects from a
dataset that are similar to a given query. This kind of tool is an elementary task in different areas, for instance pattern recognition or artificial
intelligence. To solve this problem, it is usual to use a metric index. The
permutation based index (PBI) is an unbeatable metric technique which
needs just few bits for each object in the index. In this paper, we present
a dynamic version of the PBI, which supports insertions, deletions and
updates, and keeps the effectiveness of the original technique.

1 Introduction

Similarity (or Proximity) Searching consists in retrieving the most similar elements to a given query from a dataset. This makes the proximity searching an
elementary task in many areas where the exact searching is not possible. Examples of these areas are machine learning, speech recognition, pattern recognition,
multimedia information retrieval or computational biology, to name few. The
core of such areas is precisely a searching task and the common part is a dataset
and a similarity measure among its objects.

Proximity queries can be formalized using the metric space model [3,6,8].
Given a universe of objects $\mathbb{X}$ and nonnegative distance function defined among
them $d : \mathbb{X} \times \mathbb{X} \rightarrow R^+ \cup \{0\}$, we define the metric space as a pair $(\mathbb{X}, d)$. Objects
in $\mathbb{X}$ do not necessarily have coordinates (think, for instance, in strings). On
the other hand, the function d provides a dissimilarity criterion to compare
objects from $\mathbb{X}$. In general, the smaller the distance between two objects, the
more "similar" they are. The function d satisfies the metric properties, namely:
positiveness $d(x, y) \geq 0$, *symmetry* $d(x, y) = d(y, x)$, *reflexivity* $d(x, x) = 0$, and
triangle inequality $d(x, z) \leq d(x, y) + d(y, z)$, for every $x, y, z \in \mathbb{X}$.

In practice, we are working with a subset of the universe, denoted as $\mathbb{U} \subset \mathbb{X}$,
of size n. Later, when a new query object $q \in \mathbb{X} \setminus \mathbb{U}$ arrives, its proximity query
consists in retrieving relevant objects from $\mathbb{U}$.

This work is partially funded by National Council of Science and Technology (CONA-
CyT) of México, Universidad Michoacana de San Nicolás de Hidalgo, México, and
Fondecyt grant 1131044, Chile.

© Springer International Publishing Switzerland 2015
G. Amato et al. (Eds.): SISAP 2015, LNCS 9371, pp. 97–102, 2015.
DOI: 10.1007/978-3-319-25087-8_9

There are two basic queries, namely, *range* and *k-nearest neighbor* ones. The range query (q, r) retrieves all the elements in $\mathbb{U}$ within distance r to q. The k-nearest neighbor query $NN_k(q)$ retrieves the k elements in $\mathbb{U}$ that are closest to q. Both queries can be trivially answered by exhaustively scanning the database, requiring n distance evaluations. However, as the distance function is assumed to be expensive to compute (think, for instance, when comparing two fingerprints), frequently the complexity of the search is defined in terms of the total number of distance evaluations performed, instead of using other indicators such as CPU or I/O time. Thus, the ultimate goal is to build an offline index that, hopefully, will accelerate the process of solving online queries.

2   Previous and Related Work

To solve this problem, a practical solution consists in building offline an index which is later used to solve online queries. Among the plethora of indices for metric space searching [3], the Permutation Based Index (PBI) [2] has shown an unbeatable performance. Let $\mathbb{P} \subset \mathbb{U}$ be a subset of permutants. Each element $u \in \mathbb{U}$ computes the distance towards all the permutants $p_1, \ldots, p_{|\mathbb{P}|} \in \mathbb{P}$. The PBI *does not store distances*. Instead, for each $u \in \mathbb{U}$, stores a sequence of permutant identifiers $\Pi_u = i_1, i_2, \ldots, i_{|\mathbb{P}|}$, called the permutation of u. Each permutation Π_u stores the identifiers in increasing order of distance, so $d(u, \mathbb{P}_{i_j}) \leq d(u, \mathbb{P}_{i_{j+1}})$. Permutants at the same distance take an arbitrary but consistent order. Thus, a simple implementation needs $n|\mathbb{P}|$ space. For the sake of saving space, we can compact several permutant identifiers in a single machine word. There are several improvements built on top of the basic PBI technique [1,5,7], however all of them are static indices.

The crux of the PBI is that two equal objects are associated to the same permutation, while similar objects are, hopefully, related to similar permutations. In this sense, when Π_u is similar to Π_q one expects that u is close to q. The similarity between the permutations can be measured by Kendall Tau K_τ, Spearman Footrule S_F, or Spearman Rho S_ρ metric [4], among others. As these three distances have similar retrieval performance [2], for simplicity we use S_F, defined as $S_F(\Pi_u, \Pi_q) = \sum_{j=[1,|\mathbb{P}|]} |\Pi_u^{-1}(i_j) - \Pi_q^{-1}(i_j)|$, where $\Pi_u^{-1}(i_j)$ denotes the position of permutant p_{i_j} in the permutation Π_u. For example, if we have two permutations $\Pi_u = (42153)$ and $\Pi_v = (32154)$, then $S_F(\Pi_u, \Pi_v) = 8$.

Finally, at query time, we compute Π_q and compare it with all the permutations stored in the PBI. Next, $\mathbb{U}$ is traversed in increasing permutation dissimilarity. If we limit the number of distance computations, we obtain a probabilistic search algorithm that is able to find the right answer with some probability.

3   Our Approach

We propose a dynamic scheme for the PBI. That is, we grant the PBI the capability of inserting or deleting objects in the index while preserving the searching performance. Bestowing dynamism on the PBI allows us to manage real-world applications, where the whole dataset is unknown beforehand and objects are inserted or deleted as the retrieval system evolves. At the rest of the paper, we show how to do that.

3.1   Dynamic Permutants

A dynamic permutant based index has to support object insertions and deletions, while preserving the retrieving performance. We note that when we insert a new object into the index, a new permutant can also be added. So, each object in the index has a dynamic permutation. On the other hand, we need to support the case when we delete an object which is a permutant.

In order to deal with permutations that are continuously changing, for each object we split its permutation in buckets. This way, we can limit the scope of the changes. The number of buckets is a parameter we study experimentally. All these buckets make a valid permutation and the last bucket is considered *in process*. Formally, let B be the size of a bucket, then every object has a permutation divided in pieces of size B. That is, every object $u \in U$ has a permutation Π_u divided in $\lceil \frac{|\Pi_u|}{B} \rceil = m$ pieces.

The main idea is processing small permutations of size B. Therefore, we will consider three sections: a list of bucket completed, a bucket of size B (which store the bucket in process), and a list of computed distances D of size B, to manage the distances for the bucket in process. Formally, for an object $u \in U$, we have its complete permutation Π_u divided in $m = \lceil \frac{|\Pi|}{B} \rceil$ pieces. Therefore, $\Pi_u = \Pi_u^1, \Pi_u^2, \ldots, \Pi_u^m$. Particularly, Π_u^m is the permutation in process. We also need a small array D for the distances of the bucket Π_u^m.

Inserting an Object. When inserting an object into the database we have two possibilities: it is a simple object or is also a new permutant. In first case, the object computes all the distances to the set of permutants and computes its permutation Π. The cost is $O(k)$ distances.

The interesting case is when an object v becomes a permutant (in this work, we chose the permutants at random). Firstly, v computes $d(u, v)$ for all $u \in U$. Next, u modifies both its Π_u^m and its vector of distances D. In Algorithm 1 we show details of the insertion process as permutant.

Algorithm 1.. InsertionAPermutant(p)

1: INPUT: Let p be the new permutant
2: Let $U = \{u_1, \ldots, u_n\}$ be our database
3: **for each** $u \in U$ **do**
4:     $d1 = d(p, u)$
5:     insert p in bucket Π_u^m and $d1$ at D
6:     Rebuild the small bucket Π_u^m
7: **end for**

Notice that when the bucket m is completed, it is transfered to the list of bucket completed and when a new permutant arrives, we use a new small bucket, and this is now the bucket Π_u^m.

For example: Let be u an element of the database, $B = 4$ and $\Pi_u = \{1, 2, 3, 4, 5, 6, 7, 8, 9, 10\}$, that is:

- List of bucket completed = $\{\Pi_u^1 = \{1, 2, 3, 4\}, \Pi_u^2 = \{5, 6, 7, 8\}\}$.
- Permutation in process = $\Pi_u^3 = \{9, 10\}$
- Distances $D = [d(u, p_9), d(u, p_{10})]$

Comparing Small Permutations. Using Spearman Footrule, we can compare two permutations in two ways. Let Π_u and Π_q be the permutations of an element u and a query q:

- If we consider use a sequential number for every permutant like the previous example, then we can compare all the permutation in a classic way, that is: $S_F(\Pi_u, \Pi_q) = \sum_{1 \leq i \leq k} |\Pi_u^{-1}(i) - \Pi_q^{-1}(i)|$.
- If we numerate each bucket separately, we just need numbers in $[1, B]$. So, we have $\Pi_u = \{1, 2, 3, 4, 1, 2, 3, 4, 1, 2\}$, where $\{\Pi_u^1 = \{1, 2, 3, 4\}, \Pi_u^2 = \{1, 2, 3, 4\}\}$ and $\Pi_u^3 = \{1, 2\}$. In this case we compute:

$$S_F(\Pi_u, \Pi_q) = \sum_{1 \leq i \leq m} \sum_{1 \leq j \leq B} |(\Pi_u^i)^{-1}(j) - (\Pi_q^i)^{-1}(j)|$$

Notice, that we can get the same performance that the original technique. However, this alternative allows better compaction of the permutation.

Deleting Permutants. In this case, we consider two element types. The first one is a simple object, which can be deleted without any consideration from the database. The second type is a permutant, that can also be deleted without modify the permutations of objects in $\mathbb{U}$, because the order in the rest of elements is conserved. A bucket with less than three permutants can be deleted, because is too short to help in the retrieving process.

Searching. The search process is almost identical to the basic PBI one. The only consideration is that we need to compute the query permutation according to the buckets, and then we compare the permutations as we explain above.

4   Experiments

In this section we evaluate and compare the performance of our technique in different metric spaces, such as synthetic vectors on the unitary cube and NASA images. The experiments were run on an Intel Xeon workstation with 2.4 GHz CPU and 32 GB of RAM with Ubuntu server, running kernel 2.6.32-22.

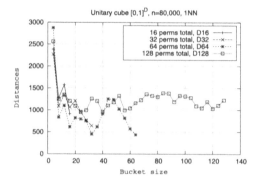

Fig. 1. Example of our technique. All points use the same amount of memory in the index. For example, line with × means 32 permutants and the first point has $B = 4$ that is $32/4 = 8$ buckets. The next point is $32/8$ we are using 4 buckets, and so on. The last point has $32/32 = 1$ bucket, that is the original idea. Notice that axe x represents the size of bucket.

4.1   Synthetic Databases

In these experiments, we used a synthetic database with vectors uniformly distributed on the unitary cube $[0, 1]^D$, in order to control the dimensionality of the space. This also allows us to define some extra parameters. We use 80,000 points in different dimensions $D = 16$, 32, 64, and 128, under Euclidean distance.

4.2   Optimal Value of B

For this experiment, different values of B (bucket size) are plotted in Fig. 1 for different dimensions. Notice that if $B = m$ then we have the original permutation based index. In axe x, we change the size of bucket, the values start at 4, and increases in values of 4. In this case, we represent distances computed for 1NN in dimension 16, 32, 64, and 128. Notice that the last point is the value of the original technique. This plot shows that we can get a better performance with a dynamic technique. For example, in dimension 128 the original technique makes 1224 distances (1 completed bucket of $B = 128$) while using $B = 24$ (that is, 5 buckets), only 948 computations of distances are required for the same query, that is a 27% less distances.

4.3   NASA Images

We use a set of 40,700 images from NASA, represented as 20-dimension feature vectors. For simplicity we compare the vectors with the Euclidean distance. This dataset is available at www.dimacs.rutgers.edu/Challenges/Sixth/software.html.

Notice that in this datase, our proposal keeps its performance. Fig 2 shows that after some B size our technique can improves the original idea. For example, 128 permutants with $B = 100$, or $B = 52$.

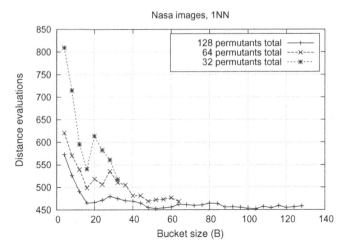

Fig. 2. Our approach keeps its performance on the real database of NASA images. All points use the same amount of memory.

5   Conclusions

In this paper we present a technique to turn the permutation based index (PBI) into a dynamic one. That is, an index that support both insertions and deletions of objects, while preserving the unbeatable performance of the original PBI. To do so, we process the complete permutation by parts.

As future work, we plan to test our technique in other metric spaces and research another alternatives to grant dynamism to the PBI strategy.

References

1. Amato, G., Savino, P.: Approximate similarity search in metric spaces using inverted files. In: Proc. 3rd Intl. Conf. Scalable Information Systems, ICST (2008), article 28
2. Chávez, E., Figueroa, K., Navarro, G.: Effective proximity retrieval by ordering permutations. IEEE Trans. on Pattern Analysis and Machine Intelligence (TPAMI) **30**(9), 1647–1658 (2009)
3. Chávez, E., Navarro, G., Baeza-Yates, R., Marroquín, J.: Searching in metric spaces. ACM Computing Surveys **33**(3), 273–321 (2001)
4. Fagin, R., Kumar, R., Sivakumar, D.: Comparing top k lists. SIAM J. Discrete Math. **17**(1), 134–160 (2003)
5. Figueroa, K., Paredes, R.: List of clustered permutations for proximity searching. In: Brisaboa, N., Pedreira, O., Zezula, P. (eds.) SISAP 2013. LNCS, vol. 8199, pp. 50–58. Springer, Heidelberg (2013)
6. Hjaltason, G., Samet, H.: Index-driven similarity search in metric spaces. ACM Transactions Database Systems **28**(4), 517–580 (2003)
7. Mohamed, H., Marchand-Maillet, S.: Quantized ranking for permutation-based indexing. In: Brisaboa, N., Pedreira, O., Zezula, P. (eds.) SISAP 2013. LNCS, vol. 8199, pp. 103–114. Springer, Heidelberg (2013)
8. Zezula, P., Amato, G., Dohnal, V., Batko, M.: Similarity Search - The Metric Space Approach. Advances in Database System, vol. 32. Springer (2006)

Finding Near Neighbors Through Local Search

Guillermo Ruiz[1], Edgar Chávez[2]([✉]), Mario Graff[3], and Eric S. Téllez[3]

[1] Universidad Michoacana de San Nicolás de Hidalgo, Morelia, Mexico
gruiz@dep.fie.umich.mx
[2] CICESE, Ensenada, Mexico
elchavez@cicese.mx
[3] INFOTEC México, Mexico, Mexico
{mario.graff,eric.tellez}@infotec.com.mx

Abstract. Proximity searching can be formulated as an optimization problem, being the goal function to find the object minimizing the distance to a given query by traversing a graph with a greedy algorithm. This formulation can be traced back to early formulations defined for vector spaces, and other recent approaches defined for the more general setup of metric spaces.

In this paper we introduce three searching algorithms generalizing to local search other than greedy, and experimentally prove that our approach improves significantly the state of the art. In particular, our contributions have excellent trade-offs among speed, recall and memory usage; making our algorithms suitable for real world applications. As a byproduct, we present an open source implementation of most of the near neighbor search algorithms in the literature.

1 Introduction

The problem of proximity search consists in identifying objects from a collection that are *near* a given query. In spaces with high intrinsic dimension it is difficult to avoid a sequential scan. For those kind of situations, the only practical option is to use *approximate indexes* with recall smaller than one.

A large portion of the literature in proximity searching is devoted to spaces with coordinates with some Minkowsky distance. In this realm, vector quantization[1] is the fastest option; although with low recall in practice. One interesting option is the *Randomized Neighborhood Graph*(RNG)[2] with vertex set the database and random edges in circular cones attached to the objects. With $O(n^2)$ construction cost and $O(n \log n)$ space, the RNG is impractical, and cannot be generalized to general metric spaces; however the ideas are appealing.

Combinatorial Approaches and Rank Cover Trees. In [3], the authors introduced the Range Cover Trees (RCT), where a tree is built using ordered rank for pruning instead of rules derived from distances and the triangle inequality. Node descendants in the tree are obtained using the rank order.

© Springer International Publishing Switzerland 2015
G. Amato et al. (Eds.): SISAP 2015, LNCS 9371, pp. 103–109, 2015.
DOI: 10.1007/978-3-319-25087-8_10

Approximate Proximity Graph. Malkov et al [4] introduced the *Approximate Proximity Graph* (APG), an index with excellent searching times. APG can trade speed and accuracy for memory usage.

The construction is incremental and consists of a simple rule. To insert the j-th element, simply find the (approximate) t-nearest neighbors among the $j-1$ elements already in the collection and link them to the newly inserted element.

In a subsequent paper [5], the searching is improved, as decribed below in Algorithm 1.

Algorithm 1. The search algorithm for k nearest neigbors of APG as described in [5]. We use our notation $\mathcal{N}(u)$ to describe the connected vertices of u, see Section 2 for more details.

Name: Search algorithm APG
Input: A transition function $\mathcal{N}$ (neighbors), the database S, and the query q, the number of restarts m
Output: The set *res* of near neighbors of q.

```
 1: Let res be an empty min-queue of fixed size k
 2: Let candidates be an empty min-queue
 3: for i = 1 to m do
 4:     Define r as current distant radius in res, empty res defines r = ∞
 5:     Randomly select c ∈ S \ visited
 6:     Append c into visited and (d(q,c),c) into candidates and res
 7:     loop
 8:         Let (r_b, best) be the nearest pair in candidates
 9:         Remove best from candidates
10:         if r_b > r then
11:             break loop
12:         end if
13:         for u ∈ N(best) do
14:             if u ∉ visited then
15:                 Add u to visited and (d(q,u),u) to candidates and res
16:             end if
17:         end for
18:     end loop
19: end for
```

---

k-Nearest References (KNR). In some indexes, every node is associated with a set of k nearest references (KNR), where the set of references is a sample of the database. The similarity between items is hinted by a similarity function over the shared neighbors. In this approach there is no navigation and proximity queries are solved using an inverted index. This structural similarity was systematically explored in [6] adding several indexes to the list, like the KNR-Cos that uses the cosine similarity. The experimental evidence suggests that it surpasses the majority of the state of the art KNR indexes, so we will use it in our experimental section.

Contribution. We introduce three new indexes for near neighbor search i) APG* an improvement over APG computing *online m* (the number of restarts), ii) APG*-R that improves APG* limiting the shared state among search steps, and iii) BS, which is a beam search based algorithm.

2   Improving APG

Let S be the database, define $\mathcal{N} : S \rightarrow S^+$ as a transition function describing the connection among items, i.e., $\mathcal{N}(u)$ returns the neighborhood of u. We can think of $\mathcal{N}$ as the edges of a graph over S. Consider the nearest neighbor (nn) problem and let $d_q(u) = d(q, u)$. We want to find $w \in S$ that is the global minimum of d_q and we only have access to $\mathcal{N}(u)$ for a given u. We only have local information. An *optimal* $\mathcal{N}$ has minimum size and allows to find the global minimum using local search. It turns out that the construction of an optimal $\mathcal{N}$ is NP-hard and in some cases, this optimal $\mathcal{N}$ defines the complete graph (equivalent to sequential scan). In order to overcome this, we need to relax our constraints allowing approximate results. For approximate search, it is accepted to locate a good enough local minimum for d_q.

Recall from Section 1 that APG's search is repeated m times to improve the expected recall. As the authors suggest in [5], m must be adjusted for each dataset. A significant overestimation of m would impact both search and construction performances.

Algorithm 2. Beam Search over S and $\mathcal{N}$ for k nearest neighbors.

Name: Beam Search
Input: A transition function $\mathcal{N}$, the database S, the query q, the size of the beam b
Output: The set res of near neighbor of q.

```
 1: Let res be an empty min-queue of fixed size k
 2: Let beam be an empty min-queue of fixed size b
 3: Let visited ← ∅
 4: Let beam ← {}
 5: for i = 1 to b do
 6:     Randomly select u from S
 7:     Add u to visited and (d(u, q), u) to res and beam
 8: end for
 9: repeat
10:     Let cov* ← cov(res)
11:     for i = 1 to σ do
12:         Let beam* ← {} {Fixed sized priority min-queue of size b as beam}
13:         for c ∈ beam do
14:             for u ∈ N(c) do
15:                 if u ∉ visited then
16:                     Add u to visited and (d(q, u), u) to res and beam*
17:                 end if
18:             end for
19:         end for
20:         beam ← beam*
21:     end for
22: until cov* = cov(res) {Stops when there is no improvement over d_q, i.e., cov(res)}
```

Let *result* be a min priority queue (min-queue) of fixed size k. If *result* is under its full capacity then $\mathrm{cov}(result)$ equals to the maximum possible distance value; if it is full, then $\mathrm{cov}(result)$ is the radius of the furthest item.

The objective of our first search algorithm is to compute m online checking the changes over the global result set, i.e., the covering radius $\mathrm{cov}(res)$. Here, we are minimizing d_q, and stopping the algorithm when our guess of d_q cannot be improved after σ tries. All our new algorithms use the same technique as stop condition. σ should be small, in the range of 2 to 4. The idea is to relax the stop condition, but not to introduce an extra parameter. This search algorithm is used to produce the index APG*.

Our second index is named APG*-R. APG*-R uses random starting points, not the best known in APG and APG*, and also, both *candidates* and *res** are local to each step. This modification helps to escape from a local minima. For lack of space we do not include the full, formal description of the algorithm.

Our final index is BS (produced by Algorithm 2), a novel index using beam search over S and $\mathcal{N}$. As in the previous algorithms, at least σ steps are taken before stopping the search. Hence, if after some steps d_q does not change, then we claim a good local minimum. The main parameter is the size of the beam b, which can be adjusted at any time. Notice that beam search does not need to restart; however, it needs to know b.

3   Experimental Results

All experiments were performed in a 24-core Intel Xeon 2.60 GHz workstation with 256GB of RAM, running CentOS 7, without using the multiprocessing capabilities in the search process. Both, the index and the database where maintained in memory. Our implementations were created in the mono framework, in C#. For the RCT [3] we used the C++ implementation, kindly shared by authors. We fixed our attention in the recall and search speed. Since comparing them by time would be unjust, we measure the time of the search compared with the sequential scan (speedup).

Our benchmarks consist of synthetic databases called RVEC which are random vectors using the L_2 distance. We tested the effect of varying the dimension and database size. First, we used datasets of $16, 32, 64, 128$ and 256 dimensions, each one with 10^5 randomly generated items. Next, the performance as the size increases was measured using 16, 32 and 64-dimensional datasets of sizes $3 \times 10^5, 10^6$, and 3×10^6 items.

The Effect of the Dimension on the Search Performance. On Figure 1, each point on the lines represents the result for a dimension. We show the recall and the speedup for each instance on the same figure. The results are grouped by size of the indexes on the figures (controlled by the parameter t).

Figure 1(a) shows the speedup for the indexes that take from 1 to 10 integers per element of the database. The APG and APG* are very fast because of the small number of distances they have to compute but their poor recall makes them useless. The APG*-R variant shows a big improvement on the recall. The different instances of the BS gain in recall but lose speed when we increase the number of searches. The KNR gets good recalls but is slower than the average of the others. Finally, the RCT has a good overall performance specially for $c = 128$.

The Figure 1(b) contains the speeds for indexes using 10 to 30 integers per element. Because of their small recall, the APG and APG* are very fast. The

APG*-R and BS with $b = 8$ are very similar, one a little faster and the other with better recall. As expected, the other configurations of BS have higher recalls.

Note how the APG and APG* are always very similar, also they need more memory to achieve good recalls. The APG*-R and BS have a great range of action with the option to get recalls very close to one with good speeds.

The Figures 1(c) and 1(d) are very similar. The APG and APG* are not very fast. The APG*-R is faster for the dimension $16, 32$, and 64 but the BS with $b = 8$ becomes faster for the big dimension.

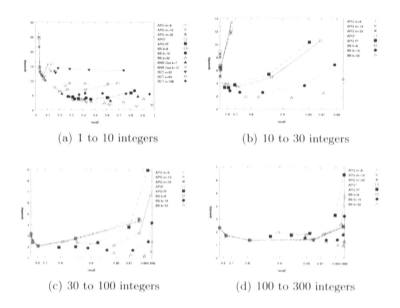

(a) 1 to 10 integers   (b) 10 to 30 integers

(c) 30 to 100 integers   (d) 100 to 300 integers

Fig. 1. Recall and speedup comparison in four different classes of memory among our indexes and several state of the art techniques for fixed $n = 10^5$ over several dimensions. Here we show performance for low to medium memory resources (i.e., 1 to 30 integers per item). Each curve corresponds to a different dimension $16, 32, 64, 128$, and 256; as a hint to review figures, large dimensions correspond to smaller recalls and scale is exponential.

Scalability. We used the randomly generated database of dimension 16 with sizes $3 \times 10^5, 10^6$, and 3×10^6. Each point in a curve is produced by a different database size; speed ups are larger as n increases.

Figure 2(a) shows the recall and the search speed for the different sizes of the database for 16 dimension. The APG and APG* scored a recall below 0.7 on all the sizes even when using 32 searches for the query. As shown in the previous experiments, the APG*-R gets a boost on the recall compared to the APG*. Note how the lines of the BS are practically vertical meaning that the index gains in speed much more than it loses in the recall. Compare this to the KNR with $k = 7$, its recall is very affected as the size increase. Also, in general the KNR are not very fast. The RCT does not get a recall as good as the BS and is considerable slower. On Figure 2(d) we see the recall and the number of distances for the queries. This general behavior is the same in the next figures.

Figure 2(b) shows the same tendency, the APG*-R being an improvement over the APG's, the BS having very good recalls and speeds, the KNR being slow, and the RCT behind the BS. The Figure 2(e) shows the same but with the distances computed. The same can be said about the Figure 2(c) and 2(f) but the interesting part is to note how the change of dimension affects more the indexes than the size of the database. For 16 dimensions, the performance is maintained for increasing database size.

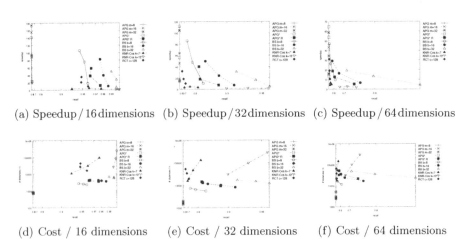

(a) Speedup / 16 dimensions (b) Speedup / 32 dimensions (c) Speedup / 64 dimensions

(d) Cost / 16 dimensions   (e) Cost / 32 dimensions   (f) Cost / 64 dimensions

Fig. 2. Speedup as function of the expected recall for our indexes and the state of the art. We use datasets with dimensions $16, 32$, and 64. Each dataset has three instances with sizes of $3 \times 10^5, 10^6$, and 3×10^6. We also limited memory to be under 8 direct neighbors per item, i.e. 16 undirected neighbors in average.

4   Conclusions

We introduced three near neighbor searching indexes, called APG*, APG*-R and BS. These indexes use an underlying graph and several variants of local search to navigate them. As part of our contribution, we compare our techniques among the current state of the art indexes. Our indexes significantly surpass the performance of the majority of the alternatives, in almost all of our benchmarks.

In the present work we used $\mathcal{N}$ as introduced by APG; however, there exists several open questions about it, such as how to determine its precise parameters online, i.e., without having any a priori knowledge about it.

References

1. Silpa-Anan, C., Hartley, R.: Optimised kd-trees for fast image descriptor matching. In: IEEE Conference on Computer Vision and Pattern Recognition, CVPR 2008, pp. 1–8, June 2008
2. Arya, S., Mount, D.M.: Approximate nearest neighbor queries in fixed dimensions. In: Proceedings of the Fourth Annual ACM/SIGACT-SIAM Symposium on Discrete Algorithms, pp. 271–280, Austin, Texas, 25–27 January 1993 (1993)
3. Houle, M.E., Nett, M.: Rank cover trees for nearest neighbor search. In: Brisaboa, N., Pedreira, O., Zezula, P. (eds.) SISAP 2013. LNCS, vol. 8199, pp. 16–29. Springer, Heidelberg (2013)
4. Malkov, Y., Ponomarenko, A., Logvinov, A., Krylov, V.: Scalable distributed algorithm for approximate nearest neighbor search problem in high dimensional general metric spaces. In: Navarro, G., Pestov, V. (eds.) SISAP 2012. LNCS, vol. 7404, pp. 132–147. Springer, Heidelberg (2012)
5. Malkov, Y., Ponomarenko, A., Logvinov, A., Krylov, V.: Approximate nearest neighbor algorithm based on navigable small world graphs. Information Systems **45**, 61–68 (2014)
6. Chávez, E., Graff, M., Navarro, G., Téllez, E.: Near neighbor searching with K nearest references. Information Systems **51**, 43–61 (2015)

Metrics and Evaluation

When Similarity Measures Lie

Kevin A. Naudé[(✉)], Jean H. Greyling, and Dieter Vogts

Computing Sciences, Nelson Mandela Metropolitan University,
Port Elizabeth, South Africa
{kevin.naude,jean.greyling,dieter.vogts}@nmmu.ac.za

Abstract. Do similarity or distance measures ever go wrong? The inherent subjectivity in similarity discernment has long supported the view that all judgements of similarity are equally valid, and that any selected similarity measure may only be considered more effective in some chosen domain. This paper presents evidence that such a view is incorrect for structural similarity comparisons. Similarity and distance measures occasionally do go wrong, and produce judgements that can be considered as errors in judgement. This claim is supported by a novel method for assessing the quality of similarity and distance functions, which is based on relative scale of similarity with respect to chosen reference objects. The method may be applied in any domain, and is demonstrated for common measures of structural similarity in graphs. Finally, the paper identifies three distinct kinds of relative similarity judgement errors, and shows how the distribution of these errors is related to graph properties under common similarity measures.

Keywords: Similarity measures · Distance measures · Similarity judgement errors · Similarity judgement quality · Information retrieval

1 Introduction

Numeric measures of similarity are versatile tools for solving information retrieval problems. They serve both in classification and similarity search, and have been used effectively in a variety of problem domains [1–6]. Similarity is typically quantified as either a notional proportion of matching (similarity functions), or as a cumulative sum of differences (distance functions).

The performance of similarity measures may be evaluated in two orthogonal dimensions: *resource* performance and *task* performance. The former of these is easy to study, as computational resource usage may be either directly observed through empirical research, or studied through theoretical models of computation. This research concerns itself with the second dimension of performance. Specifically, it examines the general efficacy with which different similarity measures are able to judge similarity between structured object representations. The most important of such discrete structures are graphs.

K.A. Naudé—This research was supported by the National Research Foundation, South Africa.

G. Amato et al. (Eds.): SISAP 2015, LNCS 9371, pp. 113–124, 2015.
DOI: 10.1007/978-3-319-25087-8_11

There are well established methods of evaluating the task performance of classification and similarity search algorithms that incorporate measures of similarity. Section 2 highlights the main processes of these conventional evaluation techniques, as they form a basis for the present work. An important problem with such methods is that the conclusions they provide are confined to a specific problem domain tested, leaving general conclusions about the embedded similarity measures hard to obtain.

There is a significant need to understand the judgement quality of similarity measures directly. The present research introduces a new evaluation technique which directly characterises the decision behaviour of similarity measures. The motivation for direct evaluation is advanced throughout Section 3, with the main processes such and evaluation set out in Sections 3.1 and 3.2. The new method focuses upon relative judgements, such as claiming that *objects A and B are more alike than objects A and C*. Consequently, there are no specific requirements about the scale of difference between similarity scores.

The main difficulty in directly assessing the judgement quality of similarity measures is the lack of justifiable ground truth data. The new evaluation method circumvents the problem by transforming a dataset of graphs into test instances that have determinable outcomes. The ground truths obtained are related to graph edit distances, but are procedurally created to ensure that test instances satisfy constraints that provide a justification for the relative decision outcomes.

The new method of evaluation is particularly valuable in that it produces new kinds of information about the decision behaviour of the tested similarity measures. In particular, it provides frequency data for three different kinds of inconsistent judgements that can be made. Section 4 demonstrate the application of the evaluation method to a selection of graph similarity measures. The first evaluated measure is the similarity derived from the maximum common induced subgraph (MCIS), a popular alternative to the maximum common subgraph (MCS), which is itself seldom computationally feasible. The second evaluated measure is attributed to Blondel *et al.*[7]. The Blondel measure is an example of a family of contemporary fixed-point graph similarity measures that are discover pair-wise vertex similarity scores. The experiments are described in Section 4, followed by the findings and general performance characteristics for each measure reported in Section 5.

2   Conventional Task Performance

The conventional approach for evaluating systems of indexing, search and classification originates with the *Aslib Cranfield* projects of the 1960s [8]. The assessment of task performance of modern similarity search and classification follows very much the same method, although present day researchers benefit from the wide variety of collected datasets. One such dataset, the Columbia Object Image Library [9], shall now be used to illustrate the method and limitations of conventional task performance evaluation.

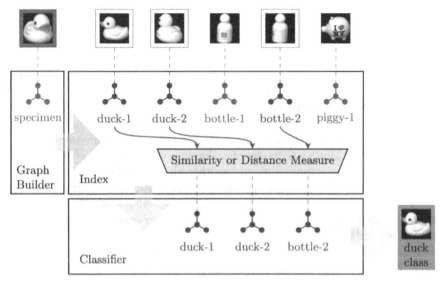

Fig. 1. The software components for object recognition, using a similarity measure.

Consider the task of identifying an ordinary object from a source photograph, based on a library of previously observed photographs. The internal organisation of a software solution for the task is abstractly represented in Fig. 1. The figure describes solutions that use an explicitly computed measure of similarity[1]. Note that there are several computational sub-tasks besides computing similarity scores. The main sub-tasks are

- building a graph representation of the object,
- querying an index for related known objects, using the similarity measure strategically to minimise comparisons, and
- using a selection of the query response to classify the specimen object.

A popular and effective representation for an object found in an image is a Region Adjacency Graphs (RAG) [10, 11]. In the present scenario, a RAG would be constructed to represent each known object, and these would be inserted into the index. During classification, the index treats the representation of the specimen object as a query, and selects a subset of the known objects with which to compare it. It is here that the distance or similarity measure comes to the foreground as a measure quantifying relatedness. However, the similarity scores still do not constitute a response to the stated task. Instead, the results of the index query are delivered to a classifier to perform object identification. A simple k nearest neighbour classifier would treat the k most relevant query responses as votes in favour of the matched classes. The recogniser identifies the object

---

[1] Other machine learning techniques could also be applied to the task, but these are not relevant to the subject of this paper.

according to the class with the greatest number of votes, and reports that class as the identification of the object.

The evaluation of classification task performance depends upon the existence of test instances, which do not appear in the index, and for which ground truth data are known. For example, Fig. 1 shows an image identified as a *duck*, which is correct. The set of test data for which expected outcomes are known is called the ground truth data for the problem task.

For classification tasks there are many suitable datasets available. Riesen and Bunke have collected ten distinct dataset specifically for investigating the application of graph similarity to classification in a variety of fields of study [12]. However, the situation is more difficult for other kinds of tasks, such as similarity search. The greatest difficulty in applying the conventional evaluation method for task performance remains obtaining relevant ground truth data. A rising solution to the deficient supply of ground truths is to extract knowledge from crowdsourced judgements [13,14].

If ground truth data are available, the conventional evaluation method is effective in providing meaningful answers about how effectively the software stack solves the problem under investigation. However, if the researcher intends to understand the behaviour of the similarity measure in particular, the results are not informative. If, for example, the object recogniser of Fig. 1 performs poorly in the recognition task, little is learned about the judgement performance of the similarity measure. The reason for this is that the prevailing evaluation techniques are holistic in nature. They produce task performance data for a complex arrangement of several software components working together. Consequently, poor task performance could be attributed to inappropriate choices in the object representation, or aspects within the indexer, or classifier. The extent to which the judgements of the similarity measure are correct remain unclear. The next section seeks to address this problem through direct evaluation of the similarity measure itself.

3   Direct Performance Assessment

The direct evaluation of similarity judgements requires ground truth data, as described for conventional evaluation. However, similarity is intrinsically subjective, so it is important that ground truths are only asserted for test cases that can be objectively motivated. Three possible scenarios for ground truth data items are now considered:

single pairs   having expected similarity scores,
                    i.e. $s(A, B)$ is provided,
independent pairs having expected relative similarity,
                    i.e. $s(A, B) - s(C, D)$ is provided, and
dependent pairs   having expected relative similarity,
                    i.e. $s(A, B) - s(A, C)$ is provided.

Two of the three possibilities are non-viable sources of ground truth. Firstly, since the scale of scores is subjective, single scores between pairs can appear anywhere on the accepted scale of values. It is therefore not usually meaningful to impose any specific similarity score outcome for any single pair of test graphs. The second suggestion is therefore more interesting as it considers relative strength of observed similarity. However, differences between independent pairs are not required to be measured on the same scale. Consider, for example, that the very well-known Jaccard measure scales scores with the inverse of the combined size of the source structures. Consequently, differences are measured relative to the scale of the pair under consideration. Ground truth data for independent pairs is therefore not recommended.

The third option is more compelling. It is reasonable to establish ground truth data for relative similarity between graph pairs, provided that the pairs are dependent. The presence of graph A common to pairs (A, B) and (A, C) makes it possible to relate their similarity scores. Furthermore, if the graphs are carefully chosen, it may be reasonable to expect sensitive similarity measures to rank the differences between the pairs in a consistent way. The next sections describe a procedure for generating such pairs, and a method for using them in task performance evaluation.

3.1   Procedurally Generated Truths

The task of obtaining ground truth similarity data can now be separated into two distinct steps: i) selecting graphs triples (A, B, C) such that the sign of $s(A, B) - s(A, C)$ has an expected value, and ii) discovering that expected value. The first step of selecting three graphs may seem easy, if some supply of graphs is available. However, it is ill-advised to select graphs that differences that are trivially ordered. For instance, if B and C have *substantially* different size, that information alone may be enough to order B and C with respect to A. For the purpose of comparing the quality of similarity judgement, it is best to present cases of graphs that are not easily distinguished. Hence, it is required that graphs A, B, and C have shared topological features. In particular, they should have equal

- number of vertices,
- number of edges,
- frequency distribution for vertex attributes, and
- frequency distribution for edge attributes.

It is clear by these requirements that differences between such graphs would not manifest in easily observable divergent features. The suggested process for constructing such graph triples is to assign graph A from a source of interesting graphs. Graphs B and C can then be derived from A by applying a chosen number of independent transformations. A suitable transformation which preserves all of the stated graph features is edge rewiring [15]. Edge rewiring is accomplished by selecting a single edge at random for removal, and then inserting a new edge at a random location, with the same edge attribute as that which was previously removed.

For the purposes of this paper, rewiring is applied once to obtain B, and twice to obtain C (but independently of B). The effect is that B is either isomorphic with A, or exactly one edge-rewiring distance away from it. C is likewise either isomorphic with A, or one at most two edge-rewiring steps from A. Hence, there are three possible outcomes:

- B is nearer to A than C is,
- the graphs B and C are equally similar to A, or
- C is the nearer graph (perhaps surprisingly).

The final step in establishing the ground truth data is to determine the expected outcome, or alternatively, the expected sign of $s(A, B) - s(A, C)$. The authors have found it quite practical to employ a fast isomorphism check to determine if A and B are equivalent. If the isomorphism check fails, then a single edge-rewiring operation separates the graphs. The minimum number of changes required to reproduce C may be determined in a similar manner.

It may be tempting to instead compute $D(A, C) - D(A, B)$, where D is some preferred graph edit distance. However, there are important reasons to avoid using a conventional edit distance to determine ground truth. The most important of these is that there are many edit distances, and they are not necessarily consistent with one another. Rather they each provide a selected set of edit operations, and are parameterised with a vector of costs associated with the chosen edit operations. Two edit distances that employ the same edit operations may still disagree considerably if they operate with different cost vectors. Since the costs assigned to edit distances are arbitrary, there is no reason to favour the judgements of one edit distance over another.

The process described above is different from the use of edit distances because it employs only a single edit operation, and it only considers graphs that differ in that specific form of change. In particular, it does not assert relative ordering between arbitrary graphs. In this way, the choice of graphs A, B, and C is considerably constrained to avoid the possibility that some competing structural similarity claim could be justifiable.

3.2   Evaluation Procedure

The ground truths described in the preceding section may be directly used for evaluation purposes. The evaluation process follows conventional task performance evaluations, by examining a chosen number of test graph triples. For each triple, the similarity measure's judgement over the triple is compared with the ground truth outcome, and tabulated in a multinomial contingency table, as illustrated in Fig. 2.

The conventional measures of task performance are *precision*, *sensitivity (recall)*, and the *recognition rate (global sensitivity)*. These may be computed directly from the contingency table. It is suggested that the sensitivity is the most useful measure of judgement quality, as it describes the proportion of orderable graphs which are correctly ordered. These performance measures may be

	Ground Truth		
	B is nearer	**neither** is nearer	**C is** nearer
B is nearer	r_1	a_1	c_1
neither is nearer	b_1	r_2	b_2
C is nearer	c_2	a_2	r_3

Similarity (left axis label)

$r = r_o + r_b$: correct cases

$r_o = r_1 + r_3$: ordered
$r_b = r_2$: balanced

$a = a_1 + a_2$, : arbitrary order
$b = b_1 + b_2$, : unordered
$c = c_1 + c_2$: inverse order

Fig. 2. Bivariate frequency distribution of similarity judgements and ground truth.

computed using the following formulae.

$$\text{precision} \quad = \quad \frac{\text{number correctly ordered}}{\text{number of order assertions}} \quad = \quad \frac{r_o}{r_o + a + c} \qquad (1)$$

$$\text{sensitivity} \quad = \quad \frac{\text{number correctly ordered}}{\text{number of orderable cases}} \quad = \quad \frac{r_o}{r_o + b + c} \qquad (2)$$

$$\text{recognition rate} \quad = \quad \frac{\text{number correct}}{\text{number of instances}} \quad = \quad \frac{r}{r + a + b + c} \qquad (3)$$

The contingency table also shows that similarity measures may make three distinct kinds of judgement errors. For purposes of discussion, the proportion of correct judgements shall be denoted ρ, while the proportions of the three kinds of errors shall be denoted α, β, and γ, respectively. These types of errors are analogous to errors that may be made during directional hypothesis testing, and so they carry similar names here. They are:

- Type I errors: ordering differences that are actually equal in scale,
- Type II errors: failing to note orderable differences, and
- Type III errors: produce an inverse ordering of orderable graphs.

4   Experiment

An experiment was performed to demonstrate the application of the new method for evaluating and characterising the decision behaviour of common structural graph similarity measures. Two similarity measures were examined. The first is the similarity derived from the size of the maximum common induced subgraph of the source graphs, giving by (4).

$$s_{\text{MCIS}}(X, Y) = 2\frac{|\text{MCIS}(X, Y)|}{|X| + |Y|} \qquad \text{where } |g| = \text{ number of vertices in } g \quad (4)$$

The second similarity measure is attributed to Blondel *et al.*[7]. It is a member of a family of measures that propagate local vertex similarity scores within a product graph iteratively, until a fixed point is reached. This family of similarity measures are an interesting contemporary approach to computing graph similarity. The computed scores are, however, between vertex pairs taken from each of the source graphs. In order to obtain a final similarity score between the graphs, the method used by Zager and Vergese [16] is followed. First the Hungarian algorithm is applied to find the optimal assignment between the vertices of the source graphs, using their local vertex similarity scores, and accounting for labels. Following this, the average vertex similarity score across the optimal assignment is taken as the final graph similarity score.

The similarity measures were examined across a collection of 555 synthetic graph datasets. All of these datasets were constructed using the Erdös-Rényi $G_{n,m}$ model for random graphs with fixed edge densities. The graph datasets were chosen to reflect a wide range of different graph properties. Two broad cases were considered: a) fixed edge density (50%) with the number of vertices varying between 10 and 30, and b) a fixed number of vertices (20) with edge density varying between 10% and 90%, in 5% increments. Therefore, there are 37 distinct combinations of graph sizing parameters. The number of distinct label categories was varied between 1 (equivalent to unlabelled graphs), and 15. The labels were assigned with uniform probability to all edges and vertices. Thus, there is a total of 555 distinct datasets generated, each having different graph properties.

Every dataset was comprised of two samples: one of 500 instances, and another of 10 000 instances. The smaller samples are sufficient for hypothesis testing, as the tests available (such as Boschloo's test) are very powerful. However, the small samples do not give narrow confidence intervals for the descriptive statistics. Thus, the large samples were provided to obtain high-precision estimates of binomial proportions.

5   Results

The results shown in Fig. 3 highlight a selection of important cases[2]. A close examination of error distribution for the MCIS similarity reveals two interesting facts. The first is that the errors distribution is dominated by Type II errors. The Type II errors are instances in which the graph triples should be strictly ordered, but the MCIS measure does not observe any particular ordering. In other words,

---

[2] All raw data is available upon request.

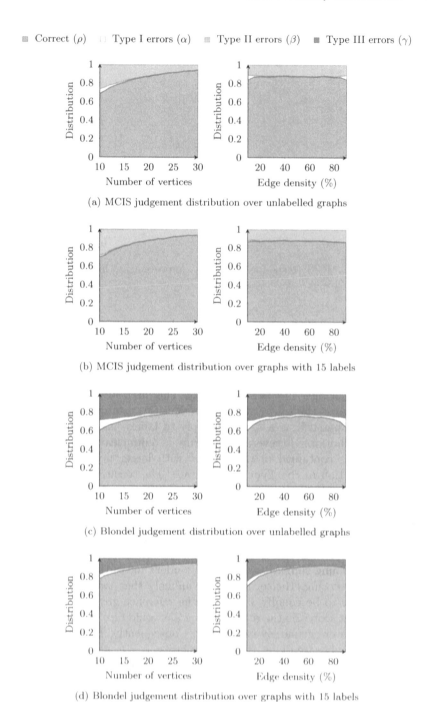

Fig. 3. Multinomial outcome distributions for MCIS and Blondel similarity.

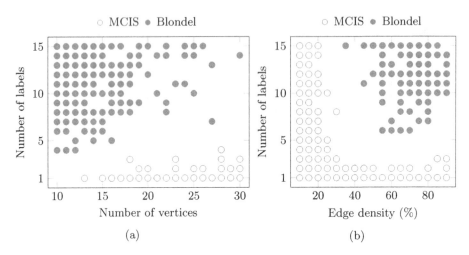

Fig. 4. Blondel vs. MCIS: tests for sensitivity improvement, highlight cases with $p < 0.05$.

the MCIS measure frequently does not observe fine detail in the graphs. Paradoxically, this leads to high precision as Type II errors do not contribute to the precision score. However, the sensitivity is adversely affected.

The second interesting detail regarding the MCIS measure is that it is not at all sensitive to changes in the edge density. The distribution of outcomes is nearly constant across the full range of edge densities examined. The effect of labels is similarly passive, as a diverse supply of labels has little effect upon the outcome distribution. However, the measure is quite clearly sensitive to the number of vertices contained in the graphs, with larger numbers of vertices leading to a reduction in the Type II error rate. The sensitivity and precision for larger graphs are therefore improved. It is unfortunate that the measure is not computationally feasible for large graphs.

The Blondel measure stands in contrast to the MCIS measure in that it makes essentially no Type II errors, but incurs a larger proportion of Type I and Type III errors. This fact is easily explained. The nature of the iterative process for determining similarity scores in Blondel's measure gives it access to very fine grained scoring. Hence, it is very unlikely that two strictly ordered graphs will appear to be equally similar to the reference graph. Achieving the precisely equivalent scores on the real number scale would be a peculiarity, unless the source graphs were genuinely isomorphic. Consequently, Type II judgement errors are exceedingly rare under Blondel similarity.

The low Type II error rate creates opportunities for other kinds of judgement errors. For unlabelled graphs, the Blondel measure incurs a large proportion of Type III errors, which are the least desirable kind: asserting order in the wrong direction. Fortunately, the case of labelled graphs shows that Blondel similarity has much reduced error rates when there is increased label diversity.

Finally, hypothesis testing was applied using Boschloo's test for binomial proportions [17], and simultaneously directional hypotheses [18] to determine the improvement direction between the two measures. Fig. 4 shows a graphical representation of the one of the two measures producing statistically improved sensitivity, with $p < 0.05$.

The hypothesis tests are consistent with the earlier observations. The evaluation shows that the MCIS measure dominates Blondel in sensitivity if the number of distinct labels is low (less than 5), or if the edge density is low (less than 30%). When neither of these conditions are present in the source graphs, Blondel's measure outperforms MCIS similarity convincingly. This is particularly encouraging, as it shows that the Blondel similarity measure is a robust substitute for MCIS similarity in larger graphs.

6   Conclusion

The empirical evidence gathered under the new evaluation method shows a variety of useful results that were not previously known. The most valuable of these is that the Blondel similarity measure is a reliable substitute for MCIS similarity when the source graphs are labelled and not sparse. In addition, the MCIS similarity has difficulty detecting fine grained differences between very similar graphs. These facts illustrate how the new evaluation method can establish general characteristics of structural similarity measures. Most importantly, the characteristics are independent of any specific application domain.

References

1. Morain-Nicolier, F., Landré, J., Ruan, S.: Binary symbol recognition from local dissimilarity map. In: 8th International Workshop on Graphic Recognition GREC 2009, pp. 143–148 (2009)
2. Boyer, L., Habrard, A., Sebban, M.: Learning metrics between tree structured data: application to image recognition. In: Kok, J.N., Koronacki, J., Lopez de Mantaras,R., Matwin, S., Mladenič, D., Skowron, A. (eds.) ECML 2007. LNCS (LNAI), vol. 4701, pp. 54–66. Springer, Heidelberg (2007)
3. Rahman, S.A., Bashton, M., Holliday, G.L., Schrader, R., Thornton, J.M.: Small Molecule Subgraph Detector (SMSD) toolkit. Journal of Cheminformatics 1(1), 12 (2009)
4. Cao, Y., Jiang, T., Girke, T.: A maximum common substructure-based algorithm for searching and predicting drug-like compounds. Bioinformatics 24(13), i366–i374 (2008)
5. Islam, A., Inkpen, D.: Semantic similarity of short texts. In: Nicolov, N., Angeliva, G., Mitkov, R. (eds.) Text, pp. 227–236. John Benjamins Publishing Company (2009)
6. Markines, B., Cattuto, C., Menczer, F., Benz, D., Hotho, A., Stumme, G.: Evaluating similarity measures for emergent semantics of social tagging. In: Proceedings of the 18th International Conference on World Wide Web, pp. 641–650. ACM, New York (2009)

7. Blondel, V.D., Gajardo, A., Heymans, M., Senellart, P., Van Dooren, P.: A measure of similarity between graph vertices: applications to synonym extraction and web searching. SIAM Review **46**(4), 647–666 (2004)

8. Cleverdon, C., Mills, J., Keen, M.: Factors Determining the Performance of Indexing Systems. ASLIB Cranfield project, Cranfield University, Cranfield, Technical report (1966)

9. Nene, S.A., Nayar, S.K., Murase, H.: Columbia Object Image Library (COIL-100). Technical report CUCS-006-96, Columbia University (1996)

10. Colantoni, P., Laget, B.: Color image segmentation using region adjacency graphs. In: Sixth International Conference on Image Processing and its Applications, vol. 2, pp. 698–702, July 1997

11. Chevalier, F., Domenger, J., Benoispineau, J., Delest, M.: Retrieval of objects in video by similarity based on graph matching. Pattern Recognition Letters **28**(8), 939–949 (2007)

12. Riesen, K., Bunke, H.: IAM graph database repository for graph based pattern recognition and machine learning. In: da Vitoria Lobo, N., Kasparis, T., Roli, F., Kwok, J.T., Georgiopoulos, M., Anagnostopoulos, G.C., Loog, M. (eds.) Structural, Syntactic, and Statistical Pattern Recognition. LNCS, vol. 5342, pp. 287–297. Springer, Heidelberg (2008)

13. Agrawal, R., Gollapudi, S., Kannan, A., Kenthapadi, K.: Similarity search using concept graphs. In: Proceedings of the 23rd ACM International Conference on Conference on Information and Knowledge Management, Shanghai, China, pp. 719–728 (2014)

14. Zafarani, R., Liu, H.: Evaluation Without Ground Truth in Social Media. Communications of the ACM **58**(6), 54–60 (2015)

15. Albert, R., Barabasi, A.L.: Topology of evolving networks: local events and universality. Physical Review Letters **85**(24), 5234–5237 (2000)

16. Zager, L., Verghese, G.: Graph similarity scoring and matching. Applied Mathematics Letters **21**(1), 86–94 (2008)

17. Boschloo, R.: Raised conditional level of significance for the 2×2-table when testing the equality of two probabilities. Statistica Neerlandica **24**(1), 1–9 (1970)

18. Shaffer, J.P.: Recent developments towards optimality in multiple hypothesis testing. Lecture Notes-Monograph Series, 16–32 (2006)

An Empirical Evaluation of Intrinsic Dimension Estimators

Cristian Bustos[1], Gonzalo Navarro[2], Nora Reyes[1], and Rodrigo Paredes[3]([✉])

[1] Departamento de Informática, Universidad Nacional de San Luis,
San Luis, Argentina
{cjbustos,nreyes}@unsl.edu.ar
[2] Department of Computer Science, Center of Biotechnology and Bioengineering,
University of Chile, Santiago, Chile
gnavarro@dcc.uchile.cl
[3] Departamento de Ciencias de la Computación, Universidad de Talca, Curicó, Chile
raparede@utalca.cl

Abstract. We study the practical behavior of different algorithms that aim to estimate the intrinsic dimension (ID) in metric spaces. Some of these algorithms were specifically developed to evaluate the complexity of searching in metric spaces, based on different theories related to the distribution of distances between objects on such spaces. Others were originally designed for vector spaces only, and have been extended to general metric spaces. To empirically evaluate the fitness of various ID estimations with the actual difficulty of searching in metric spaces, we compare one representative of each of the broadest families of metric indices: those based on pivots and those based on compact partitions. Our preliminary conclusions are that Fastmap and the measure called Intrinsic Dimensionality fit best their purpose.

1 Introduction

Similarity search in metric spaces has received much attention due to its applications in many fields, ranging from multimedia information retrieval to machine learning, classification, and searching the Web. While a wealth of practical algorithms exist to handle this problem, it has been often noted that some datasets are intrinsically harder to search than others, no matter which search algorithms are used. An intuitive concept of "curse of dimensionality" has been coined to denote this intrinsic difficulty, but a clear method to measure it, and thus to predict the performance of similarity searching in a space, has been elusive.

The similarity between a set of objects $\mathbb{U}$ is modeled using a *distance function* (or *metric*) $d : \mathbb{U} \times \mathbb{U} \mapsto \mathbb{R}^+ \cup \{0\}$ that satisfies the properties of triangle inequality, strict positivity, reflexivity, and symmetry. In this case, the pair $(\mathbb{U}, d)$ is called a *metric space* [6,21,24].

Partially funded by basal funds FB0001, Conicyt, Chile and Fondecyt grant 1131044, Chile.

G. Amato et al. (Eds.): SISAP 2015, LNCS 9371, pp. 125–137, 2015.
DOI: 10.1007/978-3-319-25087-8_12

In some applications, the metric spaces are of a particular kind called "vector spaces", where the elements consist of D coordinates of real numbers and the distance is some Minkowski metric. Many works exploit the geometric properties of vector spaces, but they usually cannot be extended to general metric spaces, where the only available information is the distance between objects. Since in most cases the distance is very expensive to compute, the main goal when searching in metric spaces is to reduce the number of distance evaluations. In contrast, vector space operations tend to be cheaper and the primary goal when searching them is to reduce the CPU cost or the number of I/O operations carried out.

Similarity queries are usually of two types. For a given database $S \subseteq \mathbb{U}$ with size $|S| = n$, $q \in \mathbb{U}$ and $r \in \mathbb{R}^+$, the *range query* $(q, r)_d$ returns all the objects of S at distance at most r from q, whereas the *nearest neighbor query* $kNN_d(q)$ retrieves the k elements of S that are closest to q.

A naïve way to answer similarity queries is to compare all the database elements with the query q and return those that are close enough to it. This *brute force* approach is too expensive for real applications. Research has then focused on ways to reduce the number of distance computations performed to answer similarity queries. There has been significant progress around the idea of building an *index*, that is, a data structure that allows discarding some database elements without explicitly comparing them to q.

In uniform vector spaces, the *curse of dimensionality* describes the well-known exponential increase of the cost of all existing search algorithms as the dimension grows. Non-uniform vector spaces may be easier to search than uniform ones, despite having the same *explicit* dimensionality. The phenomenon also extends to general metric spaces despite their absence of coordinates: some spaces are intrinsically harder to search than others. This has lead to the concept of *intrinsic dimensionality (ID)* of a metric space, as a measure of the difficulty of searching it. A reliable measure of ID has been elusive, despite the existence of several formulas.

Computing the ID of a metric space is useful, for example, to determine whether it is amenable to indexing at all. If the ID is too high, then we must just resort to brute-force solutions or to approximate search algorithms (which do not guarantee to find the exact answers). Even when exact indexing is possible, the ID helps decide which kind of index to use. For example, pivot-based methods work better on lower dimensions, whereas compact partiioning methods outperform them in higher dimensions [6].

In this work we aim to empirically study the fitness of various ID measures to predict the search difficulty of metric space searching. Some measures were specifically developed for metric spaces, based on different theories related to the distribution of distances between objects. Others were originally designed for vector spaces and have then been adapted to general metric spaces. We chose various synthetic and real-life metric spaces and two indexing methods that are representatives of the major families of indices: one based on pivots one and another based on compact partitions. Our comparison between real and estimated search difficulty yield, as preliminary conclusions, that *Fastmap* [10] and the formula by Chávez et al. [6] are currently the best predictors in practice.

2   Intrinsic Dimension Estimators for Vector Spaces

There are several interesting applications where the data are represented as D-dimensional vectors in $\mathbb{R}^D$. For instance, in pattern recognition applications, objects are usually represented as vectors [14]. So, data are embedded in $\mathbb{R}^D$, even though this does not imply that its *intrinsic* dimension is D.

There are many definitions of ID. For instance, the ID of a given dataset is the minimum number of free variables needed to represent the data without loss of information [2]. In general terms, a dataset $\mathbb{X} \subseteq \mathbb{R}^D$ has ID $M \leq D$, if its elements fall completely within an M-dimensional subspace of $\mathbb{R}^D$ [12]. Another intuitive notion is the logarithm of the search cost, as in many cases this cost grows exponentially with the dimension.

Even in vector spaces, there are many reasons to estimate the ID of a dataset. Using more dimensions (more coordinates in the vectors) than necessary can bring several problems. For example, the space to store the data may be an issue. A dataset $\mathbb{X} \subseteq \mathbb{R}^D$ with $|\mathbb{X}| = n$ requires to store $n \times D$ real coordinates. Instead, if we know that the ID of $\mathbb{X}$ is $M \leq D$, we can map the points to $\mathbb{R}^M$ and just store $n \times M$ real coordinates. The CPU cost to compute a distance is also reduced. This can in addition help identify the important dimensions in the original data.

There are two approximations to estimate the ID of a vector space [2,14], namely, *local* and *global* methods. The local ones make the estimation by using the information contained in sample neighborhoods, avoiding the data projection over spaces of lower dimensionality. The global ones deploy the dataset over an M-dimensional space using all the dataset information.

In this work we focus on global ID estimators. That is, we consider all the dataset information to estimate the ID as accurately as possible. Global methods can be split into three families: projection techniques, multidimensional scaling methods, and fractal based methods. The last two are more suitable to extend to metric spaces, so we have selected and adapted some representatives of these groups.

3   Intrinsic Dimension Estimators for Metric Spaces

In this section we analyze various methods to estimate the ID of vector spaces and others to general metric spaces. We discuss how to adapt the former to the case of general metric spaces. Note that, since multidimensional spaces are a particular case of metric spaces, our estimators can also be applied to obtain the ID of D-dimensional vector spaces.

3.1 Fractal Based Methods

Unlike other families, fractal based methods can estimate non-integer ID values. The most popular techniques of this family are *Box Counting* [17], which is a simplified version of the *Haussdorff dimension* [9,18], and *Correlation* [3].

The dimension estimation by Box Counting D_B of a set $\Omega \subseteq \mathbb{R}^D$ is defined as follows: if $v(r)$ is the number of boxes of size r needed to cover Ω, then $D_B = \lim_{r \to 0} \frac{\ln(v(r))}{\ln(\frac{1}{r})}$.

In this method, the boxes are multidimensional regions of side r on each dimension (that is, they are hypercubes of side r). Regrettably, even though efficient algorithms have been proposed, the Box Counting dimension can be computed only for low dimension datasets, because its algorithmic complexity grows exponentially with the dimension.

Estimating the dimension by Correlation is an alternative to Box Counting. It is defined as follows. Let $\Omega = \{x_1, x_2, \ldots, x_n\} \subset \mathbb{R}^D$ and the correlation integral $C_m(r) = \lim_{n \to \infty} \frac{2}{n(n-1)} \sum_{1 \leq i < j \leq n} I(\|x_j - x_i\| < r)$, where $I(\cdot)$ is the indicator function. Intuitively, $C_m(r)$ is the fraction of object pairs whose distance is lower than r. So, the dimension estimation by Correlation D_C is $D_C = \lim_{r \to 0} \frac{\ln(C_m(r))}{\ln r}$.

The most popular method to estimate the dimension by Correlation and Box Counting is the log-log plot. It consists in plotting $\ln(C_m(r))$ versus $\ln(r)$. The dimension by Correlation is the slope of the linear section of the curve. To estimate the dimension by Box Counting we replace $\ln(C_m(r))$ by $\ln(v(r))$.

In the general case of metric spaces, we do not have coordinates in general. Thus, to adapt the Box Counting method, we consider *balls* of radius r, that is, the set of objects within a distance r from a reference object o. We randomly pick the reference objects from the dataset, and count the number $B(r)$ of balls of radius r needed to cover the dataset. To do so, we use the *List of Clusters (LC)* index [5], whose code is available from SISAP [11], with the variant of fixed radius and centers chosen at random. Then the ID is just the length of the LC.

To estimate the dimension by Box Counting, which in this case is Ball Counting, we replace $\ln(v(r))$ by $\ln(B(r))$, plot $\ln(B(r))$ versus $\ln\left(\frac{1}{r}\right)$ in log-log and obtain the slope of the linear section of the curve by using linear regression with least squares over the experimental data $\left(\ln(B(r)), \ln\left(\frac{1}{r}\right)\right)$.

3.2 Distance Exponent

Traina et al. [22] discuss the problem of the selectivity estimation for range queries in real-world metric spaces, including spatial or multidimensional datasets as special cases. Their main finding is that several datasets follow the so-called *Power Law*. They call *Distance Exponent* the exponent of the power law, and show how to use it to derive formulas for estimating the selectivity of range queries. For instance, the number of objects relevant to the query, the number of I/Os to answer the query when the data is stored on disk, the amount of time needed to answer the query, and so on.

To find a formula that estimates the number of neighbors of objects within a distance r in a n-objects dataset, they introduce the following notions: (i) the *Distance Plot* of a metric set is the number of object pairs at distance at most r versus the distance r, and both axes are drawn in logarithmic scale; and (ii) the *Distance Exponent* is the slope of the line that better fits the distance plot in case it is linear for a range of scales. Using these two notions, they define the *Distance Law*.

Distance Law - *Given a dataset of n objects from a metric space with distance function $d(x, y)$, the average number of distances lower than a radius r follows a power law; that is, the average number of neighbors $\overline{nb}(r)$ within a distance r is proportional to r^D. Formally, $N \cdot \Phi(r) = \overline{nb}(r) \propto r^D$.*

If a dataset has a metric to evaluate the distance between every object pair, then this plot can always be drawn. They show that the distance plot has an almost linear behavior for many databases, both real and synthetic. Building the distance plot requires $O(n^2)$ distance computations. To reduce this cost, $\overline{nb}(r)$ is estimated using an index [22], in particular the *M-tree* [7]. Since in this work we are only interested in comparing the different ID measures, indexing the space is not necessary and we compute $\overline{nb}(r)$ directly, considering a reference object chosen at random from the dataset. We only determine the number of elements at distance r from that object. The result is averaged over various choices for the object.

3.3   Fastmap

This method arises from the proposal [13] of a fast algorithm to map objects of any metric space onto points of a k-dimensional space (k being defined by the user), so that the dissimilarities are preserved. Its goal is to speed up searches in traditional or multimedia databases.

To do so, the objects are mapped onto the k-dimensional space using k feature extraction functions, provided by domain experts [13]. The main issue is how to define such feature extraction functions. For example, in the metric case of strings with the edit distance [16], it is not clear which features can be considered.

For a domain expert, it is generally easier to provide a distance function to compare objects than to provide feature extraction functions. *Fastmap* [10] is a generalization of the original method [13], where the objects are mapped using only a distance function.

Fastmap finds, given a dataset of n objects from a metric space $(\mathbb{U}, d)$, n image points in a k-dimensional target space, such that the distances between the objects in the original space are preserved as much as possible in the target space.

For evaluating the dissimilarity preservation in the target space, a *stress* function is defined as follows, $stress^2 = \frac{\left(\sum_{i,j} (d_{ij} - \hat{d}_{ij})^2\right)}{\left(\sum_{i,j} d_{ij}^2\right)}$, where d_{ij} is the dissimilarity measure (the distance of the original space) between objects o_i and o_j, and $\hat{d}_{ij}$ is the Euclidean distance between their respective images p_i and p_j. The stress function gives the relative error that the distances in the target space suffer on average after the transformation. Fastmap begins with an estimation that is iteratively improved, until no additional improvement is possible.

In the metric case, we can assume that we have the $n \times n$ matrix of distances between all the dataset objects, and Fastmap must find n points in the k-dimensional space whose Euclidean distances are close to the original matrix of $n \times n$ distances. The crux is to assume that objects are points in some m-dimensional space, with unknown m, and to project these points over k mutually orthogonal directions. The challenge is to compute all these projections using only the distance matrix. Fastmap projects the objects over carefully selected lines. It chooses two objects o_a and o_b, and considers the "line" passing through them in the original space. The projections of the objects over this line are obtained using the *cosine law*:

Theorem 1 (*Cosine Law*). Any triangle $o_a \overset{\triangle}{o_i} o_b$ satisfies:

$$d(o_b, o_i)^2 = d(o_a, o_i)^2 + d(o_a, o_b)^2 - 2x_i' d(o_a, o_b). \tag{1}$$

Eq. 1 can be solved for x_i' to compute the projection of object o_i with the formula $x_i' = \frac{d(o_a,o_i)^2 + d(o_a,o_b)^2 - d(o_b,o_i)^2}{2d(o_a,o_b)}$.

Thus, the input of Fastmap is a set S of size n and, in each iteration, it computes the coordinates of all the n objects over the new axis. So, after k iterations, it produces a k-dimensional target space S' where each object $o_i \in S$ is mapped to a k-coordinate vector $p_i = (x_{i,1}', x_{i,2}', \ldots, x_{i,k}') \in S'$, where $x_{i,j}'$ is the j-th projection of the image p_i of the object o_i.

In our case, we want to estimate the number of projections needed so that the target space reaches a mapping with a small enough *stress*, that is, preserving the distances sufficiently well. Thus, we modify the Fastmap algorithm so that it computes the *stress* of the target space after each new dimension is added. If the difference between the current and the previous *stress* values is significative, we compute another projection (thus increasing the dimensionality of the target space). Otherwise, the current dimension of the target space is reported as the estimation of the ID of the original metric space.

3.4   Intrinsic Search Difficulty

Chávez et al. [6] introduced a measure of the intrinsic complexity of searching in general metric spaces. It is easy to estimate and independent of the search algorithm.

Several authors [1,4,8] have proposed to use the *distance histogram* to characterize the hardness of searching in arbitrary metric spaces, yet the cost was tailored to a specific index. Instead, this measure [6] depends only on the histogram and not on any assumption on the indexing method.

The intuition behind this measure is that, in random vectors in D dimensions, the histogram has a larger mean μ and a smaller variance σ^2 as D increases. In fact, it holds $D = c \cdot \mu^2/\sigma^2$ for some constant c [23]. Thus, the same formula could be used to estimate a dimension D from the mean and variance of the histogram of distances in a general metric space. We do not have the histogram of the whole universe $\mathbb{U}$, but we can approximate it using the histogram of the dataset $S \subset \mathbb{U}$.

Definition: Let μ be the mean and σ^2 be the variance of the histogram of distances of a metric space. Then, its *intrinsic search difficulty* is $\rho = \frac{\mu^2}{2\sigma^2}$.

An obvious advantage of this measure, which has contributed to its popularity, is that it is easy to compute from a reasonable sampling of pairs in S. Other techniques require more complex and expensive computations.

Pestov [19] presents a system of three axioms an intrinsic dimension function must satisfy. He proves that the intrinsic dimension measure ρ satisfies a weak version of these axioms. Later [20], he introduces a set of goals an intrinsic dimension function should fulfill, and ρ achieves many of them.

As the measure ρ has been designed for general metric spaces, we use it as is. We consider the dataset S and we compute all the distances $d(x, y), \forall\ x, y \in S$. Then we compute the average $\mu = \frac{1}{n^2} \sum_{x,y \in S} d(x, y)$ and the variance $\sigma^2 = \frac{1}{n^2} \sum_{x,y \in S} (d(x, y) - \mu)^2$. Finally, we obtain the value of $\rho = \frac{\mu^2}{2\sigma^2}$ and report it as the ID of the metric space.

4   Experimental Results

We evaluate experimentally the four ID estimators described, on general metric spaces. We consider two kinds of metric spaces, depending on the data source:

Synthetic: these are spaces generated artificially so that they present some interesting characteristic to be evaluated. For instance, uniformly distributed vectors in $\mathbb{R}^D$ with known dimension.

Real world: these are metric spaces obtained from real-world applications. For instance, a feature vectors space of images obtained from a NASA image set.

4.1   Synthetic Metric Spaces

These are vector spaces with Euclidean distance. They are treated as metric spaces, as we do not consider the coordinate information. A first set is formed by vectors with uniform distribution, so that the representational dimension matches the ID. Here we can test the estimators in a case where the ID is known. A second set is formed by vectors with Gaussian distribution, so that the representational dimension is greater than the ID (the more clustered is the space, the lower is the ID). The distance is also Euclidean. Here we aim to check whether the estimators give lower values as the ID decreases.

Uniformly Distributed Vectors with Euclidean Distance. We generate four datasets of 100,000 vectors uniformly distributed in the unitary cube $[0, 1)^D$, with $D = 5$, 10, 15 and 20. The spaces are called C5, C10, C15 and C20, respectively.

Fig. 1(a) depicts the estimations for these four metric spaces. As it can be seen, the Fractal estimator (Ball counting) is insensitive to the correct dimension. The Distance Exponent increases with D, but not proportionally. The other two estimations grow at the same rate of D, with Fastmap matching it almost perfectly and Intrinsic Search Difficulty showing a consistent factor multiplying D.

Search degradation as ID grows. To verify that the dataset ID is responsible of the search degradation, we pick C5 and extend its vectors with zeroes to produce spaces with 10, 15 and 20 representational dimensions, and study the search performance over it.

We perform 10 executions of the algorithms, building the index with 90% of the database elements, and reserving the remaining 10% (chosen at random) for the queries. So, the query objects do not belong to the index. We average the results over the 10 executions. In each execution, the objects in the metric space are permuted at random. Therefore, each of the 10 indices uses a different dataset S, and the query objects are also different.

We use a pivot index and a compact partition index. For the pivot index family, we use the generic pivot algorithm. We choose at random a set of pivots $\mathcal{P} = \{P_1, P_2, \ldots, P_k\} \subset S$ of size $|\mathcal{P}| = k$. We store the kn distances between pivots and objects, and use them to filter out candidates using the triangle inequality. For each space, we experimentally determine the number of pivots that obtains the best search performance. Thus, the results shown for each case correspond to the best possible ones for this method, in the corresponding metric space.

For the case of compact partition based algorithms, we consider the LC, which is one of the best indexes for medium and high dimensions [5]. We use the LC variant that has a maximum size for each cluster. For each metric space considered, we experimentally determine the cluster size whose perfomance is the best, and this is the result shown in the plots.

In Fig. 2, we show the cost of range queries retrieving 0.01%, 0.1% and 1% of the vector dataset per query, using the generic pivot index (Fig. 2(a)) and the LC (Fig. 2(b)). These results are compared with the ones for searching C10, C15 and C20. Both plots show that the four spaces of ID 5 overlay each other (independently of the representational dimension of the space), while the curves for C10, C15 and C20 show the usual degradation.

Gaussian Distributed Vectors with Euclidean Distance. We generate 100,000 vectors in $\mathbb{R}^D$ with mean $\mu = 1$ and variance $\sigma^2 = 0.1$, for $D = 5$, 10, 15 and 20. In these spaces, there are no, a priori, clusters of elements. These spaces are called G5, G10, G15 and G20.

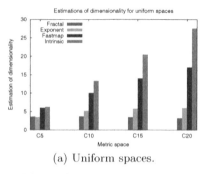
(a) Uniform spaces.

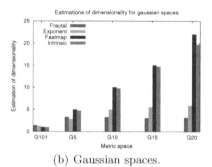
(b) Gaussian spaces.

Fig. 1. Comparison of dimensionality estimations for synthetic metric spaces.

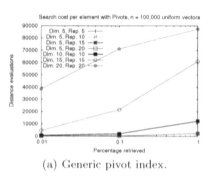
(a) Generic pivot index.

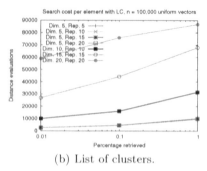
(b) List of clusters.

Fig. 2. The searching effort does not vary when the ID of the space does not change.

We also generate 100,000 vectors in $\mathbb{R}^{101}$ with mean $\mu = 1$ and variance $\sigma^2 = 0.1$ with 200 clusters (the cluster centers are uniformly distributed in the space). This space is called G101.

Fig. 1(b) shows the estimations obtained with Fractal, Distance Exponent, Fastmap, and Intrinsic Search Difficulty, for these metric spaces. Again, the Fractal estimation fails in these spaces, being insensitive to the dimension, and the Distance Exponent grows very slowly. The other two measures grow proportionally to D as they should, although Fractal is less sensitive to the fact that the distribution is not uniform. Instead, the Intrinsic Search Difficulty gives markedly lower values than in the uniform case.

4.2   Real Metric Spaces

We pick four spaces from the Metric Library [11] [1] in order to estimate their IDs with the four ID estimators. The selected spaces are varied:

Dictionary: it is a dictionary of 69,069 English words. In this space, we use a discrete function, the *Edit Distance* or *Levenshtein Distance* [16].

---
[1] Available at http://www.sisap.org/library/dbs/.

NASA: this is a set of 40,700 images from NASA, represented as feature vectors of 20 real coordinates per vector, under the Euclidean distance. They were generated from images downloaded from the NASA site.

Histograms: this is a dataset of 112,682 histograms of medical images, each one composed by 8-D color histograms of 112 real components. As any quadratic form function can be used as the distance in this case, we also have chosen the Euclidean distance, as it is the simplest alternative.

Documents: this space has 1,265 documents, represented as vectors according to the vectorial model of documents used in the Information Retrieval field. To compare documents we use the *cosine distance*. Each vector has a coordinate for each vocabulary term in the colection, and documents can be seen as points in a vector space of high representational dimension. The documents are files obtained form the TREC-3 collection.

To measure the intrinsic hardness of the searching, we consider the same two indices as before, using range queries:

Dictionary: As the metric is discrete, we use radii 1, 2, 3, and 4, retrieving on average about 0.003%, 0.037%, 0.326%, and 1.757% of the database.

NASA: In this continuous metric we use radii 0.012, 0.285, and 0.53, retrieving on average approximately 0.01%, 0.1%, and 1% of the dataset.

Histograms: This metric is also continuous. To retrieve on average approximately 0.01%, 0.1%, and 1% of the dataset, we use query radii 0.051768, 0.082514, and 0.131163.

Documents: The distance is also continuous. We use query radii 0.14, 0.15, and 0.195, which retrieve on average 0.01%, 0.1%, and 1% of the database.

Figs. 3 and 4 show the correlation between the search cost with the Pivot index and the List of Clusters, respectively, and the estimation reported for each considered ID estimator. For lack of space, we only show the results of the search that retrieve 0.01% and 0.1% of the database.

We plot the ratio between the logarithm of the search cost and the estimations of the ID. This measures how close is the logarithm of the predicted ID to the

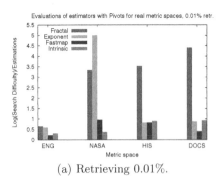

(a) Retrieving 0.01%.

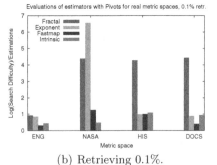

(b) Retrieving 0.1%.

Fig. 3. Comparison of ID estimators for real metric spaces, using Pivots.

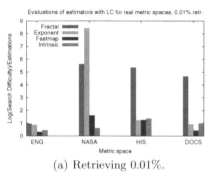
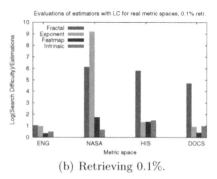

(a) Retrieving 0.01%.   (b) Retrieving 0.1%.

Fig. 4. Comparison of ID estimators for real metric spaces, using List of Clusters.

actual search costs: if the search cost is consistently $s = c^d$, where d is the predicted ID and c is a constant, then the plots should always be close to $\log c$. Thus the best methods are those that give roughly the same value regardless of the query radius and index used.

As on the synthetic spaces, Fastmap and the Intrinsic Search Difficulty turn out to be the best predictors for both types of indices. The Distance Exponent performs generally well, except for the NASA dataset.

5   Conclusions

The intrinsic dimension (ID) of metric spaces measures their search difficulty, independently of the type of index used. Computing the ID is useful to determine whether a metric space can be indexed at all (or we must resort to sequential scanning or approximate methods), which kind of index would perform better, and what search performance to expect.

In this paper we have analyzed several ID estimators in a practical perspective. Some were defined for D-dimensional coordinate spaces, and we have adapted them to the more general metric spaces. We compared their predictions with the actual search cost using various synthetic and real-life metric spaces, so as to verify which are better at predicting the search difficulty.

Although our results are preliminary, they suggest that the best performing measures in practice are *Fastmap* [10] and the simple measure proposed by Chávez et al. [6]. Instead, the Distance Exponent [22] and our generalization of Box Counting [17] did not perform so well.

The reason for the failure of Box Counting may be that it needs an extremely large number of objects to correctly estimate D. An estimation [2] is $D < 2 \log_{10} N$, which in our case implies that the method could have worked well up to $D = 10$ only. However, in our experiments the adapted method failed even in this case. It may be that our adaptation to computing it using the List of Clusters [5] is too crude (as other clustering methods may cover the dataset with fewer balls). In any case, this shows that the method is not easy to apply, but we plan to further study this issue with more points and other clustering methods. The reason for the failure of the distance exponent, which does not present issues to be adapted, is also unclear and deserves further research.

We also plan to analyze other estimators. For instance, we can study the correlation dimension [3], the concentration dimension [19], or the classical Principal Component Analysis (PCA) method [15] (which is defined on vector spaces).

References

1. Brin, S.: Near neighbor search in large metric spaces. In: Proc. 21st Conf. on Very Large Databases (VLDB 1995), pp. 574–584 (1995)
2. Camastra, F.: Data dimensionality estimation methods: a survey. Pattern Recognition **36**(12), 2945–2954 (2003)
3. Camastra, F., Vinciarelli, A.: Estimating the intrinsic dimension of data with a fractal-based method. IEEE TPAMI **24**(10), 1404–1407 (2002)
4. Chávez, E., Marroquín, J.: Proximity queries in metric spaces. In: Proc. 4th South American Workshop on String Processing (WSP 1997), pp. 21–36. Carleton University Press (1997)
5. Chávez, E., Navarro, G.: A compact space decomposition for effective metric indexing. Pattern Recognition Letters **26**(9), 1363–1376 (2005)
6. Chávez, E., Navarro, G., Baeza-Yates, R., Marroquín, J.: Searching in metric spaces. ACM Computing Surveys **33**(3), 273–321 (2001)
7. Ciaccia, P., Patella, M., Zezula, P.: M-tree: an efficient access method for similarity search in metric spaces. In: Proc. 23rd VLDB, pp. 426–435 (1997)
8. Ciaccia, P., Patella, M., Zezula, P.: A cost model for similarity queries in metric spaces. In: PODS, pp. 59–68 (1998)
9. Eckmann, J.P., Ruelle, D.: Ergodic theory of chaos and strange attractors. Rev. Mod. Phys. **57**, 617 (1985)
10. Faloutsos, C., Lin, K.-I.: Fastmap: a fast algorithm for indexing, data-mining and visualization of traditional and multimedia datasets. In: Proc. 1995 ACM SIGMOD Intl. Conf. on Management of Data, pp. 163–174. ACM Press (1995)
11. Figueroa, K., Navarro, G., Chávez, E.: Metric spaces library (2007). http://www.sisap.org/Metric_Space_Library.html
12. Fukunaga, K.: Introduction to Statistical Pattern Recognition, 2nd edn. Academic Press Professional Inc, San Diego (1990)
13. Jagadish, H.V.: A retrieval technique for similar shapes. In: SIGMOD Conference, pp. 208–217. ACM Press (1991)
14. Jain, A.K., Dubes, R.C.: Algorithms for Clustering Data. Prentice-Hall Inc, Upper Saddle River (1988)
15. Jolliffe, I.T.: Principal Component Analysis, 2nd edn. Springer Series in Statistics. Springer (2002)
16. Levenshtein, V.I.: Binary codes capable of correcting deletions, insertions, and reversals. Soviet Physics Doklady **10**(8), 707–710 (1966)
17. Mandelbrot, B.: Fractals: Form, Chance and Dimension. W. H. Freeman, San Francisco (1977)
18. Ott, E.: Chaos in Dynamical Systems. Cambridge University Press, Cambridge (1993)
19. Pestov, V.: Intrinsic dimension of a dataset: what properties does one expect? In: Intl. Joint Conf. on Neural Networks (IJCNN), pp. 2959–2964 (2007)

20. Pestov, V.: An axiomatic approach to intrinsic dimension of a dataset. Neural Networks **21**(23), 204–213 (2008). Advances in Neural Networks Research: 2007 International Joint Conference on Neural Networks (IJCNN)
21. Samet, H.: Foundations of Multidimensional and Metric Data Structures (The Morgan Kaufmann Series in Computer Graphics and Geometric Modeling). Morgan Kaufmann Publishers Inc., San Francisco (2005)
22. Traina Jr., C., Traina, A.J.M., Faloutsos, C.: Distance exponent: a new concept for selectivity estimation in metric trees. Research Paper 99–110, School of Computer Science, Carnegie Mellon University, 03/1999 (1999)
23. Yianilos, P.: Excluded middle vantage point forests for nearest neighbor search. In: DIMACS Implementation Challenge, ALENEX 1999, Baltimore, MD (1999)
24. Zezula, P., Amato, G., Dohnal, V., Batko, M.: Similarity Search: The Metric Space Approach. Advances in Database Systems, vol. 32. Springer (2006)

A Belief Framework for Similarity Evaluation of Textual or Structured Data

Sergej Znamenskij[✉]

Ailamazyan Program Systems Institute of RAS,
Pereslavl-Zalesskii, Yaroslavl Region 152021, Russia
svz@latex.pereslavl.ru

Abstract. This paper discovers the major shortcomings of the Levenshtein Distance method, the longest common subsequence (LCS) method, and other general approaches to finding common parts, including the unjustified fragmentation of selected parts, the lack of sensitivity to transposition of large blocks, and no mechanisms to prevent accidental matches. The belief function theory leads to a flexible framework for similarity evaluation. The framework is aimed on new similarity models which are free of described shortcomings and can be effectively calculated. A sketch of better sequence alignment algorithm illustrates the framework's utility.

Keywords: Levenshtein distance · Longest common subsequence · Sequence alignment · Change detection · Graph similarity · Diff utility · Fuzzy measure · Belief function

Introduction

The Levenshtein Distance, which measures the similarity of two sequences of objects, is widely used in various tools. Its quantification involves a search for a common subsequence of objects, which determines how many elements need to be deleted or inserted in order to obtain the second specified sequence from the first one.

As there may be multiple common subsequences, the longest common subsequence (LCS) is chosen among them to minimize the edit distance (also called *Damerau-Levenshtein Distance*) — the total number of objects deleted and inserted.

It has long been known that the longest subsequence may not be the best one. In many cases, the lengths of sequential deletions and (or) insertions (substitutions) series are also accounted while quantifying the edit distance. Therefore, the chosen subsequence may not be the longest one eventually; nevertheless, the practical outcome is often improved.

S. Znamenskij—This work was performed under financial support from the Government, represented by the Ministry of Education and Science of the Russian Federation, within the scope of Project No. FMEFI60414X0138.

G. Amato et al. (Eds.): SISAP 2015, LNCS 9371, pp. 138–149, 2015.
DOI: 10.1007/978-3-319-25087-8_13

Identifying the longest common subsequence can be inefficient in practical applications as well [1]. For example, the following sequences of letters and spaces

"`ineffective common efforts`"
"`self-finance comes ineffective`"

have the nonsensical longest common subsequence "`effie com efft`", instead of the obviously meaningful long common substring "`ineffective`". When tracking changes in a source code file, this approach yields (see, for example, [2]) hardly meaningful repeated strings that are either empty or contain a single character, usually a curly brace, identified as common parts.

Even one or two decades ago, it was possible to edit a macro in LATEX and use diff or another similar utility to track changes in logs in order to analyze errors in the test file via the `\tracingall` command. But as the size of logs increased, the change tracking tool virtually became ineffective because it yielded a useless long chain of short, frequently repeated fragments as a common part of long files. For exactly the same reasons, change tracking in logs generated by tracing system calls and signals (strace), which had been quite helpful during the era of short logs, also ceased to function.

This situation calls for a study in order to understand whether the length of a common subsequence should be substituted by another criterion that better reflects the infortmativeness of the selection.

A sense of the "similarity" term may differ even for textual data: one can consider similarity of character encodings, languages, styles of writing, subjects, ideas and so on. This paper treats similarity only as an expected value of most valuable common part. Both the common part identification and the identification of its value may also have different meaning. For example, have the strings "expected value" and "value expected" common part consisting of two words or not, depends on the *common part* meaning.

This paper aims moving from a choice between a too slow optimal algorithms and fast unsafe heuristic algorithms to a framework for constructing a meaningful problem formulation with fast and reliable solution algorithms. The framework should contain a general plan for the convenient formulation of similarity problem. This plan must be extremely flexible, giving a chance to select a formulation which can be solved with a fast parallel reliable algorithm.

Starting at a measuring of common subsequence value, we need a function μ which takes a subsequence and return appropriate positive number. It is the set function like the cardinality and the measure. If it is shift invariant and additive, it directly leads to LCS.

Replacing additivity by superadditivity, we comes to belief function [3], also known as fuzzy measure [4] or (non-harmonic) capacity [5].

A finite sequence appears to be a directed graph of linear ordered finite set. Next we introduce possibly generic definition of *believed similarity framework*. The section 2 discuss a problem of choice for the proper belief function and describe large application areas of Levenshtein distance (and also LCS), where it sometimes gives obviously improper solution.

The exact proper solution is known to be too expensive. The section 3 contains some explanations how the framework lead to very scalable and robust similarity algorithms. Unfortunately they are not so simple to fit this paper.

1   Framework Definition

Let's start with convenient notation of commonly used terms to define common part of two graphs which is a base for similarity.

A **graph** $X = (X_o, X_i)$ is a set of objects (vertices and edges) $X_o = X_v \cup X_e$ with the incidence relation $X_i \subset X_v \times X_e$ such that

$$\forall_{e \in X_e} \quad 1 \leqslant \left| \{ v \in V_x \mid (v, e) \in X_i \} \right| \leqslant 2 . \tag{1}$$

The number 2 in the last formula can be replaced by greater value if hypergraphs are wanted. A **map** φ of graph X to graph Y is a map $\varphi : X_o \to Y_o$ such that

$$(v, e) \in X_i \iff (\varphi(v), \varphi(e)) \in Y_i . \tag{2}$$

A map is isomorphic if it puts X_o to one-to-one correspondence onto Y_o. We note $\mathfrak{G}$ the set of all **graphs**, $\mathcal{O} = \mathcal{V} \cup \mathcal{E}$ the set of all their **objects** (vertices and edges), and $\Phi(X, Y)$ the group of all isomorphic **maps** from $X \in \mathfrak{G}$ to $Y \in \mathfrak{G}$.

A graph X is called to be a **subgraph** of Y and write $X \subset Y$ if $X_o \subset Y_o$ and $X_i = Y_i \cap (X_v \times X_e)$. The set of all **subgraphs** of graph X is usually noted as 2^X.

Definition 1. *A **common part** p of graphs X, Y means a triple of $p_1 \subset X_o$, $p_2 \subset Y_o$ and $p_3 \in \Phi(p_1, p_2)$ We note $\mathfrak{P}(X, Y)$ the set of all **common parts** for X, Y. A subset $\mathcal{P}(X, Y) \subset \mathfrak{P}(X, Y)$ is Φ-invariant if both*

$$\forall_{\psi \in \Phi(X, Z)} (p_1, p_2, p_3) \in \mathcal{P}(X, Y) \iff (\psi(p_1), p_2, \psi^{-1} \circ p_3) \in \mathcal{P}(Z, Y) \tag{3}$$

and

$$\forall_{\psi \in \Phi(Y, Z)} (p_1, p_2, p_3) \in \mathcal{P}(X, Y) \iff (p_1, \psi(p_2), p_3 \circ \psi^{-1}) \in \mathcal{P}(X, Z) . \tag{4}$$

As far as finite sequences or related objects are compared for similarity identification, the inclusion $X \subset Y$ appears to be a very special similarity case, just because $(X \subset Y) \& (Y \subset X) \iff X = Y$. We doomed to miss it if we focus only on symmetric similarity functions such as metric. For example, if we compare the versions of document, the deleting of the last half is a simple change, but the opposite isn't.

The classic idea is that the believed property is a total amount of all features of X which potentially can be regarded as common. Believed similarity is just a common features amount. Then the difference looks like distance but may be asymmetric. It satisfies the **basic similarity inequalities**:

$$S(X, X) \geqslant S(X, Y) \leqslant S(Y, Y) \geqslant S(X, Y) + S(Y, Z) - S(X, Z). \tag{5}$$

The last inequality is equivalent to triangle equality

$$d_q(X,Y) + d_q(Y,Z) \geqslant d_q(X,Z) \qquad (6)$$

for an **asymmetric similarity distance** function $d_q(X,Y) = S(X,X) - S(X,Y)$ and also for **symmetric similarity distance**

$$d(X,Y) = \frac{d_q(X,Y) + d_q(Y,X)}{2} = \frac{S(X,X) - S(X,Y) + S(Y,Y) - S(Y,X)}{2}. \qquad (7)$$

If similarity happen to distinguish objects ($d_q(X,Y) = d_q(X,X) \implies X \subset Y$), then d is a metric and d_q is a quasi-metric[1] [12,13].

Definition 2. *We call a **believed similarity framework** (BSF) any fixed triple of following grounds:*

1. *An **object similarity** function* $s : (\mathcal{V} \times \mathcal{V}) \cup (\mathcal{E} \times \mathcal{E}) \to \mathbb{R}_+$ *which satisfies basic similarity inequalities 5.*
2. *A Φ-invariant subset of **acceptable common parts** $\mathcal{P}(X,Y) \subset \mathfrak{P}(X,Y)$.*
3. *A Φ-invariant **superadditive fuzzy measure** also called **belief function** μ over any $X \in \mathfrak{G}$ which maps subgraphs to their* informativity *values $\mu : 2^X \to \mathbb{R}_+$.*

The first usually initially given only as a relation between vertex elements or as a similarity matrix [9]. It can incorporate information on local structure similarities (e.g. vertex valences or bridges).

The second implements beliefs related to structures of common parts (should they be connected or not, linearly ordered or not, embeded or minor, etc.) and may be asymmetric.

The third (μ) either reflects *a priori* beliefs and may also benefit from learning. It effects in reliability and quality of application result. A Φ-invariance of μ means $\forall_{X \in \mathfrak{G}} \forall_{\varphi \in \Phi} \mu(A) = \mu(\varphi(A))$.

Definition 3. BSF *identifies a **weight** w of common part $p \in \mathcal{P}(X,Y)$ for graphs X and Y as the Choquet integral [5] of object similarity $s_p(o) = s(o, p_3(o))$ over μ*

$$w(p) = (C)\int_{p_1} s_p \, d\mu \qquad (8)$$

and *believed similarity* $S(X,Y) = \max_{p \in \mathcal{P}(X,Y)} w(p)$.

Here the Choquet integral can be calculated by formula

$$w(p) = (C)\int_{p_1} s_p \, d\mu = \sum_{A \subset p_1} m(A) \min_{o \in p_1} s_p(o) \qquad (9)$$

where m is the the Möbius transform of μ

---

[1] Don't mix with *quasimetric* - this term by mischance has different meaning!

$$m(A) = \sum_{B \subset A} (-1)^{|A \setminus B|} \mu(B) \tag{10}$$

which is non-negative for superadditive fuzzy measure, and the original set function can be recovered from m through the zeta transform [14]:

$$\mu(A) = \sum_{B \subset A} m(B) . \tag{11}$$

Here $m(A)$ is the expected value of special information contained in A which does not appear in its subsets. For brevity we shall call $m(A)$ *significance* of A for our belief function. As a result of learning it may reflect a set of graphs $\mathcal{G}$ as a number of occurences $\mu(A) = \left| \{ G \in \mathcal{G} \mid \exists_{\varphi \in \Phi} \ \varphi : A \to G \} \right|$ for use in similarity search.

Theorem 1. *The believed similarity S satisfies basic similarity inequalities* (5).

Proof. The first two inequalities immediately follows from the definitions. Let's prove the last inequality.

Let's fix for each $A, B \in \{X, Y, Z\}$ the common part $p_{\max}(A, B)$ which has a maximal weight w. We can select their intersection in Y and construct a common part $p_c(X, Z)$ from intersection $(p_{\max}(X, Y))_2 \cap (p_{\max}(Y, Z))_1$ and their maps. By a similarity inequalities for object similarity function we get

$$w(p_{\max}(Y, Y)) \geqslant w(p_c(X, Z)) = w(p_{\max}(X, Y)) + w(p_{\max}(Y, Z)) - w(p_{\max}(X, Z)) \tag{12}$$

just from reduction of similar terms after applying the formula 9 and a basic similarity inequalities for object similarity function.

Computational complexity is the main challenge for graph similarity algorithms. Non-polinomial complexity of the search for the best common part of graphs is mentioned in a number of publications, for example in [15]. Nonetheless, the big data applications depends on algorithms of nearly linear time complexity.

To speed up the algorithm, while preserving the required accuracy and reliability, it is customary to slightly modify the definition of the problem. Instead of a precise solution, the accelerated algorithm most often provides an approximate one. "Greedy" algorithms, which are way off the mark sometimes, are also popular.

In order to use the framework in applications we have to get a special framework grounds to match the concrete practical needs in unknown fancy way. The framework needs careful preparation for use based on research of framework grounds under discovering of peculiarities of the practical problem and ideas of effective algorithm development.

2   Selection of a Belief Function

Let's apply the believed similarity framework to the diff utility problem which was already mentioned in the introduction.

We focus on the *basic probability mass assignment* to a subsets m which is known to be a general approach to define μ. The idea is to select subsets with a special meaning and identify their *significance*.

2.1 Subsets with Special Significance

Table 1. Informative parts in applications of LCS and Levenshtein distance

Applied task:	Significant part
Data comparison and synchronization of:	word, sentence, phrase,
- LaTeX documents	paragraph, or section, balanced code fragment
- XML code	tag, node, branch
- file systems	file, directory, block
- source code	procedure, function, block
- system logs	function call trace
Similarity search for:	
- fuzzy matching of records in textual database	geographic name, standard term or combination
- melodic theme in a music records database [6]	particular fragment of the composition
- recognizing human speech and voices of birds and animals [7,8]	phoneme, syllable, word, phrase
Plagiarism detection	text fragment containing an idea
Clustering textual data	simple or compound term
Analyzing genetic information	RNA complementary fragments ...

The Table 1 outlines typical examples of informative parts from various areas of applications. They leads to important observations:

1. Exact identification of each subset significance seems to be impossibly hard.
2. All identified significant subsets are substrings.
3. Significant substrings may be long.

We may limit for mentioned in the Table 1 consideration of significant subsets by the class of substrings as finite linear ordered set graphs. In order to simplify identifying of substring significance, we consider m to be a function of its length $m(A) = p(|A|)$ and compare different approaches.

2.2 Longest Common Subsequences as a Belief Function

The classic LCS approach obviously fits into the simplified scheme. If m is equal to 1 for a singletons and vanishes on other sets, then $\mu(A) = |A|$. It explains why

LCS works correctly only when contributions of long matches can be neglected, which occurs more frequently when short sequences are compared.

If the cost of deleting or inserting an element is equal to half the cost of replacing the element, then the Levenshtein Distance is explicitly expressed in terms of the LCS distance and therefore, represents the sum of contributions (equal to 1) of substrings containing 1 character.

2.3   Linear Gaps Accounting as a Belief Function

The Needleman–Wunsch algorithm [9] targets approximate matches and adds the quantified similarity of the values in the identified common subsequence to the sum of the values of a linear[2] (affine) function of the gap size.

Remark 1. Let the sequences to align A and B consist of common part C and indels and n_2 is a total number of strings in C of length 2. Suppose that there is no replacement, insertions and deletions are separated. Then

1. The common part contains $|C|$ strings of lengths 1.
2. The total amount of gaps (new-indel cost) is equal to number of substrings in C, which is equal to $|C| - n_2$.
3. The total number of gaps (extend-indel cost) is equal to $|A| + |B| - 2|C|$ - n_2.

Therefore the Levenshtein score for $|C|$ with linear gap function is equal to linear combination of $n_1 = |C|$, n_2 and $|A| + |B|$.

We see that the score of linear gap function depends linearly on the sum of the contributions of the strings with the lengths of 1 and 2. Therefore, the enhancement of LCS method by the linear gap function accounting, could also lead to unjustified fragmentation of the common subsequence.

For instance, for the strings

"mathematical informatics"
"informatics for mathematics"

the common subsequence, "matic formatics", as compared to "informatics", has more common substrings with the lengths of 1 to 3, but fewer common substrings with the lengths of 5 to 11. One does not need to have a dictionary or know the language in order to reasonably select "informatics" based on these numbers. It is sufficient to be aware of the scarcity of English words shorter than four letters among major parts of speech, and compare the chances of finding a common word.

---

[2] Use of nonlinear dependencies does not produce any major quality improvement [10,11].

2.4   Longest Common Substrings as a Belief Function

If the dependence of the contribution on the length of strings grows very quickly, then it is often possible to speed up the search for the most informative common part. For example, SCM Revlog and Mercurial [2] feature a heuristic "greedy" algorithm that first finds the longest substrings and then, if their choice is unique, these substrings are definitely part of the final solution. The algorithm is in use and often helps find precise solutions to practical optimization tasks.

Striving to avoid fragmented common subsequences, the algorithm goes to the opposite extreme: searching for the sole longest common substring sometimes leads it away from a much more informative common subsequence consisting of an arbitrary large number of substrings that are almost as long. For example, given the strings

"31, 29, 28, 27, 26, 25, 24, 23, 22, 21,_20"
"31,_29,_28,_27,_26,_25,_24,_23,_22,_21, 20"

it will identify the following common subsequence "1,_2", instead of much more informative "31,29,28,27,26,25,24,23,22,21,20".

The algorithm yields the optimal subsequence for a problem when the substring of any length has a contribution greater than any sequence of shorter substrings. A sufficient condition for this is obviously a rapid growth of contributions with increasing length, accelerating along with the growth of $|S_1|$ and $|S_2|$. It is sufficient, for example, that the following inequality hold true: $p(n + 1) > (1 - \frac{1}{n})p(n)|S_1|$. So a simple definition $m(A) = p(|A|) = |S_1|^{|A|}$ is sufficient.

2.5   The Belief Function for All Substrings Accounting

As we have seen, classic sequence comparison algorithms can be divided into two groups. The first one focuses on short common substrings and finds fragmented common subsequences. The second one, to the contrary, focuses on the longest common substrings, and sometimes the resulting match is too scarce.

In order to account for the substrings of all lengths, let us simply assume the significance of all the substrings to be equal $m \equiv 1$. Then, the subsequence containing the maximum possible number of common substrings (NCS) will be the optimal one [1].

The number of all substrings in a subsequence consisting of disjoint substrings with lengths $l_1, \ldots l_k$ is equal to

$$n = \sum_{j=1}^{k} \frac{l_j(l_j + 1)}{2} .$$

(13)

As the number of strings increases quadratically along with the size of the graph, one can scale the similarity measure to some comparable with string lengths value. The value may be equal to the size of the string with the obtained $S = n$:

$$l(S) = \sqrt{2S + 1} - 1 . \tag{14}$$

Be aware that this non-linear scale may disrupt the basic similarity inequalities.

The Table 2 illustrate the NCS method of quantifying similarity of sequences is free from the above-noted shortcomings of the classic methods (n_1 and n_2 were described in subsection 2.3).

Table 2. Scores for sample selections

Common part selection	n_1	n_2	S	$l(S)$
"effie com efft"	13	7	25	6.1
"ineffective"	10	9	55	10.0
"matic formatics"	14	11	52	9.3
"informatics"	10	9	55	10.0
"1,_2"	3	2	6	3.0
"31,29,28,27,26,25,24,23,22,21,20"	36	24	72	11.0

Remark 2. In the absence of longest common substrings with the length between l and $2l - 1$, when identifying strings of the length $2l - 1$ and above, contained in the NCS-optimal common part, shorter strings can be skipped without effect to total weight.

This statement points to the possibility of first identifying a sufficient number of the longest strings, and then updating the solution found by accounting for shorter ones. Its proof directly follows from the lemma below.

Lemma 1. *Let's assume that, when comparing two sequences, there were no common substrings in the length range between l and $2l - 1$ for some natural number $l > 0$. Then, for any NCS-optimal subsequence S and any common substrings longer than l there is a substring $s' \subset S$ longer than l, contained entirely in S and having a non-empty intersection with s.*

Let's prove the lemma by contradiction. Let only strings of length l or less intersect with s. Then all strings intersecting with s are contained in an enveloping string of the length $4l - 3$. Because the length of each of the intersecting strings does not exceed l, each of their characters provides a contribution of not more than $\frac{l+1}{2}$, and their joint contribution does not exceed $(4l - 3)\frac{l+1}{2}$, which is less than the contribution of s, equal to $(2l - 1)l$. The contradiction proves the lemma.

2.6   Semi-structured Data and Block Transposition

Even without a dictionary, contributions of substrings could be assigned in a more distinctive way. For example, when comparing ordinary text, a substring containing spaces and finishing in the middle of a word, or a substring finishing right before a space and clearly containing parts of two different sentences (a capital letter after a lower-case word followed by a period and a space), is not significant. It seems not to be so difficult to write simple and fast algorithm which discounts unfinished words and sentences from a large weight fragments.

The same idea is applicable to comparing software source code files, texts in the XML format or other poorly structured information, relatively simple rules could be used to identify balanced substrings — that is, the substrings with both opening and closing brackets as well as both opening and closing tags. The difficulty of accounting for regular expressions that are masked or contained in comments, character substrings or text could be partially overcome by assignment of fractional contribution values in doubtful cases.

This approach to assignment of contributions could help to improve the quality of the algorithms applied to identify common parts. Particularly noticeable improvements could be derived with respect to the usability of visual text comparison tools; moreover, there would be an opportunity to account for transpositions of large blocks of text.

The longer string accounting will help to quantify similarity of subsequences in such a way that transposition of large blocks would have a smaller negative impact on similarity than removal of one and insertion of a completely different substring of the same size in another location. For example, the string "next run" is more similar in this sense to "run next" than to "new next".

3   Acceptable Parts and Algorithm Performance

Transposition of large blocks is one of the basic capabilities of text editors, yet the current edit distance method cannot deal with it. The detection of large block transposition largely depends on appropriate acceptable common part selection. The commonly used restriction of order preserving by isomorphism p_ϕ on acceptable common parts suppress block transposition detection and simplifies calculations. Remembering a dominant significance of longer substrings, suppose most valuable common parts to be identified in three stages also in the alternative case $\mathcal{P}(X,Y) = \mathfrak{P}(X,Y)$:

1. Searching for all the long common substrings, including calculation of believed significance m and rough sorting by m.
2. Selecting the optimal subset of non-conflicting significant long strings in the order: the subset with more strings of large significance is considered earlier.
3. Extending the identified common part by adding common strings in its close proximity.

The last item means reducing the set of acceptable parts to only ones without far isolated short common substrings. This provide both algorithm accelerating and random noise filtering.

The only known way to reflect dominant significance of longer substrings is the NCS solution which has too expensive for modern applications algorithm [1] of cubic worst case complexity. Practical problems on block transposition detection or graph similarity are well known to have non-polynomial complexity. The big data needs sublinear or near to linear algorithms. There are at least three way that usually combined to get a fast algorithm:

1. Get approximate solution, sufficiently close to optimal.
2. Get a solution with a probability very close to 1.
3. Modify problem formulation by extra constrains usual in practical applications.

In case with graphs, stage 1 will involve identification of matching vertices, which could be performed randomly, and for each identified pair, it will be necessary to check whether the neighboring vertices match and so on, until a sequence that cannot be further extended is identified. Sorting could be based on lists of identified strings, differing in length by a factor of two.

If a graph has a known simple structure, the process can be simplified. For a sequence, the complexity of matching long substrings is certainly limited by the cubic and could likely be brought down to quadratic complexity or in practically useful cases, even to less then linear complexity (for example, when the number of possibly matching strings and the lower boundary of their relative size and the chance of allowed error are all fixed).

For example, for two huge files (e.g. terabyte), it should be possible to detect if they have huge common substrings (e.g. 100 gigabytes). It should be done in a few minutes without total string processing with a probability of $1 - 10^{100}$. The algorithm for solving this problem will be a key to our first stage.

The search tree depth required at stages 2 and 3 could be proportionally limited. Even if we take the first possible option and get result immediately, we will get a "greedy" heuristic algorithm, which could turn out to be sufficiently effective in some applications.

The following resource restrictions should be defined: the total number of long strings or their minimum length, the number of searched versions of subsets of long strings, maximum length of an adjacent chain of short strings. If the task complexity for specific graphs compared is low relative to the allocated resources, the resulting solution will be precise; otherwise, it will be an approximate one and it is possible to estimate a chance to get an exact result with increased search depth. Any of the stages could be performed in parallel.

These capabilities are also valuable for a database search. A combination of speed, accuracy, and completeness could be achieved by utilizing a fast search with subsequent slow updating on a positive forecast. Double-sided estimates for similarity and execution completeness could be fixed and utilized for each of the stages.

Conclusion

A new believed similarity framework for sequences and graphs produced an idea of comparison technique promising a new level of speed, accuracy and reliability. This looks like a way to nearly linear execution time algorithm which may practically solve problem expected to be NP-complete.

Practical implementation and experimental performance evaluation will come soon for diff utility application.

Acknowledgments. This work was performed under financial support from the Government, represented by the Ministry of Education and Science of the Russian Federation (Project ID FMEFI60414X0138).

References

1. Znamenskij, S.V.: A model and algorithm for sequence alignment. Program Systems: Theory and Applications **6**(1), 189–197 (2015)
2. Mackall, M.: Towards a better SCM: revlog and mercurial. In: Proceedings of Linux Symposium, vol. 2, pp. 83–90 (2006)
3. Shafer, G.: A Mathematical Theory of Evidence. Princeton University Press (1976)
4. Sugano, M.: Fuzzy measure and fuzzy integrals. Trans. of the Soc. of Instrument and Control Engineers **8**(2) (1972) (in Japanese)
5. Choquet, G.: Theory of capacities. In: Annales de l'Institut Fourier, vol. 5, pp. 131–295 (1953)
6. Rho, S., Hwang, E.: FMF: Query adaptive melody retrieval system. Journal of Systems and Software **79**(1), 43–56 (2006)
7. Wieling, M., Bloem, J., Mignella, K., Timmermeister, M., Nerbonne, J.: Measuring foreign accent strength in English. Validating Levenshtein Distance as a Measure, The Mind Research Repository (beta) **1** (2013). http://openscience.uni-leipzig.de/index.php/mr2/article/view/41/30
8. Wu, X., Wu, Z., Jia, J., Meng, H., Cai, L., Li, W.: Automatic speech data clustering with human perception based weighted distance. In: ISCSLP 2014, September 12–14, Singapore, pp. 216–220. IEEE (2014)
9. Needleman, S.B., Wunsch, C.D.: A general method applicable to the search for similarities in the amino acid sequence of two proteins. Journal of Molecular Biology **48**(3), 443–453 (1970)
10. Cartwright, R.A.: Logarithmic gap costs decrease alignment accuracy. BMC Bioinformatics **7**, 527 (2006)
11. Wang, C., Yan, R.X., Wang, X.F., Si, J.N., Zhang, Z.: Comparison of linear gap penalties and profile-based variable gap penalties in profile-profile alignments. Computational Biology and Chemistry **35**(5), 308–318 (2011)
12. Künzi, H.-P.A.: Nonsymmetric distances and their associated topologies: about the origins of basic ideas in the area of asymmetric topology. In: Handbook of the History of General Topology, vol. 3, pp. 853–968. Kluwer Academic Publisher, Dordrecht (2001)
13. Stojmirović, A., Yi-Kuo, Y.: Geometric Aspects of Biological Sequence Comparison. Journal of Computational Biology **16**(4), 579–610 (2009)
14. Chateauneuf, A., Jaffray, J.-Y.: Some characterizations of lower probabilities and other monotone capacities through the use of Mobius inversion. Mathematical Social Sciences **17**, 263–283 (1989)
15. Sippl, M.J., Wiederstein, M.: A note on difficult structure alignment problems. Bioinformatics **24**(3), 426–427 (2008)

Similarity of Attributed Generalized Tree Structures: A Comparative Study

Mahsa Kiani[(✉)], Virendrakumar C. Bhavsar, and Harold Boley

Faculty of Computer Science, University of New Brunswick, Fredericton, NB, Canada
{mahsa.kiani,bhavsar,harold.boley}@unb.ca

Abstract. In our earlier attributed generalized tree (AGT) structures, vertex labels (as types) and edge labels (as attributes) embody semantic information, while edge weights express assessments regarding the (percentage-)relative importance of the attributes, a kind of pragmatic information. Our AGT similarity algorithm has been applied to e-Health, e-Business, and insurance underwriting. In this paper, we compare similarity computed by our AGT algorithm with the similarities obtained using: (a) a weighted tree similarity algorithm (WT), (b) graph edit distance (GED) based similarity measure, (c) maximum common subgraph (MCS) algorithm, and (d) a generalized tree similarity algorithm (GT). It is shown that small changes in tree structures may lead to undesirably large similarity changes using WT. Further, GT is found to be not applicable to AGT structures containing semantic as well as pragmatic information. GED and MCS cannot differentiate AGT structures with edges having different directions, lengths, or weights, all taken into account by our AGT algorithm.

Keywords: Tree similarity · Attributed generalized tree · Generalized tree similarity · Weighted tree similarity

1 Introduction

We introduced earlier an attributed generalized tree (AGT) similarity algorithm, which matches pairs of AGT structures [4].

Generalized trees [1] are hierarchical and directed graphs, which were introduced as an extension of rooted tree structure. The edge set of a generalized tree is unconstrained, since edges in a generalized tree, as in a general graph, can form cycles. Generalized trees can model more contextual information compared to trees, and as a result, they are able to represent complex relational objects in practical applications. While non-attributed generalized trees (GT) [2] are only defined based on the topological structure (i.e., the vertex and edge sets), our attributed generalized trees (AGT) [4] can represent semantic and pragmatic information using vertex labels, edge labels, as well as edge weights. The root vertex of an AGT carries a class label, which types the main object; therefore, the root has the highest importance. The importance of vertices decreases as

© Springer International Publishing Switzerland 2015
G. Amato et al. (Eds.): SISAP 2015, LNCS 9371, pp. 150–161, 2015.
DOI: 10.1007/978-3-319-25087-8_14

their depth increases. While our attributed generalized tree structures can have arbitrary size, they are always finite.

In this paper, Weighted tree similarity algorithm (WT), generalized tree similarity algorithm (GT), Graph Edit Distance (GED) based similarity measure, and maximum common subgraph (MCS) algorithm for matching AGT structures are examined, and the computational results are compared with our AGT similarity algorithm.

The remainder of this paper is organized as follows. Section 2 briefly reviews various algorithms and outlines our methodology for computing similarities using the algorithms. In Sections 3 and 4, similarity approaches with the weighted tree similarity algorithm (WT) and the generalized tree similarity algorithm (GT) are compared to our AGT similarity algorithm. In Section 5, graph edit distance (GED) based similarity measures and the maximum common subgraph (MCS) approach are again compared to the AGT similarity. For each method, its application for matching of attributed generalized tree structures is analyzed, and its computational results are compared with the AGT similarity algorithm. Finally, we conclude with Section 6.

2   Methodology

WT algorithm is a recursive similarity algorithm for comparing vertex-labeled edge-labeled edge-weighted trees [6]. In order to compute structure similarity, the algorithm visits all vertices of two trees starting from their roots. If root vertex labels of two (sub)trees are identical, their similarity is computed by top-down traversal of the (sub)trees accessible through identical edge labels. Vertex attributes as well as edge attributes are compared using exact string comparison resulting in either 0.0 or 1.0. The algorithm calculates arithmetic mean of the weights of corresponding edges; and the results are considered while integrating the similarities of vertex attributes, edge attributes, as well as structures. The pair of structures being compared could have different sizes. If the concepts represented by roots of two trees are the same, the missing substructure could have existed in the structure. The effect of missing substructures in the overall similarity is considered using simplicity. Simplicity of a substructure is computed by calculating its similarity to a corresponding empty substructure. In WT approach, breadth, depth, and weight factors are considered in simplicity computation. Simplicity is a value in real interval $[0, 1]$. The WT approach has a linear time complexity.

In GT algorithm [7], structures of two generalized trees are compared by transforming each tree to property strings. Property strings are formed based on out-degree and in-degree sequences of vertices on each level of hierarchy. Optimal sequence alignments of the corresponding property strings are determined; and the structure similarity is computed by aggregating the local similarity for out-degree and in-degree alignments on each level of generalized trees. The time complexity of GT approach is $O(|\hat{V_1}| \cdot |\hat{V_2}|)$ were $|\hat{V_1}|$ and $|\hat{V_2}|$ are the matrices representing the number of vertices in various levels of generalized trees.

Graph Edit Distance (GED) [9] [10] [11], a transformation-distance app-roach, is an extension of the string edit distance [12] to the domain of graphs. In this paper, GED transforms the corresponding generalized trees using insertion, deletion, and substitution operations on vertex, edge, and weights. As the gen-eralized trees are based on a common schema, vertices in corresponding levels are matched. Also, the structure similarity measure for generalized trees has the symmetric property; therefore, the costs of deletion and insertion of vertices are equal. Similarly, the costs of deletion and insertion of edges are equivalent. In GED based similarity measure, the distance value obtained from GED approach is transformed into a similarity value.

Also, the Maximum Common Subgraph (MCS) [13] approach is applied to generalized trees; a pair of generalized trees are given and the purpose is finding the largest induced substructure common to both of them; therefore MCS finds a substructure of both generalized trees such that there is no other substruc-ture with more vertices. Note that the complexity of the *GED based similarity measure* as well as the *MCS* approach is linear.

Before similarity computation using our AGT algorithm, AGT structures are transformed to an internal representation, a weighted extension of Object Ori-ented RuleML (OO RuleML) [5], which preserves all structural information of generalized trees. Then, our AGT similarity algorithm [4] recursively traverses a pair of OO RuleML representations top-down and computes the similarity bottom-up based on matching corresponding pairs of edge labels, edge weights, and vertex labels. A depth-first search traversal with a cycle-detection strategy is used to handle cycles in the generalized trees. Labels attached to both ver-tices and edges are strings, and both vertex labels and edge labels are compared using an exact string matching approach. When we compute the similarity of two generalized trees, a substructure in one generalized tree might be missing in the other one. The contribution of missing substructures to the overall structure similarity is computed using a recursive simplicity algorithm. In computing sim-plicity, the numbers of vertices and edges are considered, as wider and deeper substructures have smaller simplicity values. Edges in a substructure can have different lengths, directions, positions, and weights; therefore, these factors are taken into account when computing simplicity of edges as well. We found that the complexity of our AGT similarity algorithm is linear in the number of vertices and edges [4].

3   WT Versus AGT Algorithm

The structures of generalized trees are more complex than the structures of trees, as edges in a generalized tree can form cycles. A rooted tree is a special case of a generalized tree. WT can only consider direct hierarchical relations (i.e., edges) between vertices in rooted trees; as a result, this approach is not suitable for comparing generalized trees [2].

Some shortcomings of WT have been discussed in [3]. WT considers edge weights only as scaling factors to ensure that the overall similarity value is in the real interval $[0, 1]$. As a result, the similarity of a pair of trees with identical structures but different edge weights would always be considered equal to 1.0, while in some applications, these structures are not considered identical.

In addition, the simplicity approach in WT does not support the simplicity of edges with different lengths, and directions (i.e., cross and back edges in generalized tree structure). Also, the effect of hierarchical levels in all the structural factors (e.g., branch factor) are not considered. WT provides lower accuracy in the ranked results compare to our AGT similarity algorithm.

In order to evaluate the results of WT, a dataset containing 25 metadata having tree structures is represented in Figure 1. While H, A, and B represent vertex labels, edge labels are expressed as la and lb. If there is only one outgoing edge, the edge weight would be equal to 1.0. When a pair of outgoing edges exist, the weight of each edge is considered to be 0.5.

G_{25} represents the full binary tree, whereas other trees are subtrees of this tree. Therefore, to compare the performance of the AGT algorithm and the WT algorithm, we compute similarities of G_1 through G_{24} compared to G_{25} (see Figure 2). Note that complexity of the tree structures slowly increases (or remains the same) as we consider trees from G_1 to G_{25}. Therefore, we expect the similarity to be monotonic w.r.t. G_{25}.

As it is illustrated in Figure 2, compared to our AGT similarity, trend of the computed similarity values using the WT algorithm is not monotonic. Furthermore, small changes in a tree structure can cause large changes in the similarity value.

We have also found similarities of the two subsets of metadata, (G_8, G_9, and G_{25}) as well as (G_{18}, G_{19}, and G_{25}). Using WT, similarity between G_{25} and G_8 trees, $Sim(G_{25}, G_8) = 0.664$, is found to be greater than the similarity between G_{25} and G_9, $Sim(G_{25}, G_9) = 0.376$, structures; this result does not agree with intuition. In addition, the computed similarity value between G_{25} and G_8 structures (i.e., 0.664) is a higher value than expected. However, using our AGT similarity algorithm, the similarities of G_{25} to G_8 and G_9 are found to be 0.2722, and 0.6554, respectively (i.e., $Sim(G_{25}, G_8) < Sim(G_{25}, G_9)$). Furthermore using WT, the computed similarity value between G_{25} and G_{18} trees is greater than the similarity value between G_{25} and G_{19} structures. However, using our AGT similarity algorithm, similarity of G_{25} to G_{18} and G_{19} is 0.5293 and 0.6031, respectively, $Sim(G_{25}, G_{18}) < Sim(G_{25}, G_{19})$, which is what we expect.

In addition, WT cannot differentiate all the trees with different structures. For instance, while structures of G_{21} and G_{25} are not identical, the computed similarity value between G_9 and G_{21}, $Sim(G_9, G_{21}) = 0.6456$, (not shown in the figure) is found to be equal to the similarity value of G_9 and G_{25}. However, using our AGT similarity, the similarity of G_9 to G_{21} and G_{25} is 0.4312 and 0.3775, $Sim(G_9, G_{21}) \neq Sim(G_9, G_{25})$. Also, WT uses edge weights only as scaling factors, and as a result it cannot differentiate trees with different edge weights having identical structures, vertex labels, as well as edge labels.

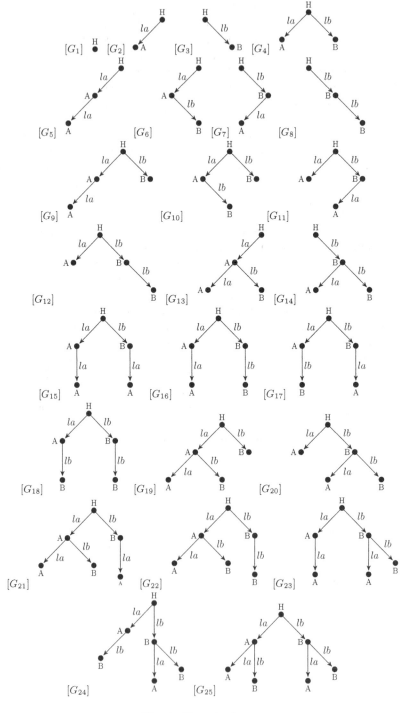

Fig. 1. Trees in dataset.

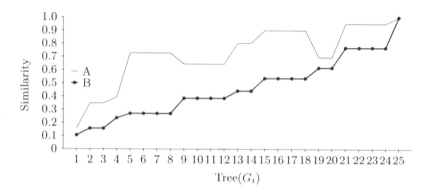

Fig. 2. Similarity of G_{25} with G_1 through G_{25} shown in Figure 1. A: weighted tree similarity algorithm (WT) [6] and B: AGT similarity algorithm [4].

4   GT Versus AGT Algorithm

In the GT algorithm, generalized trees are considered to have homogeneous vertices which are connected by homogeneous edges. Vertex-labels and edge-labels of attributed generalized tree structure are not considered in the matching process by GT; as a result, this approach only takes into account the similarity of structures, and it is not applicable in domains containing semantic and pragmatic information.

Consider the generalized trees corresponding to Figure 3 defined in Table 1 having identical edge labels and edge weights. Vertex labels of G_1 and G_2, and G_3 are illustrated in Table 1. The similarity of G_1 to G_2 and G_3 is computed using the GT and AGT similarity algorithms. While the similarity values computed using GT similarity algorithm are equivalent ($Sim(G_1, G_2) = Sim(G_1, G_3) = 1.0$); our AGT similarity algorithm could differentiate G_2 and G_3 ($Sim(G_1, G_2) = 1.0; Sim(G_1, G_3) = 0.6994$).

Fig. 3. Structure of AGTs having different vertex labels.

Now we show that the GT similarity algorithm cannot differentiate AGTs having different edge weights. Generalized trees G_1, G_2 and G_3 corresponding to Figure 4 given in Table 2 have identical vertex labels and edge labels; however, their edge weights are different (Table 2). While GT similarity algorithm ignores the difference between their edge weights ($Sim(G_1, G_2) = Sim(G_1, G_3) = 1.0$),

Table 1. Vertex labels of G_1, G_2, and G_3 with the structure given in Figure 3

AGT	$l(v_1)$	$l(v_2)$	$l(v_3)$	$l(v_4)$
G_1	H	A	B	C
G_2	H	A	B	C
G_3	H	A	B	D

Fig. 4. Structure of $AGTs$ having different edge weights.

Table 2. Edge weights of generalized trees G_1, G_2, and G_3 with the structure given in Figure 4

AGT	$w(e_1)$	$w(e_2)$	$w(e_3)$
G_1	0.8	0.2	1.0
G_2	0.5	0.5	1.0
G_3	0.1	0.9	1.0

our AGT similarity considers the edge weights in similarity computation, and as a result G_2 and G_3 could be differentiated ($Sim(G_1, G_2) = 0.5916$; $Sim(G_1, G_3) = 0.2606$).

Generalized trees G_2 and G_3 defined in Table 3 (see also Figure 5) have identical vertex labels and edge weights; however, their edge labels are different. GT similarity algorithm is used to compute similarity of G_1 to G_2 and G_3. As this approach does not consider edge labels in the similarity computation, it generates identical similarity values (i.e., $Sim(G_1, G_2) = Sim(G_1, G_3) = 1.0$). The AGT similarity algorithm takes into account edge labels and it is able to distinguish G_2 and G_3 (i.e., $Sim(G_1, G_2) = 1$; $Sim(G_1, G_3) = 0.7349$).

5   GED and MCS Versus AGT Algorithm

Intuitively, two generalized trees are not identical, if they have edges with different directions, lengths, or weights. Also, two generalized trees are not considered identical if the level in which their edges exist are different. *GED based similarity measure* and *MCS* cannot differentiate attributed generalized trees having edges with different directions, lengths, or weights as discussed below. Also, *GED based similarity measure* and *MCS approach* cannot be used to differentiate generalized trees having edges in different levels. The computational results from our experiments are illustrated in this section.

Fig. 5. Structure of *AGTs* having different edge labels.

Table 3. Edge labels of G_1, G_2, and G_3 with the structure given in Figure 5

AGT	$l(e_1)$	$l(e_2)$	$l(e_3)$
G_1	lb	lc	ld
G_2	lb	lc	ld
G_3	la	lc	ld

In order to evaluate the results of GED on generalized trees, *GED based similarity measure* is integrated with Depth-First Search. Two versions of *GED based similarity measure*, GED_1 and GED_2, are developed based on two sets of costs for edit operations. In GED_1, the cost of deletion\insertion of one edge\vertex as well as the cost of substitution of one edge weight is equal to 1. While in GED_2, the cost of deletion\insertion of a vertex is 2. The cost of deletion\insertion of an edge weight is equal to 0.25; however, the cost of deletion\insertion of an edge equals 1.

MCS is integrated with Depth-First Search as well; and two versions of MCS approach are defined. In MCS_1 only the difference between vertices of two generalized trees are considered; the cost of insertion and deletion of a vertex is equal to 1. MCS_2 considers difference between both vertices and edges of two generalized trees; the cost of insertion\deletion of a vertex\edge are equal to 1.

Generalized trees in the following datasets are compared using GED_1, GED_2, MCS_1, and MCS_2, and the results are compared with our AGT similarity algorithm.

Generalized trees G_2 and G_3 in Figure 6 have identical number of missing edges compared to G_1; however, they appear in different levels of hierarchy. Their vertex labels, edge labels, and edge weights are considered to be identical. The similarity of G_1 to G_2 and G_3 is illustrated in Table 4. Using the AGT similarity algorithm, G_2 and G_3 are differentiated; however, using GED_1, GED_2, MCS_1, and MCS_2, the similarity of vertices and edges in different levels of hierarchy have the same contribution in similarity and therefore these approaches generate identical similarity values.

GED_1, GED_2, MCS_1, and MCS_2 cannot differentiate edges with different lengths as well. Generalized trees G_2 and G_3 in Figure 7 have two missing edges compare to G_1. Note that length of back edges are different in G_2 and G_3.

These structures are considered to have identical vertex labels, edge labels, and edge weights. G_1 is compared to G_2 and G_3 and the results are illustrated in Table 5. In our AGT similarity, edges between concepts in different levels are differentiated; however, GED_1, GED_2, MCS_1, and MCS_2 only consider the

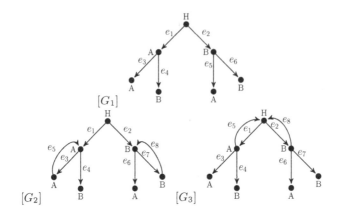

Fig. 6. *AGTs* having edges in different levels.

Table 4. Similarity of G_1 to G_2 and G_3 in Figure 6

AGT	AGT-Sim	GED_1	GED_2	MCS_1	MCS_2
G_2	0.7736	0.9047	0.9450	1.0	0.8666
G_3	0.6692	0.9047	0.9450	1.0	0.8666

number of missing edges and ignore the difference of edges between concepts in different levels of hierarchy.

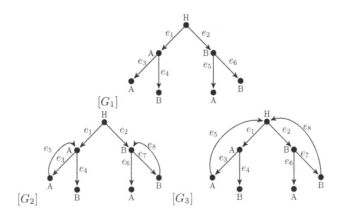

Fig. 7. Structures of *AGTs* having edges with different length.

Consider the generalized trees in Figure 8. These structures have identical vertex labels, edge labels, and edge weights. Missing edges in generalized trees G_2 and G_3 compared to G_1 have different directions (forward edges and back edges).

Table 5. Similarity of G_1 to G_2 and G_3 in Figure 7

AGT	AGT-Sim	GED_1	GED_2	MCS_1	MCS_2
G_2	0.7736	0.9047	0.9450	1.0	0.8666
G_3	0.7401	0.9047	0.9450	1.0	0.8666

Table 6. Similarity of G_1 to G_2 and G_3 in Figure 8

AGT	AGT-Sim	GED_1	GED_2	MCS_1	MCS_2
G_2	0.6766	0.7142	0.8571	1.0	0.8666
G_3	0.6624	0.7142	0.8571	1.0	0.8666

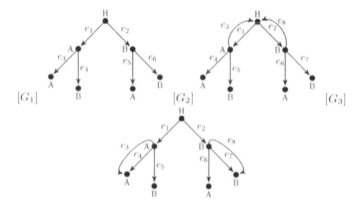

Fig. 8. Structures of AGTs having edges in different directions.

Similarity of G_1 to G_2 and G_3 in Figure 8 is illustrated in Table 6. Although the directions of the edges in pairs of structures are not identical, the similarity value computed using GED_1, GED_2, MCS_1, and MCS_2 approaches are equivalent. However, our AGT similarity algorithm could differentiate G_2 and G_3.

Also, using GED_1, GED_2, MCS_1, and MCS_2 approaches, structures with different edge weights cannot be differentiated. In G_1 (see Figure 9), edge weights of substructures beneath the root are different; Weights of e_1, e_2, and e_3 are 0.1, 0.2, 0.7 respectively. Generalized tree G_1 is compared to generalized trees G_2 and G_3 in Figure 9 and the results are illustrated in Table 7.

While the AGT similarity algorithm generates different similarity values for G_2 and G_3, MCS_1 and MCS_2 approaches fail to distinguish G_2 and G_3. The reason is that MCS_1, and MCS_2 do not consider the edge weights in similarity computation. While GED_1 and GED_2 consider the cost of difference in edge weights between the generalized trees, edge weights in a higher level of hierarchy

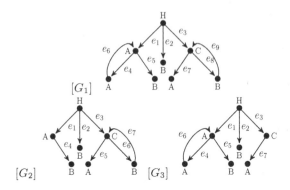

Fig. 9. Structures of *AGTs* having edges with different weights.

Table 7. Similarity of G_1 to G_2 and G_3 in Figure 9

AGT	AGT-Sim	GED_1	GED_2	MCS_1	MCS_2
G_2	0.5384	0.8936	0.9024	0.875	0.8823
G_3	0.3859	0.8936	0.9024	0.875	0.8823

do not effect the similarity of substructures beneath them, as in these approaches the similarity of each level is computed independently. This is in contrast to the AGT similarity algorithm in which difference in substructures reaching through edges with higher weights has more effect in the similarity value. Therefore, the AGT similarity algorithm generates different similarity values for comparing G_1 with G_2 and G_3, while GED_1 and GED_2 are unable to differentiate them.

Thus, GED_1, GED_2, MCS_1, as well as MCS_2 approaches could not differentiate G_2 and G_3 in Figures 6, 7, 8, and 9, as edge factors (i.e., length, direction, position, and weight) are not considered in the definitions of insertion, deletion, and substitution costs.

We can clearly see that our AGT similarity algorithm can achieve higher accuracy (i.e., higher degree of agreement with intuition) by considering attributes (i.e., vertex label, edge label, as well as edge weights) in similarity computation process. The AGT similarity algorithm is applicable in domains having concepts with complex relations as well as rich semantic and pragmatic information. In [4] and [8], we have demonstrated application of the AGT similarity algorithm for the retrieval of Electronic Medical Records (EMRs) as well as life-insurance application underwriting.

6 Conclusion

In this paper, we have compared similarity computed by our AGT algorithm with the similarities obtained using various inexact similarity measures: a weighted tree similarity algorithm (WT), graph edit distance (GED) based similarity measure, maximum common subgraph (MCS) algorithm, and a non-attributed generalized tree similarity algorithm (GT). We have illustrated that applicability of existing similarity measures for comparing AGTs is limited and our AGT similarity algorithm could achieve higher accuracy by integrating the similarity of attributes with structure similarity as well as considering directions, lengths, and weights of edges.

References

1. Mehler, A., Dehmer, M., Gleim, R.: Towards logical hypertext structure-a graph-theoretic perspective. In: Böhme, T., Larios Rosillo, V.M., Unger, H., Unger, H. (eds.) IICS 2004. LNCS, vol. 3473, pp. 136–150. Springer, Heidelberg (2006)
2. Dehmer, M., Emmert-Streib, F., Mehler, A., Kilian, J.: Measuring the Structural Similarity of Web-based Documents: a Novel Approach. J. Computational Intelligence **3**, 1–7 (2006)
3. Kiani, M., Bhavsar, V.C., Boley, H.: Combined structure-weight graph similarity and its application in e-health. In: 4th Canadian Semantic Web Symposium, Montreal, pp. 12–18 (2013)
4. Kiani, M., Bhavsar, V.C., Boley, H.: Structure similarity of attributed generalized trees. In: 8th IEEE International Semantic Computing, pp. 100–107. IEEE Press, Newport Beach (2014)
5. Boley, H.: Object-oriented RuleML: User-level roles, URI-grounded clauses, and order-sorted terms. In: Schröder, M., Wagner, G. (eds.) RuleML 2003. LNCS, vol. 2876, pp. 1–16. Springer, Heidelberg (2003)
6. Bhavsar, V.C., Boley, H., Yang, L.: A Weighted-tree Similarity Algorithm for Multi-agent Systems in e-Business Environments. J. Computational Intelligence **20**, 584–602 (2004)
7. Dehmer, M., Emmert-Streib, F., Kilian, J.: A Similarity Measure for Graphs with Low Computational Complexity. J. Applied Mathematics and Computation **182**, 447–459 (2006)
8. Kiani, M., Bhavsar, V.C., Boley, H.: A fuzzy structure similarity algorithm for attributed generalized trees. In: 13th IEEE International Conference on Cognitive Informatics and Cognitive Computing, pp. 203–210. IEEE Press, London (2014)
9. Bunke, H.: What is the Distance between Graphs?. J. Bulletin of the European Association for Theoretical Computer Science, 35–39 (1983)
10. Sanfeliu, A., Fu, K.S.: A Distance Measure between Attributed Relational Graphs for Pattern Recognition. IEEE Transactions on Systems, Man and Cybernetics **3**, 353–362 (1983)
11. Bunke, H.: On a relation between Graph Edit Distance and Maximum Common Subgraph. J. Pattern Recognition Letters **18**, 689–694 (1997). Elsevier Science
12. Wagner, R.A., Fischer, M.J.: The string-to-string correction problem. JACM **21**, 168–173 (1974). ACM
13. Bunke, H., Shearer, K.: A Graph Distance Metric based on the Maximal Common Subgraph. J. Pattern Recognition Letters **19**, 255–259 (1998)

Evaluating Multilayer Multimedia Exploration

Juraj Moško[1]([⊠]), Jakub Lokoč[1], Tomáš Grošup[1], Přemysl Čech[1],
Tomáš Skopal[1], and Jan Lánský[2]

[1] SIRET Research Group, Department of Software Engineering,
Faculty of Mathematics and Physics, Charles University in Prague,
Prague, Czech Republic
{mosko,lokoc,grosup,cech,skopal}@ksi.mff.cuni.cz
[2] Department of Computer Science and Mathematics,
University of Finance and Administration, Prague, Czech Republic
lansky@mail.vsfs.cz

Abstract. Multimedia exploration is an entertaining approach for multimedia retrieval enabling users to interactively browse and navigate through multimedia collections in a content-based way. The multimedia exploration approach extends the traditional query-by-example retrieval scenario to be a more intuitive approach for obtaining a global overview over an explored collection. However, novel exploration scenarios require many user studies demonstrating their benefits. In this paper, we present results of an extensive user study focusing on the comparison of 3-layer Multilayer Exploration Structure (MLES) structure with standard flat k-NN browsing. The results of the user study show that principles of the MLES lead to better effectiveness of the exploration process, especially when searching for a first object of the searched concept in an unknown collection.

Keywords: Similarity search · Multimedia exploration · Content-based retrieval · Exploration operation · Multimedia browsing · User study

1 Introduction

Traditional similarity retrieval scenarios based on single *query-by-example* approaches are not sufficient in applications where advanced system/retrieval features are required. To fulfill this gap, when progressive factors like a restricted GUI (e.g., smart phones) or the need of interactivity/entertainment directly affect a retrieval process, novel retrieval scenarios were recently investigated. The typical example of such scenario is *multimedia exploration* [2]. The *multimedia exploration* is an advanced retrieval scenario where users want to explore and get an idea of an unknown multimedia collection. Usually, the only thing a user has in the beginning of the exploration process is an idea of the result in her/his mind and thus a single *query-by-example* similarity search does not have to be sufficient for effective retrieval. On the other hand, the *multimedia exploration* systems often use the *query-by-example* principles as a basic supportive

© Springer International Publishing Switzerland 2015
G. Amato et al. (Eds.): SISAP 2015, LNCS 9371, pp. 162–169, 2015.
DOI: 10.1007/978-3-319-25087-8_15

task for more complex exploration operations. Usual tasks that the users expect from the *exploration system* are iterative navigation, browsing and visualization of the collection, therefore we can see an analogy between operations used in *multimedia exploration* and navigating in a map (zooming, panning)[1].

In this paper, we evaluate properties of a multilayer multimedia exploration structure that natively supports zoom in, zoom out and pan exploration operations and compare the structure to simulated flat k-NN browsing in an extensive user study.

2   Multilayer Multimedia Exploration

In the following, we present a structure for multimedia exploration that utilizes similarity indexes for efficient *horizontal browsing* and employs multiple layers enabling *vertical browsing* [4]. We assume that the exploration process always starts with a limited number of representative objects displayed on a screen. Starting in the initial view, users can consecutively zoom in to specific parts of the view, pan to other groups of objects, or zoom out if the actual view is filled with undesired objects. Note that the users see the same number of objects all the time. Given such exploration use case, a good exploration structure should provide representative distinct objects for earlier phases of the exploration, while objects more similar to selected examples should be retrieved in later stages. Based on these assumptions, we define a Multilayer Exploration Structure (Figure 1):

Definition 1. *(Multilayer Exploration Structure). Given a dataset $\mathbb{S}$ of objects (descriptors of the respective multimedia objects), the Multilayer Exploration Structure, $MLES(\mathbb{S}, m, v, \phi)$, is a system of $m + 1$ subsets $\mathbb{L}_i$ (layers), where the subset condition holds:*

$$\mathbb{L}_i \subset \mathbb{L}_{i+1}, \forall i = \{0, .., m - 1\}$$

The smallest subset $\mathbb{L}_0$ represents the first depicted v objects (i.e., page zero view) and the proper subset $\mathbb{L}_m = \mathbb{S}$ represents the whole database. Selection of objects for each layer is determined by a selection function $\phi : \mathbb{N} \to 2^{\mathbb{S}}$ that has to comply with the subset condition.

The Multilayer Exploration Structure enables retrieval on different levels of detail if the subsets (e.g., randomly selected) correspond to representative samples of the dataset. Furthermore, each layer can be indexed independently by a similarity index suitable for a specific layer (e.g., memory-based pivot table [6] for upper layers, or disk-based PM-Tree [7] for lower layers).

In the following text, we present formal definitions of basic exploration operations over the Multilayer Exploration Structure. Note that a popular k-NN query is considered as a supportive task for exploration operations.

Definition 2. *Given a Multilayer Exploration Structure $E = MLES(\mathbb{S}, m, v, \phi)$, a query object $q \in \mathbb{L}_i$ and a parameters k, i, the operation $Zoom\text{-}In(E, q, k, i)$ on a layer $\mathbb{L}_i, \forall i = \{0, .., m - 2\}$ returns a set of objects being the k nearest neighbors to the query object q from objects in layer $\mathbb{L}_{i+1}$.*

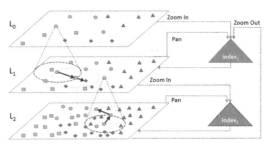

Fig. 1. MLES with indexes $Index_1$ for layer L_1 and $Index_2$ for layer L_2.

The *Zoom-In* operation enables users to select a query object from actually investigated layer L_i to query for objects from more populated layer L_{i+1}, thus seeing more details in the new view.

Definition 3. *Given a Multilayer Exploration Structure, $E = MLES(\mathbb{S}, m, v, \phi)$, a query object $q \in \mathbb{L}_i$ and a parameters k, i, the operation Zoom-Out(E, q, k, i) on a layer $\mathbb{L}_i, \forall i = \{1, .., m-1\}$ returns a set of objects being the k nearest neighbors to the query object q from objects in layer $\mathbb{L}_{i-1}$.*

The *Zoom-Out* operation enables users to hide details by moving from more populated layer L_{i+1} to less populated layer L_i. Since L_i does not contain all objects from L_{i+1}, the selected query object does not necessary have to be in the new view.

Definition 4. *Given a Multilayer Exploration Structure, $E = MLES(\mathbb{S}, m, v, \phi)$, a query object $q \in \mathbb{L}_i$ and a parameters k, i, the operation Pan(E, q, k, i) on a layer $\mathbb{L}_i, \forall i = \{0, .., m-1\}$ returns a set of objects being the k nearest neighbors to the query object q from objects in layer $\mathbb{L}_i$.*

The *Pan* operation employs the k-NN query just for objects in the same layer, enabling users to reach objects not accessible by the *Zoom-In* operation.

3   User Study

In this section, we present our user study focusing on the effectiveness of multi-layer exploration. We have investigated 3-layer MLES and compared it to simple 2-layer MLES corresponding to a standard flat k-NN browsing baseline.

In order to fully compare different exploration approaches, various exploration tasks have to be utilized to challenge the performance of the structures. In our study, we have focused on known-item search tasks, where users receive a simple textual description of a searched class. Then, starting from the initial view, users have to find as much objects of the searched class as possible using just limited number of exploration operations (in our study we have used 15 operations). Let us note that the initial view was the same for all search tasks and that the initial view did not contain objects of the searched classes. For each search task, we have measured the number of clicks to find a first object of the searched class and the cumulative number of found objects for each step. Although it was not the objective of the search tasks, we have measured also the cumulative number of visited classes for each step.

In the following, we present results of our extensive user study with 94 participants from different countries (47 men and 47 women) that altogether performed 1661 search tasks. The participants are students of the University of Finance and Administration attending different study programs (IT, economics). The participants were from different groups. The participants of one group have received the instructions from one person. In order to diversify the groups even more, some groups were motivated to find as much images of the search class as possible and others have received few more keywords describing the searched class. In each group, the tasks were distributed uniformly for both compared indexes. In the following, we describe the *Find the image* application used for testing, test settings, results of the tests and then we discuss the results of the user study.

3.1 Find the Image Application

The user study was performed using the *Find the image* web application [5]. It is an open platform for image exploration user studies that can be used also as a web service. In the application, users are presented with the current search task and their progress on the top of the screen. The application shows and monitors wall-clock time, number of remaining exploration steps and number and percentage of found objects fulfilling the current task. The main part of the screen is dedicated to results of exploration operations. The results are visualized using a force-directed layout which uses images as nodes and image similarities as weights for edges between nodes. Depending on the current level of the multi-layer index structure, each node provides interactive zoom in, zoom out and pan operations. Zooming out can be performed by double-clicking the right mouse button, buttons for zoom-in and/or pan operations are offered when hovering over a node. The bottom of the screen shows the images found so far in a grid of thumbnails. When a user spends 15 exploration operations, the application saves the exploration statistics and offers another search task.

3.2 Settings

As a dataset, the PROFIMEDIA test collection [3] comprising 21993 small thumbnail images divided into 100 classes was employed. The position-color-texture feature signatures and the signature quadratic form distance were used as a similarity model. In the study, 2-layer and 3-layer variants of MLES were compared. Both variants of the structures have shared the same set of images for the initial view with 50 images and the last layer comprising the whole dataset. The three layer structure used middle layer with one eighth of the objects from the third layer. The PM-Tree index [7] was used to process k-NN queries.

The query classes for search tasks were selected from classes not present in the initial view. More specifically, 10 homogeneous (objects of the same class are visually similar) and 10 heterogeneous (objects of the same class can be less visually similar) query classes were manually selected. Each participant has received an ID, where for each ID there was a sequence of 20 search tasks. The sequence always consisted of permuted 10 homogeneous query classes followed by permuted 10 heterogeneous query classes. Using such sequences, the participant always started with easier tasks and then continued with more complex tasks. The participant could access the next search task only if he finished the actual search task. Before starting the sequence of tasks, each participant had also one test task for each MLES structure with simple query class that appeared in the initial view. Results of these test tasks were not considered in this study.

3.3   Results

In the first part of this section, we focus on the number of exploration operations to find first objects of the query class (the initial view did not contain objects from query classes). In some search tasks, users were not able to find any object of the query class, as depicted in Figure 2. We may observe that for query classes grain, pizza, running track and bee, it was difficult for the users to find objects of the searched class given just a short textual description. Finding objects of these four classes was difficult for both compared MLES structures, however, the 2-layer MLES resulted in more unsuccessful searches (in case of bee, only 30% of searches using 2-layer MLES were successful). Except five query classes, the 3-layer MLES outperformed the 2-layer MLES. Overall, there was 10% of unsuccessful searches for the 3-layer MLES and 18% of unsuccessful searches for the 2-layer MLES. Men participants were slightly more successful than women participants for both indexes.

In the top part of Figure 3, the average number of exploration operations to find first object of a searched class is depicted. Since the number of exploration operations to find first object is unknown for unsuccessful searches, we have used number 16 as the minimal number of operations required for the unsuccessful searches. We may observe that in most cases five exploration operations were sufficient to find first object of the class. Except few cases the number of exploration operations to find first object was lower for the 3-layer MLES.

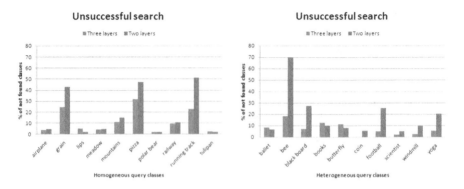

Fig. 2. Unsuccessful search.

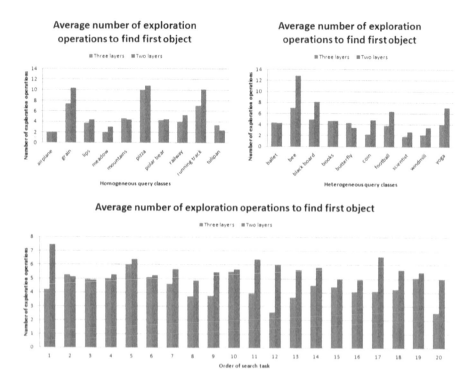

Fig. 3. Average number of exploration operations to find first object of the searched class.

As the users started always from the same initial view, the learning effect could take significant part of the exploration effectiveness. In the bottom part of Figure 3, we may observe the average number of exploration operations to find first object of a searched class for each search task. Although the numbers are slightly lower for tasks solved later, there is not a significant evidence of the learning effect. This is caused probably by the fact that the query classes were presented only after the previous search task was finished and that the description of the classes was textual. Therefore, any remembered visual information was not often connected to the actual search task. On the other side, in some cases the participants have reported that the previous experience has helped them to find quickly some searched classes.

In the second part of this section, we present graphs depicting how many objects of the searched classes were found using 15 exploration operations. In the top part of Figure 4, we may observe that searching objects of the homogeneous query classes resulted in higher percentage of found objects (even though the heterogeneous query classes could benefit from the learning effect). This is probably caused by the fact that by finding a first object of the query class the participant accesses a large cluster of visually similar objects which can be simply explored by k-NN queries. On the other side, a heterogeneous query class

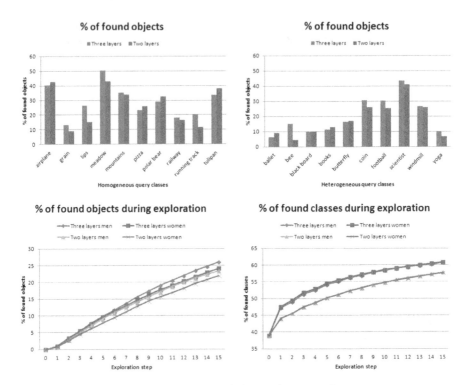

Fig. 4. Found objects and classes during exploration.

consists of more visually dissimilar clusters and thus the participant often gets stuck in just one of them during 15 exploration steps.

In the bottom left part of Figure 4, the average percentage of found objects from searched classes is depicted for each exploration step. We may observe that from the second step the 3-layer MLES starts to outperform the 2-layer MLES. We may also observe that in later steps men participants were able to find more objects of the searched class than women participants for both structures. In the bottom right part of Figure 4, the average percentage of found classes is depicted for each exploration step. We may observe that in each exploration step, the user visited more classes when using the 3-layer MLES. This behavior is caused by the middle layer of the 3-layer MLES that provides higher variability of classes already after first exploration operation (which is always Zoom in).

3.4 Discussion

The results of the user study show that using more layers in multimedia exploration could bring better effectiveness to the exploration process, especially in tasks focused on finding a first object of the searched class. On the other side, when the first object of the searched class is found, then simple k-NN queries are utilized to explore the cluster of visually similar objects. In this study, we did not primarily focus on searching of more visually dissimilar clusters of the same class and therefore we cannot conclude if more layers can help with such task. Considering the results presented in Figure 4, there is not a significant performance gain of the three layer exploration structure for heterogeneous query classes. We can just intuitively guess from results of Figure 3 that when using more than 15 exploration operations, the users could find a first object of another visually homogeneous cluster of the searched class sooner. The results would show bigger differences between both compared structures if the utilized database contained an order of magnitude higher number of objects and classes. For example, given 1 billion images and 1 million query classes, it would be probably much more complicated to find the searched class when using just 2-layer MLES.

4   Conclusion

In this paper, we have performed an extensive user study on MLES, a general structure for multilayer multimedia exploration. The MLES proved to be a more suitable concept for multimedia exploration than standard flat k-NN querying (simulated by 2-layer MLES). So far, the defined operations are based just on exact k-NN queries, however, further enhancements like approximate k-NN queries could be used for more efficient retrieval in lower layers of MLES, or multiple k-NN queries or similarity models could be combined in one exploration operation, to make the exploration process performed on the MLES more scalable.

Acknowledgments. This research has been supported in part by Czech Science Foundation project 15-08916S, by Grant Agency of Charles University projects 201515 and 910913 and by project SVV-2015-260222.

References

1. Barthel, K.U., Hezel, N., Mackowiak, R.: ImageMap - visually browsing millions of images. In: He, X., Luo, S., Tao, D., Xu, C., Yang, J., Hasan, M.A. (eds.) MMM 2015, Part II. LNCS, vol. 8936, pp. 287–290. Springer, Heidelberg (2015)
2. Beecks, C., Driessen, P., Seidl, T.: Index support for content-based multimedia exploration. In: International Conference on Multimedia, pp. 999–1002. ACM (2010)
3. Budikova, P., Batko, M., Zezula, P.: Evaluation platform for content-based image retrieval systems. In: Gradmann, S., Borri, F., Meghini, C., Schuldt, H. (eds.) TPDL 2011. LNCS, vol. 6966, pp. 130–142. Springer, Heidelberg (2011)
4. Heesch, D.: A survey of browsing models for content based image retrieval. Multimedia Tools Appl. **40**(2), 261–284 (2008)
5. Lokoč, J., Grošup, T., Čech, P., Skopal, T.: Towards efficient multimedia exploration using the metric space approach. In: Content-Based Multimedia Indexing (2014)
6. Navarro, G.: Searching in metric spaces by spatial approximation. The VLDB Journal **11**(1), 28–46 (2002)
7. Skopal, T., Pokorný, J., Snášel, V.: PM-tree: pivoting metric tree for similarity search in multimedia databases. In: Advances in Databases and Information Systems (2004)

Semantic Similarity Between Images: A Novel Approach Based on a Complex Network of Free Word Associations

Enrico Palumbo[1]([⊠]) and Walter Allasia[2]([⊠])

[1] Physics University of Torino, via Giuria, 1, 10025 Torino, Italy
enrico.palumbo@edu.unito.it
[2] EURIX, via Carcano, 26, 10153 Torino, Italy
allasia@eurix.it

Abstract. Several measures exist to describe similarities between digital contents, especially for what concerns images. Nevertheless, distances based on low-level visual features embedded in a multidimensional linear space are hardly suitable for capturing semantic similarities and recently novel techniques have been introduced making use of hierarchical knowledge bases. While being successfully exploited in specific contexts, the human perception of similarity cannot be easily encoded in such rigid structures. In this paper we propose to represent a knowledge base of semantic concepts as a *complex network* whose topology arises from free conceptual associations and is markedly different from a hierarchical structure. Images are anchored to relevant semantic concepts through an annotation process and similarity is computed following the related paths in the complex network. We finally show how this definition of semantic similarity is not necessarily restricted to images, but can be extended to compute distances between different types of sensorial information such as pictures and sounds, modeling the human ability to realize synaesthesias.[1]

Keywords: Semantic similarity · Complex networks · Free word associations · Image analysis

1 Introduction

Content-based image retrieval is a well established research branch, working on low-level visual features, such as color or texture, that can be automatically extracted from digital contents [7,19]. Unfortunately, it is often the case that purely visual features do not encode similarities regarding high-level concepts. Smeulders et al. define the *semantic gap* as "the lack of coincidence between the information that one can extract from the visual data and the interpretation that the same data have for a user in a given situation" [19]. Image annotation

---

[1] This work was partially funded by the European Commission in the context of the FP7 ICT project ForgetIT (under grant no: 600826)

G. Amato et al. (Eds.): SISAP 2015, LNCS 9371, pp. 170–175, 2015.
DOI: 10.1007/978-3-319-25087-8_16

attempts to fill the semantic gap by mapping low-level visual features into high-level concepts, either manually or through machine learning algorithms such as Support Vector Machines (as done in [13], possibly combined with more structured hierarchical knowledge bases [7,21]). After the annotation, a picture is represented as a Bag of Words, namely a vector whose elements indicate the presence (or the absence) of the concepts utilized in the annotation process and distances are evaluated in multidimensional L^p Lebesgue spaces or more generalized topological spaces [17].

2   Graph-Based Similarity

The representations of images as vectors in a metric space all rely on the assumption of independence between the words used in the annotation process [8]. As also argued in [12], this is rarely true. Let us suppose to make use of three concepts for the annotation, tree, leaf and window, and to have three pictures, one containing only a tree, one only a leaf and one only a window. The vector representation would be: $tree = (1,0,0)$, $leaf = (0,1,0)$, $window = (0,0,1)$ and the distance between the images would be the same, even if intuitively the concept of tree should be closer to the concept of leaf than to that of window. Indeed, the natural semantic correlations among the concepts used in the annotation make inadequate the euclidean representation. To comply with the necessity of a structure which well expresses relations, it is common to make use of *semantic graphs*. A semantic graph is a pair of sets $G = (C, A)$ where C is the set of nodes and A is the set of edges, i.e. links between nodes. In semantic graphs the nodes are concepts (or words) and the edges represent either logical relations between them, e.g. 'is-a', 'has-a', or simple conceptual associations. Notable examples of semantic graphs are the models of semantic memory developed by Collins and Quillian [5] and Collins and Loftus [4] or networks of words such as WordNet [14], Roget's Thesaurus [18], Word Association networks or network of tags such as those of [3] and the like. Ontologies as well may be seen as semantic graphs whose structure must be logically consistent and is often hierarchical and in which formal rigor is added by means of logical axioms and inference rules [11]. In the last years, many have tried to overcome the limitations of the euclidean representation utilizing semantic graphs [8,9,12]. In [8], for instance, the authors use ImageNet, a logically organized database of images, analogous to WordNet, to evaluate semantic similarities between images. These methods, however, only account for logical similarities, namely for shared taxonomical categories. Oppositely, humans can analyze images at different semantic levels [21] and can establish more complex relations between them, which cannot be easily encoded in a hierarchical structure (Fig. 1). A pair of images can be considered to be related because the objects represented often occur together, because they evoke similar feelings or belong to the same context. Statistical evidence of this fact is presented in [10]. The authors compare the associations of the Word Association Network of the Human Brain Cloud [22], a web-based "massively interplayer word association game", which they have validated for scientific purposes, with the logical relations of WordNet. They map the word association

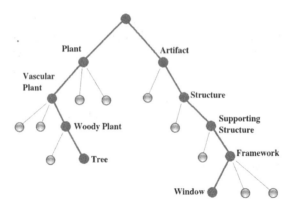

Fig. 1. *Sketch of the structure of ImageNet. The only way to reach Window from Tree is to go up and down the hierarchy: 'Tree-WoodyPlant-VascularPlant-Plant-Root-Artifact-Structure-SupportingStructure-Framework-Window'. On the Word Association Network built from the data of [16] the path is 'Tree-Shade-Window'.*

network, which completely lacks of semantics, onto WordNet and what they observe is that "human beings often construct associations with probabilities that could strongly deviate from what would be the pure statistical structure of WN", entailing that conceptual associations are often based on other criteria than pure logic.

3   Complex Networks

A further limitation of the hierarchical models is that they rarely exhibit *complex* structures. In fact, in the last years, starting from the article of Steyvers and Tenenbaum [20], many have pointed out the strong analogies between semantic graphs and *complex networks*. The study of *complex networks* is a new and emerging field, born in the late 90s as a consequence of the discovery that many real networks (WWW, Internet, science collaboration graph, the web of human social contacts,...) are *small world*, i.e. the distance between two nodes scales logarithmically with the number of nodes N, highly *clustered* and *heterogeneous*, i.e. the degree distribution is considerably different from binomial, poissonian or gaussian distributions, since it is markedly right-skewed and fat-tailed, often well approximated by a power-law distribution [1,2].

In [20] the authors have shown that Wordnet, the Roget's thesaurus and the Word Association network built from the experiment done by the University of South Florida [16] exhibit a small average path length, a high clustering coefficient and a power-law distribution of degree. Word associations are obtained through a simple experiment: subjects are asked to write down the first word that comes to their mind which is meaningfully related to a cue word, provided by the experimenters. A network can be built by identifying the words as nodes and the edges as associations, which can be weighted by the frequency of that particular association. In [10] this analysis is extended to another network of word association, the Human Brain Cloud [22], an online multiplayer word association game. In [15] similar results are obtained, without aggregating data from different individuals. In [3] the topological properties of the semantic networks spontaneously emerging from co-occurring tags of digital resources on website such as del.icio.us also exhibit the typical properties of complex networks. These independent studies, obtained from semantic graphs of diverse nature and origin, yielding similar conclusions suggest that *complexity* is a fundamental property of the structure of semantic networks. This fact has remarkable consequences on the shortest paths, therefore on similarities. Scale-free networks are more than *small world*, with average shortest path $< l > \simeq \frac{\log N}{\log \log N}$ [2] with N vertices. This is due to the hubs of the networks which act as bridges between "distant" nodes, providing shortcuts across the web.

4   The Model

To account for the role of complexity and to encompass the possibility of free conceptual association, we propose a model for evaluating semantic similarities between images based on a Complex Network of Free Word Associations. Both the Word Association Network built from the experiment of the University of South Florida [16] and the one built from the web-based experiment of Human-BrainCloud, have been proven to be *complex networks* and to share a similar structure [10, 20]. Therefore, in the following, we shall generically refer to a Word Association Network (WAN). The model works as follows (Fig. 2):

1) Build a Word Association Network
2) Annotate the images I_i and I_j with words of the WAN: f_i and f_j are the vectorial representations of I_i and I_j, whose component $f_i^k \in [0, 1]$ is a confidence score of the word k in I_i
3) Turn the most relevant words into weighted links and anchor the images to the WAN
4) The distance is the shortest weighted path length [6], namely $d(I_i, I_j) = min_{\gamma_{i,j}} \sum_{l \in \gamma_{i,j}} l$, where $l = \frac{1}{w}$ is the length of a link (the stronger the association, the closer the nodes) and $\gamma_{i,j}$ is a generic weighted path connecting I_i and I_j

Note that "most relevant" is vague and needs to be further specified. Different criteria may be applied to determine what number of words should be turned into links, but a robust method is to normalize f_i, sort the components by magnitude and select the first k concepts containing a fixed percentage α of the total norm. In the demonstration of Fig. 2, we have used $\alpha = 0.9$, but different values may be selected. We suggest that this free parameter could be set optimally through a learning process onto a training set of images whose similarity have been already evaluated by well established methods.

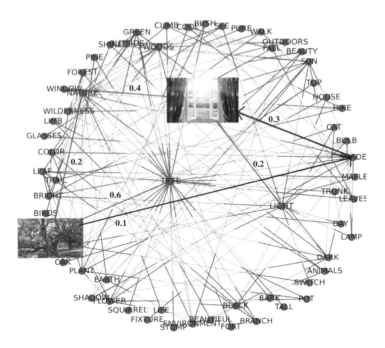

Fig. 2. *Suppose that the annotation yields* $f_1 = (window = 0.4, shade = 0.3, light = 0.2, nature = 0.05, sheep = 0.05)$ *and* $f_2 = (tree = 0.6, nature = 0.2, shade = 0.1, branch = 0.1)$. *Setting* $\alpha = 0.9$ *we select the first three attributes and turn them into weighted links (represented in red and black). The shortest path (in black) between the two images is Img1 → Shade → Img2, hence the similarity is* $d = 1/0.1 + 1/0.3 \approx 13.3$ *in the WAN built from [16].*

5   Conclusions

In this position paper, we have highlighted two possible weak points of the state-of-the-art measures of semantic similarity between images: the excessive rigidity of purely *hierarchical structures* and the *absence of complexity*. Therefore, we have proposed a model which could possibly solve these issues. We want to underline the fact that this definition of similarity can be extended to digital contents of diverse nature, such as images, sounds and more generally media objects. Once the objects are semantically annotated, the proposed algorithm allows to measure distances between different sensorial information, modeling the natural human ability to associate sensations. The model still necessitates a thorough evaluation and its performances have to be compared with the available and consolidated IR techniques in order to confirm its proximity to human behaviors. However, if our intuition is right, it can provide a general method to evaluate relations between digital contents in a way more comprehensible to humans. This approach could have a vast array of applications in information management such as retrieval, clustering and the like.

References

1. Albert, R., Barabási, A.L.: Statistical mechanics of complex networks. Reviews of Modern Physics **74**(1), 47 (2002)
2. Barrat, A., Barthélemy, M., Vespignani, A.: Dynamical processes on complex networks, pp. 116–135. Cambridge University Press (2008)
3. Cattuto, C., Barrat, A., Baldassarri, A., Schehr, G., Loreto, V.: Collective dynamics of social annotation. Proceedings of the National Academy of Sciences **106**(26), 10511–10515 (2009)
4. Collins, A.M., Loftus, E.F.: A spreading-activation theory of semantic processing. Psychological Review **82**(6), 407–428 (1975)
5. Collins, A., Quillian, M.: Retrieval time from semantic memory. Journal of Verbal Learning and Verbal Behavior **8**, 240–248 (1969)
6. Dall'Asta, L., Barrat, A., Barthélemy, M., Vespignani, A.: Vulnerability of weighted networks, March 2006. arXiv:physics/0603163v1
7. Datta, R., Li, J., Wang, J.Z.: Content-based image retrieval: approaches and trends of the new age. In: Zhang, H., Smith, J., Tian, Q. (eds.) Multimedia Information Retrieval, pp. 253–262. ACM (2005)
8. Deselaers, T., Ferrari, V.: Visual and semantic similarity in imagenet. In: Proceedings of the 2011 IEEE Conference on Computer Vision and Pattern Recognition, CVPR 2011, pp. 1777–1784. IEEE Computer Society, Washington, DC (2011)
9. Fang, C., Torresani, L.: Measuring image distances via embedding in a semantic manifold. In: European Conference on Computer Vision, pp. 402–415, October 2012
10. Gravino, P., Servedio, V.D.P., Barrat, A., Loreto, V.: Complex structures and semantics in free word association. Advances in Complex Systems **15**(3–4) (2012)
11. Guarino, N., Oberle, D., Staab, S.: What is an ontology? In: Staab, S., Studer, R. (eds.) Handbook on Ontologies, 2nd edn. Springer (2009)
12. Kurtz, C., Beaulieu, C.F., Napel, S., Rubin, D.L.: A hierarchical knowledge-based approach for retrieving similar medical images described with semantic annotations. J. of Biomedical Informatics **49**(C), 227–244 (2014)
13. Markatopoulou, F., Mezaris, V., Kompatsiaris, I.: A comparative study on the use of multi-label classification techniques for concept-based video indexing and annotation. In: Gurrin, C., Hopfgartner, F., Hurst, W., Johansen, H., Lee, H., O'Connor, N. (eds.) MMM 2014, Part I. LNCS, vol. 8325, pp. 1–12. Springer, Heidelberg (2014)
14. Miller, G., Beckwith, R., Fellbaum, C., Gross, D., Miller, K.: Introduction to wordnet: an on-line lexical database. Int. J. Lexico. **3**, 235–244 (1990)
15. Morais, A.S., Olsson, H., Schooler, L.: Mapping the structure of semantic memory. Cognitive Science **37**, 125–145 (2012)
16. Nelson, D.L., McEvoy, C.L., Schreiber, T.A.: The university of south florida word association norms. http://w3.usf.edu/FreeAssociation
17. van Rijsbergen, K.: The Geometry of Information Retrieval. Cambridge University Press (2004–2007)
18. Roget, P.: Roget's thesaurus of English words and phrases. TY Crowell Co. (1911)
19. Smeulders, A.W.M., Worring, M., Santini, S., Gupta, A., Jain, R.: Content-based image retrieval at the end of the early years. IEEE Trans. Pattern Anal. Mach. Intell. **22**(12), 1349–1380 (2000)
20. Steyvers, M., Tenenbaum, J.B.: The large scale structure of semantic networks: Statistical analyses and a model of semantic growth. Cognitive Science **29**, 41–78 (2005)
21. Tousch, A., Herbine, S., Audibert, J.: Semantic hierarchies for image annotation: a survey. Pattern Recognition **45**, 333–345 (2012)
22. Gabler, K.: The human brain cloud. http://www.humanbraincloud.com

Applications and Specific Domains

Vector-Based Similarity Measurements for Historical Figures

Yanqing Chen, Bryan Perozzi, and Steven Skiena[✉]

Department of Computer Science, Stony Brook University,
Stony Brook, NY 11794, USA
{cyanqing,bperozzi,skiena}@cs.stonybrook.edu

Abstract. Historical interpretation benefits from identifying analogies among famous people: Who are the Lincolns, Einsteins, Hitlers, and Mozarts? We investigate several approaches to convert approximately 600,000 historical figures into vector representations to quantify similarity according to their Wikipedia pages. We adopt an effective reference standard based on the number of human-annotated Wikipedia categories being shared and use this to demonstrate the performance of our similarity detection algorithms. In particular, we investigate four different unsupervised approaches to representing the semantic associations of individuals: (1) TF-IDF, (2) Weighted average of distributed word embedding, (3) LDA Topic analysis and (4) Deepwalk embedding from page links. All proved effective, but Deepwalk embedding yielded an overall accuracy of 91.33% in our evaluation to uncover historical analogies. Combining LDA and Deepwalk yielded even higher performance.

Keywords: Vector representations · People similarity · Deepwalk

1 Introduction

Historical interpretation benefits from identifying analogies among famous people: Who are the Lincolns, Einsteins, Hitlers, and Mozarts? Effective analogies should reflect shared personality traits, historical eras, and domains of accomplishment, but usually only particular facets of these individuals are captured. Analogies are of course highly subjective, and hence rest at least partially in the eyes of the beholder: "there are a thousand Hamlets in a thousand people's eyes". For instance, Figure 1 gives closest analogies on different aspects of *Isaac Newton*:

Detailed similarity quantification cannot create a perfect ranking to satisfy everyone, especially for pairs of people that sit on the same "level" of similarity. However, people on different "level" are definitely comparable. We are interested in developing a generalized model to identify analogous figures of a reasonably high similarity level, based on semantics in text and connections of history. It could be very evocative when correctly identified examples like: *Martin Luther King* and *Nelson Mandela*; *George Washington* and *Mao Zedong*; *Babe Ruth* and *Sachen Tendlukar*.

© Springer International Publishing Switzerland 2015
G. Amato et al. (Eds.): SISAP 2015, LNCS 9371, pp. 179–190, 2015.
DOI: 10.1007/978-3-319-25087-8_17

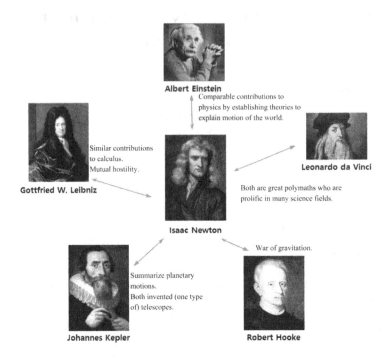

Fig. 1. Sample analogous historical figures of Isaac Newton and corresponding explanations of similarity. Analogies are highly subjective, making it impossible to find perfectly fair and objective gold standards.

In this paper, we propose methods for identifying historical analogies through the large-scale analysis of Wikipedia pages, as well as a reference standard to judge the effectiveness of our methods. The most obvious applications of this are in historical interpretation and education, but we believe that the problem runs considerably deeper since being able to identify similar individuals goes to the heart of algorithms for suggesting friends in social networks, or even matching algorithms pairing up roommates or those seeking romantic partners.

Specifically, our work makes the following contributions:

- We propose to use information extracted from Wikipedia categories to be as reference standards to solve this task. Though not perfect, these human labeled features imply relationships and are shared between similar people. We generated in total 3,000,000 triples of variable and prescribed difficulty, providing an effective standard to evaluate the performance of our similarity measurement algorithms.
- We investigate four different unsupervised approaches to extract semantic associations from Wikipedia. All proved effective and our best approach of Deepwalk yielded an overall accuracy of 91.33% in agreement with human annotated Wikipedia categories. We provide an interactive demonstration of

our historical analogies at *http://peoplesimilarity.appspot.com/*, where you can identify the most similar historical figures to any individual you query.
- We did a careful search to find the best distance function for each vector model. All these approaches yield good qualities, but may focus on different aspects of feature vectors. We also generated a model using linear combination of previously mentioned models to get a better tradeoff between graph structures and text semantics.

The rest of this paper proceeds as follows: In Section 2 we describe related work. Section 3 talks about data collection and resource processing. Section 4 focuses on model constructions. In Section 5 we introduce how we setup our experiment. Finally in Section 6 we show results and evaluations.

2   Related Work

Similarity measurements on documents are gaining popularity [6,8]. Traditional methods of TF-IDF and LDA [3] are proved to be valuable in topic similarity measurements. Topics from LDA highly agree with real tags when finding most important feature words of a page [11]. However, such as an excellent approach still has some defects. One is that the probability distribution is not deterministic, especially when there are many closely-related topics. The other is that LDA focus only on co-occurrence of words but a robust sematic grouping of topics needs more language dependent resources, like stemming and synonyms [7], which makes the procedure semi-supervised. We will show later that unsupervised models can perform as well as, or even better than LDA in our tasks.

On the other hand, researchers have developed network-based similarity measurements instead of pairwise comparisons. Adopting networks in this tasks is valuable. It generalizes classification and local graph features on large scale document collections [10], provides unsupervised hierarchic similarity structure [18] and also benefits visualization of the task to better understand "similarity" [13].

Learning vector representation of words or articles can help convert semantic features into high-dimensional spaces for easy quantification. Word-level representations are proved to be useful [2] and arouse significant interests. SENNA [5] shows that embedding are able to perform well on several NLP tasks in the absence of any other features. Embedding with local information can lead to better and more precise clustering of words [9]. It is even possible to extract specific analogies of words from embedding [4,14]. Additionally, word embedding can be easily constructed for many other languages [1], making it perfectly extensible in multilingual world.

On sentence or article level, proposed vector model in [12] demonstrates the possibility as well as the potential usage in related NLP tasks. Deepwalk [15] seeks vector representations using recent advancements in language modeling and unsupervised feature learning (or deep learning) from sequences of words to graphs. Deepwalk uses truncated random walks to learn latent representations by encoding social relations in a continuous vector space that is easily exploited by statistical models. Applying Deepwalk on Wikipedia will create vector representation for each page thus provides statistical comparison between entries in this huge graph.

3   Data Collection

We start the whole corpus processing from English Wikipedia dumps. We first split each page into content part and reference part: Content part contains main body of text while Reference part includes Wikipedia category information and links to supplemental materials. We then collect links between Wikipedia pages to create a huge adjacent list for all pages. We keep only links in content part that point to another main page (not supplemental materials and categories links). This procedure helps us collect adjacent list of totally 4,517,721 pages which will be used in Deepwalk training.

We search each remaining page in Freebase to check whether this page falls in the category of people. In order to make the similarity comparison more reasonable, we ignore pages that have less than 50 hits which may probably indicate insignificant people. We parse and tokenize all remaining pages with reasonable length and keep track of all raw texts. We finally record 557,965 people's Wikipedia text. In our experiment we use these Wikipedia raw texts to refer to corresponding people when calculate similarities.

At last that we extract Wikipedia category information from reference part to generate standards called "category description" of people. Such human-annotated labels are good for summarizing the history of famous people and will be used to create our experiment test bed.

4   Model Description

In this section we will describe four candidate models which convert the content part of Wikipedia article – a neutral description of one's history – to feature vectors to fuel similarity detection methods.

These models are TF-IDF, distributed word embedding, LDA probability of topics and Deepwalk embedding.

4.1   TF-IDF Model

TF-IDF is a frequently used bag-of-words processing in NLP tasks to reflect importance of words in a target document. One advantage of TF-IDF model is that rare words are usually emphasized due to their uniqueness in the corpus, which makes it easy to handle words that do not appear in pre-defined dictionary. Calculation of TF-IDF basically involve all words appeared in the corpus collection. We use standard TF-IDF vector representation of documents in Gensim [16] to convert each Wikipedia article into feature vector. Pairwise similarity is calculated using cosine similarity.

4.2   Distributed Word Embedding Model

Distributed word embedding model represents an article using a weighted average embedding of words in it. By using TF-IDF value as weight of each word and combining SENNA word embedding, feature vector of an article will be pulled to the barycenter of the most important and descriptive words in the article. Since word embedding cluster syntactically and semantically similar words together in embedding space, potential synonyms can be detected easily to improve the performance of very basic bag-of-words processing, thus to provide a better estimate of similarity. However, word embedding model has a disadvantage that it can only handle words with an existing embedding – words not appear in SENNAs dictionary will be discarded. Distance measurement in our experiment could be Euclidean distance, Manhattan distance or cosine similarity. Figure 2 is a simplified 2-D projection illustrating party leaders, musicians and physicists in our embedding space.

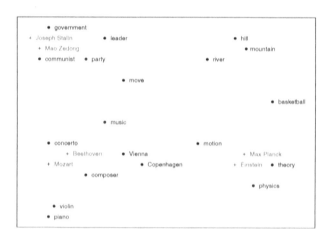

Fig. 2. Sample entities and words in projected embedding space. Entities will be attracted by the most descriptive words in embedding space and locate at the "barycenter" of these words, which can later on be used in distance calculation to show relevance between pairs.

4.3   LDA Model

LDA model provides probability distribution of possible topics. It is based on co-occurrence of words and has an advantage that the output of each topic would be easy for human beings to read and understand. We calculate LDA model for the corpus using 500 topics, converting each Wikipedia page in the corpus into a probability distribution of possible topics. Figure 3 shows an example of top related topics for some entities. Pages have higher probability to fall into same

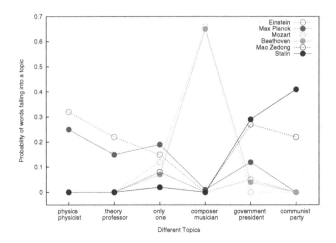

Fig. 3. Examples of entities and top six related topics in LDA method. We display top two representative words in each topic. The distribution of topics for Wikipedia people differs a lot but historical figures with comparable contributions or related professional fields are much more similar than the others.

topics should be considered more similar and the probability of topic distribution differs a lot for party leaders, musician and physicists.

We try several metrics to measure similarity distance of the final LDA vector, including previously used distance functions of Euclidean distance, Manhattan distance and cosine similarity. Additionally, KullbackLeibler divergence and JensenShannon divergence can be applied since feature vectors from LDA are probability distributions.

4.4   Deepwalk Embedding Model

Deepwalk is an online algorithm that creates statistical representation of graph by learning via random walks in the graph. Walks are considered as sentences metaphor and generate latent dimensions according to adjacent list. With a hierarchical Softmax layer these latent dimensions will be finally converted into vector representations. In our experiment, we propose that Wikipedia pages sharing more common links will sit closer in Deepwalk embedding space since random walks in corresponding pages visit through very similar paths. Groups with large fraction of links between corresponding Wikipedia people will indicate stable relationships on similarity as random walks have lower chances to step out of the group.

We use the package described in [15] with 128 output dimensions to train Deepwalk embedding on adjacent lists of all Wikipedia pages. Distance function between two Wikipedia people can be either Euclidean distance or Manhattan distance between Deepwalk embedding output of corresponding pages.

5   Experimental Setup

Wikipedia pages usually contain category information as human-annotated labels. In peoples Wikipedia pages, categories include eras, nationalities, occupations, awards and honors, educations, important historical events and other summaries of their lives. Categories usually have strong signals to distinguish people, such as "Nobel laureates in Physics" and "Presidents of the United States". We assume categories can construct memorable labels to indicate "similarities" and to remind us of the images of historical figures, thus we adopt Wikipedia categories as reference standards to make distinction between groups of people.

However, some categories provide less valuable information in our task, such as "Living people" and "Born in 1957". In order to create better test bed, we did a manual pre-processing to eliminate category patterns that are too broad to be representative, including "living people", "year of birth / death" and "Birthplace". We show in Figure 4 the distribution of Wikipedia categories for all peoples pages after preprocessing. "Famous level" is measured according to Wikipedia hits, article length and links, which is described as "ranking of significance" in [17]. Famous people usually have many Wikipedia categories but most non-significant people only have less than 5 representative categories – that's why we cannot measure people similarity based only on these category descriptions. We build up our similarity references according to number of categories shared between people, the more categories being shared, the more similar they are."

Let $F(X, Y)$ be the number of shared categories between X and Y. $F(X, Y)$ can be calculated for any pair (X, Y) according to extracted category information. We then randomly select tuples of people (X, Y, Z) with condition $F(X, Y) > F(X, Z)$, indicating X shares more common Wikipedia categories with Y than with Z and Y is more similar to X than Z is.

According to distribution of Wikipedia category numbers, we consider pairs (X, Y) have high similarity if $F(X, Y) > 3$ and pairs (X, Y) have low similarity if $0 < F(X, Y) \leq 3$. Additionally, $F(X, Y) = 0$ indicate zero similarity. We sample 500,000 tuples of (X, Y, Z) in each of the following 3 cases sorted in difficulty level:

- **Case I: High similarity VS Zero**: $F(X, Y) > 3$ and $F(X, Z) = 0$
- **Case II: Low similarity VS Zero**: $0 < F(X, Y) \leq 3$ and $F(X, Z) = 0$
- **Case III: High similarity VS Low similarity**: $F(X, Y) > 3$ and $0 < F(X, Z) \leq 3$ and $F(X, Y) - F(X, Z) \geq 2$

Notices that Case III has an extra constraint to guarantee statistic difference between two similarity levels. We construct such test bed for all 557, 965 people and most famous 50,000 people. Examples of our test bed are shown in Table 1.

To evaluate, we try to measure how well our feature vectors agree with these test bed entries. We calculate distances between X, Y and X, Z for each tuple entry (X, Y, Z) in our test bed. For a single tuple test, our feature vector passes if distance between X, Y is closer than distance between Y and Z. The final performance of each model is reported as the percentage of passed tuples tests among 3 test cases as well as the overall percentages.

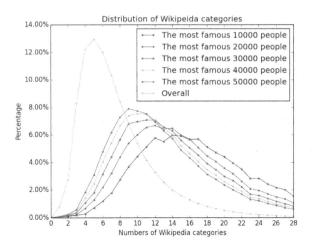

Fig. 4. Distribution of Wikipedia categories on pages of people. As expected, we usually have more detailed information on famous historical figures and category comparisons could be more precise on these people. Overall average number of categories between people lies between 8 and 9.

Table 1. Tuple examples (X, Y, Z) in 3 different cases. Pair (X, Y) is always closer than Pair (X, Z) in our tuples. Cases are sorted according to difficulty so making correct judgements on Case III is harder.

Case	(X, Y, Z)
	(Einstein, Oppenheimer, Michael Jackson)
I	(Lincoln, Reagan, Gaddafi)
	(Mozart, Brahms, Michael Phelps)
	(Einstein, Aristotle, Celine Dion)
II	(Lincoln, Bill Clinton, Heath Ledger)
	(Mozart, Charlie Chaplin, Larry Bird)
	(Einstein, Richard Feynman, John Jacob Abel)
III	(Lincoln, Ulysses Grant, George W. Bush)
	(Mozart, Beethoven, Dmitri Shostakovich)

6   Results and Analysis

Table 2 gives the accuracy of 4 models under 5 distance metrics, on tests over all people in Wikipedia and then restricted to the 50,000 most famous people.

TF-IDF model can answer approximately 4 out of 5 questions correctly, undoubtedly better than random guess (50% accuracy), which shows a quick glance at the most important words can induce similarity well. However, TF-IDF considers no syntactic changes of words (e.g. great vs greatest) and synonyms (e.g. emperor vs monarch). Study on error cases shows that TF-IDF focus too much on locations and names (e.g. the last name of James Simons ranked highly

Table 2. Accuracy performance of candidate models with different distance function. LDA model has extra metrics since the output feature vectors are probability distributions. Deepwalk slightly outperforms LDA.

			Overall		
Model	Distance	Case I	Case II	Case III	Overall
TF-IDF	Cosine	87.01%	76.23%	77.98%	80.08%
Word embedding	L2	96.95%	84.40%	74.97%	85.44%
	L1	96.56%	84.26%	75.57%	85.46%
	Cosine	96.43%	84.13%	75.45%	85.34 %
LDA	L2	98.70%	88.22%	75.39%	87.43 %
	L1	98.17%	88.35%	77.26%	87.92 %
	Cosine	98.43%	88.60%	77.53%	88.18 %
	KL	97.57%	87.86 %	76.10%	87.16 %
	JS	97.69%	87.98 %	76.22%	87.29 %
Deepwalk	L2	**99.51%**	**89.50%**	**84.97%**	**91.33%**
	L1	99.11%	89.13%	84.59%	90.98%
			Most Famous 50,000 people		
Model	Distance	Case I	Case II	Case III	Overall
TF-IDF	Cosine	87.88%	77.06%	78.37%	81.10%
Word embedding	L2	96.89%	82.92%	90.23%	90.01%
	L1	96.51%	82.90%	89.04%	89.48 %
	Cosine	96.31%	82.68%	88.85%	89.29 %
LDA	L2	97.68%	83.71%	80.99%	87.46%
	L1	97.95%	83.31%	81.26%	87.51%
	Cosine	97.63%	83.65%	80.94%	87.41 %
	KL	96.86%	83.58 %	80.84%	87.11 %
	JS	97.15%	83.89 %	81.14%	87.40%
Deepwalk	L2	**98.73%**	**85.47%**	**91.59%**	**91.93%**
	L1	98.11%	84.85%	90.79%	91.24%

in TF-IDF), which reduces the ability to recognize more important words in similarity measurement. Word embedding model clusters synonyms in embedding space thus allow feature vectors to capture various topic words with similar semantic meaning, resulting in better performance compared against original TF-IDF.

LDA model performs well in Case I and Case II, comparing to other models. However in Case III where entities have shared categories in our references, LDA failed to capture the similarity from topic distributions due to the fact that detailed topics might not benefit similarity measurement. For instance, Yao Ming is recognized as "the famous Chinese basketball player in NBA" and so does Jeremy Lin. However, it is hard to find supporting evidence from text since their positions, styles and even images as a basketball player are different. This phenomenon indicates that we may need to control topic numbers to limit the detail level in our task. Additionally, we found that distance metrics does not lead

Table 3. Examples of 10 closest neighbors we find using our vector based models and corresponding human evaluations. *C* column represents count of common Wikipedia categories between pairs of people and *HE* column shows human evaluation after reading their bibliography with ++, + or - rating according to general knowledge.

Model	Albert Einstein			Yao Ming			Larry Page		
	Candidates	C	HE	Candidates	C	HE	Candidates	C	HE
LDA	Wolfgang Pauli	5	++	Yi Jianlian	4	++	Simson Garfinkel	0	++
	Emmy Noether	1	-	Chris Bosh	2	++	Robert Metcalfe	1	++
	Erwin Schrdinger	2	++	Sun Yue	3	++	John Mashey	0	
	Eugene Wigner	4	++	Luol Deng	1	+	Ray Tomlinson	0	-
	Norbert Wiener	1	-	Bob Cousy	1	-	M. J. Dominus	0	-
	Esther Lederberg	0	-	Steve Nash	2	+	R. Piquepaille	0	-
	David Hilbert	0	+	Herschel Walker	0	-	Ellen Spertus	1	-
	Felix Ehrenhaft	0	+	R. Tomjanovich	2	+	Jon Lebkowsky	0	+
	Paul Ehrenfest	1	+	A. Kavaliauskas	1	+	R. P. Garbriel	2	++
	Ralph Kronig	1	+	Mengke Bateer	5	++	D. Giampaolo	1	+
Deepwalk	Richard Feynman	4	++	Yi Jianlian	4	++	Sergey Brin	12	++
	Max Planck	3	++	Jeremy Lin	0	++	Eric Schmidt	6	++
	Freeman Dyson	3	+	Kobe Bryant	2	+	Bill Gates	6	++
	David Bohm	1	+	Wang Zhizhi	5	++	Marc Andreessen	2	++
	Stephen Hawking	2	++	Michael Jordan	1	+	Mark Zuckerberg	6	++
	David Hilbert	0	+	Deron Williams	2	+	Esther Dyson	0	-
	Oppenheimer	4	++	Mengke Bateer	5	++	John Doerr	3	+
	Werner Heisenberg	4	++	Dwyane Wade	3	+	John Battelle	0	++
	Hermann Bondi	1	-	LeBron James	3	+	Joi Ito	0	-
	Erwin Schrdinger	2	++	Steve Francis	2	+	Jimmy Wales	1	+
Linear comb of Deepwalk and LDA	Max Planck	3	++	Yi Jianlian	4	++	Bill Gates	6	++
	Erwin Schrdinger	2	++	Mengke Bateer	5	++	Eric Schmidt	6	++
	Richard Feynman	4	++	Chris Bosh	2	++	Simson Garfinkel	0	++
	Freeman Dyson	3	+	Michael Jordan	1	+	Sergey Brin	12	++
	Wolfgang Pauli	5	++	LeBron James	3	+	Robert Metcalfe	1	++
	David Bohm	1	+	Jeremy Lin	0	++	Marc Andreessen	2	++
	Eugene Wigner	4	++	Charles Barkley	2	++	Mark Zuckerberg	6	++
	R. Millikan	2	+	Tony Parker	1	+	John Battelle	0	++
	Stephen Hawking	2	++	Steve Francis	2	+	Marissa Mayer	4	+
	George Gamow	2	+	Juwan Howard	2	++	Steve Jobs	5	++

to huge performance improvements in LDA, showing LDA feature vectors are robust and well-shaped in embedding space, reflecting good hierarchic similarity structure.

Deepwalk embedding yielded an overall accuracy of 91.33% on all people test and an even higher 91.93% on 50,000 most famous people test, winning all other vector based model. Noticeable drop in Case II accuracy is caused by "weak" categories that do not provide strong support on similarity measurement. Another discovery is that important historical event will usually pull people closer in Deepwalk embedding space, for instance, Ward Hill Lamon is considered close enough to Abraham Lincoln due to the famous assassination. However, Lamon was no match of Lincoln since he is actually not a politician.

Though we consider Case III (quantification of similarities) to be more difficult than Case II (detecting minor similarities), TF-IDF model (as well as Deepwalk and Distributed word embedding model when targeting most famous 50,000 Wikipedia figures) answers Case III questions more accurately.

We did a manual human evaluation on random selected 200 people using LDA, Deepwalk and linear combination of both. Result shows that text-based methods like LDA plays complementary roles to Deepwalk since these two approaches emphasize different part in similarity. Agreement from both methods undoubtedly increase the confidence level of quantifying similarity. Table 3 shows 10 examples and corresponding analogous historical figures, as well as human evaluations using single or combined distance measurement from our previous experiments.

7   Conclusion

We have proposed models for constructing feature vectors and measuring the similarity between historical figures, and demonstrated that it works effectively over a representative evaluation environment. We tested our models on approximately 600,000 historical figures from Wikipedia pages, and our Deepwalk embedding yielded an overall accuracy of 91.33% in our evaluation, showing high agreement with human annotated Wikipedia categories. We show that linear combination of Deepwalk and LDA make results even more reasonable.

These models naturally extending to analyzing figures in different languages, and also to extend to other classes of entities like locations (i.e. cities and countries) and organizations (companies and universities) and we are able to identify similar individuals for suggesting friends in social networks, or even matching algorithms pairing up roommates or those seeking romantic partners.

Parameterizing can capture different tradeoffs between personality, temporal, and topic-based analogies. An inspection of our closest matches suggests that topic-based analogies dominate the nearest matches when considering text only, but more revealing analogies may result from restricting the analyzed word features to particular parts of speech or sentiment polarity.

Finally, explain or naturally represent the reasons for the observed similarity or analogy require finding human-interpretable names for the dimensions/topics obtained using our learning procedures, or at least better understanding the meaning of particular strong or overrepresented features in our analysis which is an advantage for LDA model. We believe that properly defined weights in combination of different models could better generalize the definition of "similarity" and greatly improve the performance.

Acknowledgments. This research was partially supported by NSF Grants DBI-1355990 and IIS-1017181, and a Google Faculty Research Award.

References

1. Al-Rfou, R., Perozzi, B., Skiena, S.: Polyglot: Distributed word representations for multilingual NLP. In: Proceedings of the Seventeenth Conference on Computational Natural Language Learning, pp. 183–192 (2013)
2. Bengio, Y., Ducharme, R., Vincent, P., Jauvin, C.: A neural probabilistic language model. Journal of Machine Learning Research **3**, 1137–1155 (2003)
3. Blei, D.M., Ng, A.Y., Jordan, M.I.: Latent dirichlet allocation. The Journal of Machine Learning Research **3**, 993–1022 (2003)
4. Chen, Y., Perozzi, B., Al-Rfou, R., Skiena, S.: The expressive power of word embeddings. In: ICML 2013 Workshop on Deep Learning for Audio, Speech, and Language Processing (2013)
5. Collobert, R., Weston, J., Bottou, L., Karlen, M., Kavukcuoglu, K., Kuksa, P.: Natural language processing (almost) from scratch. The Journal of Machine Learning Research **12**, 2493–2537 (2011)
6. Elsayed, T., Lin, J., Oard, D.W.: Pairwise document similarity in large collections with mapreduce. In: Proceedings of the 46th Annual Meeting of the Association for Computational Linguistics on Human Language Technologies (2008)
7. Fellbaum, C.: WordNet. In: Theory and Applications of Ontology: Computer Applications, pp. 231–243 (2010)
8. Huang, A.: Similarity measures for text document clustering. In: Proceedings of the Sixth New Zealand Computer Science Research Student Conference (2008)
9. Huang, E.H., Socher, R., Manning, C.D., Ng, A.Y.: Improving word representations via global context and multiple word prototypes. In: Proceedings of the 50th Annual Meeting of the Association for Computational Linguistics (2012)
10. Kim, M., Zhang, B.T., Lee, J.S.: Subjective document classification using network analysis. In: 2010 International Conference on Advances in Social Networks Analysis and Mining (ASONAM), pp. 365–369. IEEE (2010)
11. Krestel, R., Fankhauser, P., Nejdl, W.: Latent dirichlet allocation for tag recommendation. In: Proceedings of the third ACM conference on Recommender Systems, pp. 61–68. ACM (2009)
12. Le, Q.V., Mikolov, T.: Distributed representations of sentences and documents. In: Proceedings of the 31st International Conference on Machine Learning (ICML-2014), pp. 1188–1196 (2014)
13. Maiya, A.S., Rolfe, R.M.: Topic similarity networks: visual analytics for large document sets. In: 2014 IEEE International Conference on Big Data (Big Data), pp. 364–372. IEEE (2014)
14. Mikolov, T., Chen, K., Corrado, G., Dean, J.: Efficient estimation of word representations in vector space (2013). arXiv preprint arXiv:1301.3781
15. Perozzi, B., Al-Rfou, R., Skiena, S.: Deepwalk: online learning of social representations. In: Proceedings of the 20th ACM SIGKDD International Conference on Knowledge Discovery and Data Mining, pp. 701–710 (2014)
16. Řehůřek, R., Sojka, P.: Software framework for topic modelling with large corpora. In: Proceedings of the LREC 2010 Workshop on New Challenges for NLP Frameworks, pp. 45–50 (2010)
17. Skiena, S., Ward, C.B.: Who's Bigger?: Where Historical Figures Really Rank. Cambridge University Press (2013)
18. Wang, C., Yu, X., Li, Y., Zhai, C., Han, J.: Content coverage maximization on word networks for hierarchical topic summarization. In: Proceedings of the 22nd ACM international conference on Conference on Information & Knowledge Management, pp. 249–258. ACM (2013)

Efficient Approximate 3-Dimensional Point Set Matching Using Root-Mean-Square Deviation Score

Yoichi Sasaki[1]([✉]), Tetsuo Shibuya[2], Kimihito Ito[3], and Hiroki Arimura[1]

[1] IST, Hokkaido University, Sapporo, Japan
{ysasaki,arim}@ist.hokudai.ac.jp
[2] University of Tokyo, Tokyo, Japan
tshibuya@hgc.jp
[3] CZC, Hokkaido University, Sapporo, Japan
itok@czc.hokudai.ac.jp

Abstract. In this paper, we study approximate point subset match (APSM) problem with minimum RMSD score under translation, rotation, and one-to-one correspondence in d-dimension. Since this problem seems computationally much harder than the previously studied APSM problems with translation only or distance evaluation only, we focus on speed-up of exhaustive search algorithms that can find all approximate matches. First, we present an efficient branch-and-bound algorithm using a novel lower bound function of the minimum RMSD score. Next, we present another algorithm that runs fast with high probability when a set of parameters are fixed. Experimental results on real 3-D molecular data sets showed that our branch-and-bound algorithm achieved significant speed-up over the naive algorithm still keeping the advantage of generating all answers.

Keywords: 3D point set matching · RMSD · Geometric transformation · One-to-one correspondence · Branch and bound · Probabilistic analysis

1 Introduction

1.1 Background

The approximate point set matching (APSM) is one of the fundamental problems in computer science, while it plays important roles in many application areas including molecular biology, image retrieval, pattern recognition, music information retrieval, and geographic information systems [12]. For every $d \geq 1$, the d-dimensional approximate point set matching problem considered in this paper can be described as follows. An input consists of a data set T and a pattern set P of n and k points in $\mathbb{R}^d$, respectively, and a positive integer $r > 0$, called distance threshold. The task is finding some point subset $Q \subseteq T$ of k data points

© Springer International Publishing Switzerland 2015
G. Amato et al. (Eds.): SISAP 2015, LNCS 9371, pp. 191–203, 2015.
DOI: 10.1007/978-3-319-25087-8_18

that are similar to $P \subseteq \mathbb{R}^d$ w.r.t. a given distance measure under some transformation f, such as translation and rotation, and under some correspondence π between the elements of two sets.

In molecular biology, for instance, such an algorithm for solving 3-D APSM can be used to predict the unknown function of a given target protein with known structure. To do this, we search a database of proteins with known functions for structurally similar proteins that may indicate the unknown function. In real molecular databases, individual data entry of molecules in a database may contain measurement errors, and may have different origin, coordinate, and numbering of data points from data to data. Therefore, a point set matching algorithm should have capability of finding approximation matches, and moreover should be tolerant under translation, rotation, and also one-to-one correspondence between points. For this reason, we focus on point set matching of this kind.

1.2   Related Work

APSM and its variants have been extensively studied for many years (See survey [12,17]). Our version of APSM problem in this paper is equipped with full of invariance requirements, that is, all of invariance under translation, rotation, and one-to-one correspondence, that is, *rigid motion with one-to-one correspondence*. However, this invariance makes the problem computationally much harder than the previously studied APSM problems. Most of previous theoretical results on efficient APSM problems seem to fall into two categories: one is an APSM problem [2] for finding data subset *under translation only*, and another is the congruence problem with rigid motion [3] that detects the congruence between two point sets *of the same size*.

In the first category, for point subset matching, de Rezende and Lee [8] presented $O(kn^d)$ time exact matching algorithm for $d \geq 2$ under rigid motion and global scaling. However, it seems difficult to extend their algorithm for approximate matching. For approximate point subset matching, Goodrich *et al.* [10], and later Cho and Mount [6], presented simple constant approximation algorithms for $d = 2, 3$ under directed Hausdorff distance based on aligning pairs of points. Although this algorithm has quadratic time complexity in n, it only has constant approximation ratio larger than three [6]. On the hardness, Akutsu [1] showed that the APSM problem under rigid motion is NP-hard when d is not bounded In the second category, Alt *et al.* [3] presented an algorithm for the congruence problem under rigid motion that runs in $O(n^{d-2} \log n)$ time in d dimension. Akutsu [1] improved this by presenting $O(n^{(d-1)/2} \log n)$ time Monte-Carlo type randomized algorithm. We note that all of above results were obtained for Hausdorff distance which allows many-to-one correspondence between points.

The selection of distance score gives another dimension to the APSM problem. In this paper, we consider the *minimum root-mean-square deviation score* (*minimum RMSD score*) between point sets, which is widely used similarity score in molecular biology [5]. This score requires that there is a one-to-one correspondence π between a transformed set $f(P)$ and Q such that the sum of the squares of Euclidean distances between each point p in $f(P)$ and its counterpart q in Q is within a given threshold r, that is,

$$r \geq \min_{\pi} \min_{f} \sqrt{\frac{1}{n} \sum_{i=1}^{n} ||f(p_i) - q_{\pi(i)}||^2} \tag{1}$$

for some candidate subset $Q \subseteq T$. This minimum RMSD score resembles the *directed approximate congruence* [3] between two k-point sets except that the latter requires that there is a one-to-one correspondence that maps each point in $f(P)$ to its counterpart q in Q within distance ε. In other words, the distance between $f(P)$ and Q is measured in L_2-norm in our problem, while it is measured in L_∞-norm in [3]. Overall, from the above discussions, most of the previous results on APSM do not apply to our problem.

1.3   Research Goal

In this paper, we consider the approximate 3-D point set matching problem with respect to the minimum RMSD score for sets under both of translation and rotation and under one-to-one correspondence (APSM(trans, rot, 1-1; RMSD), for short). In addition, we are also interested in finding *all* approximate matches rather than *just one* or *some* matches. A straightforward approach to solve this problem is to use exhaustive search, which enumerates all candidate point subsets of T and all one-to-one correspondences between a pattern and each of them. Then, for each enumerated candidate, we test the minimum RMSD score under translation and rotation by some matrix computation such as [15]. However, in practice, one serious problem with this exhaustive search method is its exact exponential time complexity. Since the method must exactly enumerate all of $\binom{n}{k} = n^{\Theta(k)}$ combination of k data points in T, it always visits the leaves of the search tree regardless of the content of input sets.

1.4   Main Results of this Paper

To overcome this difficulty, we study a branch-and-bound algorithm for APSM(trans, rot, 1-1; RMSD) problem. This algorithm finds all approximate matches within threshold r by systematically enumerating all one-to-one correspondences between candidate subsets of k data points from smaller to larger prefixes based on recursive computation. At each iteration of the search, it tests if the current candidate prefix of size $i < k$ satisfies a given lower bound function, and then prune the search if the test failed. As a key of the algorithm, we show that the proposed lower bound function is sound in the sense that it cannot prune any successful search branches. Although the time complexity of the obtained algorithm is still $n^{O(k)}$ in n and k in the worst case, the algorithm can make early pruning depending on the content of input data.

We also present a *fixed-parameter-tractable* style algorithm [9] that runs particularly fast in terms of the data set size n when the parameters, namely, the size k and radius ℓ of a pattern set, and the distance threshold r are fixed. For each data point, the algorithm forms a small sample set consisting of all data points within some fixed distance, and then applies any APSM algorithm to the sample set. Assuming the spatial Poisson process [14] in $\mathbb{R}^d$, we show that the algorithm runs in linear time in expected case, and runs in $O(n \log^k n)$ time and $O(\log n)$ space with high probability for fixed parameters.

Finally, the experimental results on real data sets of 3-D molecules showed that the proposed branch-and-bound algorithm was one to two order of magnitude faster than the straightforward exhaustive search algorithm. Hence, our lower bound function and pruning technique achieve significant speed-up of approximate 3-D point subset matching.

1.5   Organization of This Paper

In Sec. 2, we introduce the basic definitions and notations related to the approximate point subset matching problem with the minimum RMSD under translation, rotation, and 1-1 correspondence. First in Sec. 3, we present the branch-and-bound algorithm using lower bound function over candidate prefixes. In Sec. 4, we present the fixed-parameter-style algorithm and gives analysis of its complexity. Finally, Sec. 6 concludes the paper.

2   Preliminaries

We give brief review of basic concepts and notations in geometry [7]. Then, we introduce our point subset matching problem.

2.1   Basic Definitions

We denote by $\mathbb{R}$ and $\mathbb{N}$ the sets of all real numbers and integers, respectively. For a real-valued k-vector $Q = (q_1, \ldots, q_k) \in \mathbb{R}^k$, we denote its L_2- and L_∞-norms by $L_2(Q) = \{\frac{1}{k} \sum_i |q_i|^2\}^{1/2}$ and $L_\infty(Q) = \max_i |q_i|$. For any k-vector Q, we have the inequality $\frac{1}{k^{1/2}} L_\infty(Q) \leq L_2(Q) \leq L_\infty(Q)$. For a matrix or a vector A, $A^\top$ denotes the *transpose* of A.

Let $d \geq 1$ be the dimension of the space. In this paper, we consider the 3-D space $\mathbb{R}^3$, but all the results also apply to the d-dimensional space for every fixed $d \geq 1$. An element $P = (p_1, \ldots, p_d)^\top \in \mathbb{R}^d$ of the space $\mathbb{R}^d$ is called a *point*. The *Euclidean distance* (or *distance*, for short) between points p and q is given by the L_2-norm $||p - q|| = L_2(p - q)$.

For a *point sequence* $S = (s_1, \ldots, s_k) \in (\mathbb{R}^d)^k$ of length k in $\mathbb{R}^d$ and any $0 \leq i \leq k$, we define the *i-prefix* of S as the subsequence $S[1..i]$ consisting of the first i points of S. A *point set* is an unordered collection $T = \{t_1, \ldots, t_n\}$ of points in the space. We define the *size* of T by the number of its elements $|T| = n$. In what follows, we represent a set as a point sequence by assuming some fixed ordering of the elements. For any $k \geq 0$, a point set P is a *k-point set* (or *k-set*, for short) if $|P| = k$.

2.2   The Minimum RMSD Score for k-point Sets

From now on, we introduce the distance score *MinRMSD* between two k-point sets under translation, rotation and one-to-one correspondence.

Let $P = \{p_1, \ldots, p_k\}$ and $Q = \{q_1, \ldots, q_k\}$ be two k-point sets in $\mathbb{R}^3$. In what follows, we assume an arbitrary fixed ordering over the indices since the following discussion does not depend on this choice of ordering. Based on this, we often regard P and Q as point sequences by assuming underlying ordering. We first try to *align* the points in P with the points in $\pi(Q) = \{Q_{\pi(1)}, \ldots, Q_{\pi(k)}\}$ permuted by a *one-to-one correspondence* π between P and Q, that is, any one-to-one mapping π over indices so that p_i and $q_{\pi(i)}$ correspond each other. We denote by $\mathcal{O}$ the class of all one-to-one correspondences over $\{1, \ldots, k\}$.

A geometric transformation is any one-to-one mapping $f : \mathbb{R}^d \to \mathbb{R}^d$ over d-dim space [3,7]. A *rigid motion* is a transformation generated by any combination of translation, rotation, and reflection. A rigid motion is a ransformation that does not change the distance. It is known that matching for P under rigid motion can be reduced to matching for P and its reflection under translation and rotation [3]. We denote by $\mathcal{RT}$ the class of all compositions of rotations and translations. It is well known that any transformation f in $\mathcal{RT}$ is obtained by application of one d-dim rotation matrix $R \in \mathbb{R}^{d \times d}$ and one d-dim translation vector $v \in \mathbb{R}^d$ such that $f(p) = Rp + v$ for any point $p \in \mathbb{R}^d$ (See [3]). For a k-set $P = \{p_1, \ldots, p_k\}$, we extend f by $f(P) = \{Rp_1, \ldots, Rp_k\} = RP$ assuming the correspondence. In what follows, we identify a transformation f in $\mathcal{RT}$ and the associated pair (R, v) if it is clear from context.

Assuming a one-to-one correspondence π between k-sets P and Q, the *root-mean-square deviation* (*RMSD*) between P and $\pi(Q)$ is defined as the average of the squared Euclidean distances between the corresponding pairs of points $\|p_i - q_{\pi(i)}\|$ for $i = 1, \ldots, k$. Then, the *minimum root-mean-square deviation* under the class $\mathcal{RT}$ of rotations and translations, denoted by $MinRMSD(P, Q)$, is defined as the minimum value of the RMSD score between $f(P)$ and $\pi(Q)$ over all transformations in RT given by

$$MinRMSD(P, Q) = \min_{\pi} \min_{f} \sqrt{\frac{1}{k} \sum_{i=1}^{k} \|f(p_i) - q_{\pi(i)}\|^2}, \qquad (2)$$

where π ranges over all one-to-one correspondences over $\{1, \ldots, k\}$, and f ranges over all transformations $f_{R,v}$ in $\mathcal{RT}$ specified by a rotation R and a translation v in d-dim space. If π is already specified, it is known that the optimal transformation f in $\mathcal{RT}$ minimizing $RMSD(P, \pi(Q))$ can be computed in linear time in fixed $d \geq 1$ [11,16] by using singular value decomposition (SVD) after aligning the centroid of P and $\pi(Q)$. Such a linear time procedure does not seem to be known for Hausdorff distance.

2.3   Approximate Point Subset Matching Problem

Let $1 \leq k \leq n$ be any positive integers. A *data set* and a *pattern set* are sets $T = \{t_1, \ldots, t_n\}$ and $P = \{p_1, \ldots, p_k\}$ of points in $\mathbb{R}^d$, respectively.

Algorithm 1. A naive algorithm for solving the enumeration version of 3-D APSM(trans, rot, 1-1; RMSD) problem using exhaustive search

---

1: **procedure** MATCHNAIVE(P, T, r)
   Input: A text point set $T[1..n]$, a pattern point set $P[1..k]$,a real number $r > 0$.
   Output: All matchings of P in T with minimum RMSD score no larger than r.
2:    FINDNAIVE$((), 0, |P|, |T|, P, T)$;

3: **procedure** FINDNAIVE($Q = (q_1, \ldots, q_i), i, k, n, P, T$)
4:    **if** $i = k$ **then**   ▷ Q becomes a k-subset
5:        **if** $MinRMSD(P, Q) \leq r$ **then**
6:            Report Q as a match;
7:            **return**;
8:    **for** $j = 1, \ldots, n$ **do**
9:        **if** $T[j] \notin Q$ **then**
10:            FINDNAIVE$((q_1, \ldots, q_i, T[j]), i + 1, k, n, P, T)$;   ▷ Recursive call

---

Each member of P (Q, resp.) is called a *data point* (a *pattern point*, resp.). A *distance threshold* is any positive real number $r > 0$. A *match* for pattern set P w.r.t. r is any k-subset Q of T such that $MinRMSD(P, Q) \leq r$ holds.

Now, we state our approximate pattern matching problem as follows. A *candidate k-subset* in T is any k-subset of T.

Definition 1 (APSM(trans, rot, 1-1; RMSD) problem). The *approximate point set matching problem with MinRMSD score under translation, rotation, and one-to-one correspondence*, abbreviated as APSM(trans, rot, 1-1; RMSD), is defined as follows: Given a data set $T \subseteq \mathbb{R}^d$ of n points, a pattern set $P \subseteq \mathbb{R}^d$ of k points, and a positive real number $r > 0$, find all match $Q \subseteq T$ of k data points that satisfy the condition $RMSD(f(P), \pi(Q)) \leq r$,

The above definition is the *enumeration version* of APSM problem. In the *decision version* of the APSM problem, given T, P, and r, an algorithm must decide if $MinRMSD(P, Q) \leq r$. In the *optimization version*, given T and P, an algorithm must find some one-to-one correspondence π and transformation f in $\mathcal{RT}$ that minimizes $MinRMSD(f(P), \pi(Q))$. In what follows, we present algorithms for the enumeration version of APSM. It is not hard to convert these algorithms to solve the decision and optimization versions in the same time and space complexity though the converse is not true in general.

2.4   A Naive Algorithm for Approximate Point Subset Matching

As the basis of our discussion, in Algorithm 1, we show the naive exhaustive search algorithm for 3-D APSM(trans, rot, 1-1; RMSD) problem in $kn^{\Theta(k)}$ time. In this algorithm, the subprocedure FINDNAIVE starts with the *empty prefix* $Q = ()$ and $i = 0$. Then, it recursively traverses the search space of *i-candidate prefixes* $Q = Q[1..i]$ from smaller to larger for all $i = 0, \ldots, k$, where each $Q[1..i]$

represents an ordered set of i data points $\{Q_{\pi(1)}, \ldots, Q_{\pi(i)}\}$ with correspondence π. At each iteration, it grows the current candidate prefix by appending a new data point from $T \setminus Q$. Whenever the condition $|Q| = k$ holds, it tests if the condition $MinRMSD(P[1..i], Q[1..i]) = \min_f MinRMSD(f(P[1..i]), \pi(Q)) \leq r$ in linear time in k assuming π as mentioned in Sec. 2.2. Since each iteration takes $O(k)$ time, the total running time is $O(ks) = kn^{O(n)}$ time, where $s = n^{O(k)}$ is the number of all i-prefixes with $i \leq k$. One problem with this algorithm is that it cannot make early termination, and thus always takes $kn^{\Theta(n)}$ time regardless of the data content.

3   A Faster Point set Matching Algorithm with Pruning

In this section, we discuss speed-up of the naive algorithm in the previous section. We present an efficient point set matching algorithm MATCHFAST based on branch-and-bound search with early pruning.

The basic idea of our algorithm is using a lower bound function LB of $MinRMSD$ score to make early pruning of unsuccessful branches. We design the lower bound function LB such that for any candidate i-prefix $Q = Q[1..i]$ consisting of $i \leq k$ data points, $LB(P[1..i], Q[1..i]) > r$ implies $MinRMSD(P, R)$ for any k-subset $R = R[1..k]$ of T that is an extension of Q such that $R[1..i] = Q[1..i]$. Based on this idea, we present a simple lower bound function for $MinRMSD$ that our branch-and-bound algorithm uses.

Theorem 1 (sound lower bound function). *Let P and Q be k-point sets as sequences. For any integer $1 \leq i \leq k$, it holds that*

$$MinRMSD(P, Q) \geq \left(\tfrac{i}{k}\right)^{1/2} MinRMSD(P[1..i], Q[1..i]) \qquad (3)$$

Proof. Consider the sum of squared distances $SSD(P, Q) = \sum_{i=1}^{k} ||f_{R,v}(P[i]) - Q[i]||^2$. Suppose that we append a pair of new points $P[i]$ and $Q[i]$ to P and Q. Then, we see that the minimum of $SSD(f(P[i]), Q[i])$ over all transformations f is larger than or equal to the sum of the squared distance $||P[i] - Q[i]||^2$ and the minimum of $SSD(f'(P[i-1]), Q[i-1])$ over all transformations f'. Since the minimum of $SSD(f(P[i]), Q[i])$ over all f equals $k \cdot (MinRMSD(P, Q))^2$, we have the equality $k \cdot MinRMSD(P[1..k], Q[1..k])^2 \geq i \cdot MinRMSD(P[1..i], Q[1..i])^2$ $(*)$. By taking the square root of the both side of $(*)$, the theorem follows. □

From the above Theorem, we propose $(i/k)^{1/2} MinRMSD(P[1..i], Q[1..i])$ as the lower bound function for $MinRMSD(P, Q)$. Based on the proposed lower bound function, in Algorithm 2, we present the MATCHFAST with the modified subprocedure FINDFAST. At each iteration, it searches for all i-prefixes as in the same manner as the naive algorithm does except that it makes the test the current candidate i-prefix $Q[1..i]$ at Line 10 based on the lower bound function LB, and prunes the search when the test failed. From Theorem 1, this pruning is sound without eliminating any successful branches. Furthermore, the test at Line 8 ensures to avoid duplicated enumeration of the same one-to-one correspondence. From the above discussion, we have the main theorem of this section.

Algorithm 2. A faster branch-and-bound algorithm for solving the enumeration version of 3-D APSM(trans, rot, 1-1; RMSD) problem using exhaustive search

---

1: **procedure** MATCHFAST(P, T, r)
2:    FINDFAST$((), 0, |P|, |T|, P, T)$;

3: **procedure** FINDFAST$(Q = (q_1, \ldots, q_i), i, k, n, P, T)$ ▷ candidate prefix Q
4:    **if** $i = k$ **then**   ▷ Q becomes a k-subset
5:       **if** $MinRMSD(P, Q) \leq r$ **then**
6:          Report Q as a match;
7:    **for** $j = 1, \ldots, n$ **do**
8:       **if** $T[j] \notin Q$ **then then** continue;
9:          $R := (q_1, \ldots, q_i, T[j])$;   ▷ Append a new data point
10:         **if** $(i/k)^{1/2} MinRMSD(P[1..i+1], R) > r$ **then** continue;   ▷ Pruning
11:         FINDFAST$(R, i+1, k, n, P, T)$;   ▷ Call itself recursively

---

Theorem 2. *The algorithm* MATCHFAST *in Algorithm 2 solves the enumeration version of APSM(trans, rot, 1-1; RMSD), the approximate 3-D point set matching problem with minimum RMSD score under translation, rotation, one-to-one correspondence, in* $O(\binom{n}{k}k)) = n^{O(k)}k$ *time.*

Although the worst case time complexity of MatchFast still remains $n^{O(k)}k$ time, it can make early termination depending on the content of an input data.

4   A Fixed-parameter-like Algorithm Using Spatial Constraint

In this section, we present the second modified algorithm MATCHFP, which is inspired by fixed-parameter tractable algorithms [9], that is particularly fast for small patterns on uniformly distributed data points. In the followings, let $d = 2, 3$ be a fixed dimension, and $\theta = (k, r, \ell)$ be the tuple of parameters such that k and ℓ is the size and radius of pattern P, $r > 0$ is a distance threshold, and $\varepsilon > 0$ is a positive number explained later.

4.1   Basic Idea

For any positive number $\ell > 0$, we define the *ball* $B_{c,\ell} = \{ q \in \mathbb{R}^d \mid ||q - c|| \leq \ell \}$. with radius $\ell > 0$ centered at a point $c \in \mathbb{R}^d$, whose volume is $|B_{c,\ell}| = (4\pi/3)\ell^3$ for $d = 3$. Then, the *radius* of a point set P, denoted by $radius(P)$, is the minimum radius of the ball containing all points of P. The *maximum neighbor distance* within a given ball $B \subseteq \mathbb{R}^d$ is the maximum of the distance to the nearest neighbor of each data point in B defined by $\varepsilon = \max_{p \in B} \min_{q \in T} ||p - q|| > 0$. The next lemma says that if there is a match between P and some k-subset $Q \subseteq T$, such a candidate can be found in a small ball around P when the size and radius of P, threshold r, and the maximum neighbor distance are bounded.

Algorithm 3. A fixed-parameter algorithm MATCHFP that solves the enumeration version of 3-D approximate PSM with RMSD score under translation, rotation, and one-to-one correspondence.

Given a data point et T, a pattern point set P, distance threshold $r > 0$, and real number $0 < \delta \leq 1$, the algorithm MATCHFP executes the following steps for each data point t in T:

- **Step 1**: Compute the set T_t of all data points in the ball $B_{t,L}$ with radius $L = L(k, r, \ell) := 2(k^{1/2}r + \ell)$.
- **Step 2**: Apply the algorithm MATCHFAST(T_t, P, r) in Sec. 3 to the restricted data set T_c centered at t to find and output all matchings $Q \subseteq T_t$ within T_t.

Theorem 3 (Locality of match). *Let P be any pattern set with the center $c \in \mathbb{R}^d$ and radius $\ell > 0$, and $r > 0$ be any number. If $RMSD(f(P), Q) \leq r$ holds for some transformation f in $\mathcal{RT}$ and some candidate k-point set $Q \subseteq T$, then Q must be contained within the ball centered at $c' = f(c) \in \mathbb{R}^d$ with radius*

$$L = L(k, r, \ell) := 2(k^{1/2}r + \ell) \tag{4}$$

Proof. Let $P = (p_i)_{i=1}^k$ and $Q = (q_i)_{i=1}^k$. It is sufficient to show $\|f(c) - q_i\| \leq L$ holds for any i. Let $1 \leq i \leq k$ be any index. Note that $\|f(c) - q_i\| \leq \|f(c) - f(p_i)\| + \|f(p_i) - q_i\|$ (*) from the triangular inequation on L_2-norm. First, we see that $\|f(c) - f(p_i)\| \leq \ell$ since $\|c - p_i\| \leq radius(P) = \ell$ by assumption. Next, by assumption, $r \geq RMSD(f(P), Q) = L_2(f(P) - Q)$ holds. Since $L_2(X) \geq (1/k)^{1/2}L_\infty(X)$ holds for any k-vector X, we have $L_2(f(P) - Q) \geq (1/k)^{1/2}L_\infty(f(P) - Q) \geq (1/k)^{1/2}\|f(p_i) - q_i\|$(**). Multiplying the both side of (**) by $k^{1/2} \geq 0$, we have $\|f(p_i) - q_i\| \leq k^{1/2}r$. By combining above arguments with (*), we have $\|f(c) - q_i\| \leq \ell + k^{1/2}r$. Applying this formula again, we also see that $f(c)$ has the nearest data point $t_c \in T$ such that $\|t_c - f(c)\| \leq \varepsilon$ for $\varepsilon := \ell + k^{1/2}r$. Thus, again from triangular inequality, we have the result $\|t_c - f(p_i)\| \leq \|t_c - f(c)\| + \|f(c) - q_i\| \leq \varepsilon + (\ell + k^{1/2}r) = L$.   □

In Algorithm 3, we present the algorithm MATCHFP for APSM(trans, rot, 1-1; RMSD) based on Theorem 3 that runs particularly fast on uniformly distributed data points for fixed parameter values $\theta = (k, r, \ell)$. This algorithm first computes the parameter $L = L(k, r, \ell)$ according to Theorem 3. Then, it works on iterations with each t among n data points in T. In Step 1, it computes the *local data set* T_t in $O(polylog(n) \times |T_t|)$ time using, e.g., the range tree index [7]. In Step 2, the algorithm computes matchings on T_t using MATCHFAST in Sec. 3 in $t = O(N(B_{t,L})^k k)$ time and $s = O(N(B_{t,L}))$ working space. Then, it repeats the above process for all of n data point t.

Lemma 1. *If $L \geq L(k, r, \ell)$, then the algorithm MATCHFP in Algorithm 3 solves the 3-D APSM(trans, rot, 1-1; RMSD) problem in $t = O(\sum_{t \in T} N(B_{t,L})^k k)$ time and $s = O(\max_{t \in T} N(B_{t,L}))$ working space.*

4.2   Probabilistic Analysis

The *spatial Poisson process* (SPP, for short) with mean parameter $\lambda > 0$ is a model of uniform distribution of random points in $\mathbb{R}^d$ [14] having density λ. In SPP, (S1) for any ball B, the distribution of count $N(B)$ obeys *Poisson distribution with mean* $\lambda|B| > 0$, i.e., $Pr(N(B) = k) = ((\lambda|B|)^k/k!)e^{-\lambda|B|}$. Moreover, (S2) for any disjoint regions $A_1, \ldots, A_m$, $N(A_1), \ldots, N(A_m)$ are independent. Now, we show the main theorem of this section which says that MATCHFP runs particularly fast when parameter k, r, and ℓ are small constant over uniformly generated data sets.

Theorem 4. *Suppose that data points are generated by SPP with density $\lambda > 0$. We fix the following parameters $\theta = (k, \ell, r)$: the maximum size $k > 0$ and radius $\ell > 0$ of a pattern, a distance threshold $r > 0$. Then, for any $\delta > 0$, the following conditions holds:*

> *For any data set T of n points of arbitrary radius and a pattern set P of k points with radius at most ℓ, if we set the radius of the local data set T_t to be $L = L(k, r, \ell)$ in Theorem 3, then* MATCHFP *in Algorithm 3 solves the 3-D* APSM *(trans, rot, 1-1; RMSD) problem in $O(n \log^k n)$ time and $O(\log n)$ working space with probability at least $1 - \delta$.*

Proof. We give a proof sketch. By assumption of SPP, for any ball $B_L = B_{t,L}$, $Pr(N(B_L) = k)$ is given by Poisson distribution with mean $\lambda|B_L|$ regardless of the location of t. Let $c > 0$ be a number that will be specified later. From Lemma 1, we see that the algorithm has the claimed complexity if the following situation does not happen; err_*: the local set size $|T_i|$ exceeds $c\lambda|B_L|$. Assuming SPP, we can show upper bounds of the failure probability $Pr(err_*)$ as follows. By applying tail bound for Poisson distribution to the union bound, we can show that $Pr(err_*) \leq \sum_{t \in T} Pr\{N(B_{t,L}) > c\lambda|B_{t,L}|\} = n \cdot Pr\{N(B_L) > c\lambda|B_L|\} \leq \delta$ for some $c \geq (1/\lambda|B|)\{-\ln(\delta) + \ln(n) + \ln(2)\} = const(\delta) + O(\ln(n))$. From the above discussion, the local set size satisfies $|T_i| \leq c\lambda|B| = O(\ln(n))$ with probability at least $1 - \delta$. Hence, the theorem follows from Lemma 1.   □

5   Experiments

In this section, we give experimental results on real point data sets to evaluate the efficiency of the proposed algorithms in Sec.3.

5.1   Data and Method

As a real data set, we used the molecular 3-D point set of one variation [1] of the protein called *Hemagglutinin HA1 chain of influenza A virus* (H10N8) from RCSB Protein Data Bank (PDB) [2]. In the followings, the units of length and

---

[1] The variation with structure ID 4XQ5 of H10N8 in PDB.
[2] http://www.rcsb.org/pdb/

distance are Å($=0.1$nm $= 1.0 \times 10^{-10}$m). For each parameter n up to 100, among 3722 atoms including 477 C_α atoms in the original data, we formed a data set of size n by extracting a subset of n locations of C_α carbon atoms, which give the approximated skeleton of a part of the molecule. The radius and the average distance between neighbor atoms were 45.31 unit and 2.2 unit, respectively. For each k up to 50, a pattern set of size k was formed by randomly selecting k points from the data set.

We implemented the naive algorithm in Sec.2 (naive) and the modified algorithm in Sec.3 (pruned)in C++ with Eigen linear algebra package [3]. As the experimental environment, we used a PC (CPU Intel Core i5 2.6 GHz,8GB memory) and compiler (g++, Apple LLVM version 6.0, clang-600.0.54) with -O3 option. In the experiments, we measured the average of running time over four trials as well as the number of matches by varying input size $n = |T|$, pattern size $k = |P|$, and distance threshold r. We used default values of $n = 100$, $k = 3$, and $r = 0.1$ otherwise stated.

5.2 Results

We show the experimental results in of (a)–(d) of Fig. 1.

Exp 1: Running Time and Number of Visited Prefixes Varying Input Data Size: First, in (a) and (b) of Fig. 1, we show the number of visited candidates and running time by varying the data set size n from 10 to 150. From (a), we observed that the proposed branch-and-bound algorithm pruned using the lower bound function in Sec. 3 could effectively reduce the number of visited candidates to the 1/4 to 1/4800 of the original naive. From (b), we also observed that by this reduction of the number of visited candidates, pruned successfully achieved 50 to 600 times speed-up over naive,

Exp 2: Running Time Varying Pattern Size: In (c) of Fig. 1, we show the running time by varying the pattern size k from 1 to 50, where $n = 100$ and $r = 0.1$. Note that we could run naive up to $k = 3$ because its running time exceeded the upper bound of 120 seconds for $k > 3$. From the figure, we see that the running time of naive showed exponential growth in k as expected from theoretical upper bound $n^{\Theta(k)}$, while that of pruned quickly grew up to $k = 2$ as same as naive and was slowly increasing after $k = 3$. Hence, we can conclude that the proposed pruning method with lower bound function is effective for matching with large pattern size k.

Exp 3: Running Time Varying Distance Threshold: In (d) of Fig. 1, we show the running time by varying distance threshold r from 0.01 to 10.0. Note that the average neighbor distance and radius of the data set is 2.2 and 45.3 units, respectively. From the figure, we observed that the time reduction ratio of pruned to naive gets larger when r goes smaller, while the ratio approaches almost one when r goes larger. Thus, our technique is more effective for smaller r.

---

[3] http://eigen.tuxfamily.org/

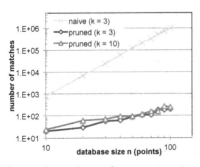

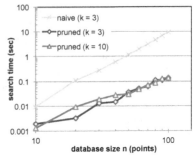

(a) Number of matches varying input size n, where $r = 0.1$, and $k = 3$ or 10.

(b) Running time varying input size n, where $r = 0.1$, and $k = 3$ or 10.

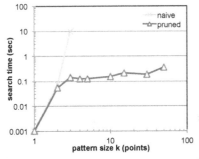

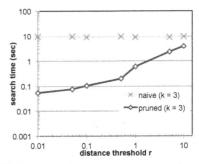

(c) Running time varying pattern size k, where $n = 100$ and $r = 0.1$.

(d) Running time varying distance threshold r, where $n = 100$ and $k = 3$.

Fig. 1. Comparison of the naive and proposed algorithms, naive and pruned, resp., where we set $n = 100$, $k = 3$, and $r = 0.1$ unless they are explicitly specified.

6   Conclusion

In this paper, we considered the approximate 3-D point set matching problem with the MinRMSD score under rotation, translation, and one-to-one correspondence, and then presented an efficient branch-and-bound algorithm based on a lower bound function. We also presented a FPT-style algorithm for fixed parameters. Experimental results showed that the first algorithm was one to two order of magnitude faster than the naive algorithm.

It will be a future work to compare the proposed algorithms and the existing constant approximation algorithms such as [6, 10] to study trade-off between the time and accuracy. We also plan to apply the proposed algorithms to real world data sets in bioinformatics, 3D-modeling, and spatio-temporal data. *Point subset mining* [4, 13] is a problem of finding point subsets from a point data set that meet a given criterion. Hence, it will be an interesting research problem how to use the proposed technique to speed-up *point subset mining* [4, 13].

Acknowledgments. The authors are grateful to anonymous reviewers for their comments which significantly improved the correctness and the presentation of this paper, and also to Takeaki Uno, Kunihiko Sadakane, Koji Tsuda, Shin-ichi Minato, and Yutaka Akiyama for their comments on this work. This research is supported in part by MEXT Grant-in-Aid for Scientific Research (A), 24240021, and the second author is also supported in part by CREST, JST, *"Foundations of Innovative Algorithms for Big Data"*.

References

1. Akutsu, T.: On determining the congruence of point sets in d dimensions. Computational Geometry **9**(4), 247–256 (1998)
2. Alt, H., Guibas, L.: Discrete geometric shapes: Matching, interpolation, and approximation, pp. 121–153. Elsevier Science Publishers B.V. North-Holland (1999)
3. Alt, H., Mehlhorn, K., Wagener, H., Welzl, E.: Congruence, similarity and symmetries of geometric objects. Discret. Comput. Geom. **3**, 237–256 (1988)
4. Arimura, H., Uno, T., Shimozono, S.: Time and space efficient discovery of maximal geometric graphs. In: Corruble, V., Takeda, M., Suzuki, E. (eds.) DS 2007. LNCS (LNAI), vol. 4755, pp. 42–55. Springer, Heidelberg (2007)
5. Carpentier, M., Brouillet, S., Pothier, J.: Yakusa: a fast structural database scanning method. Proteins **61**(1), 137–151 (2005)
6. Cho, M., Mount, D.M.: Improved approximation bounds for planar point pattern matching. Algorithmica **50**(2), 175–207 (2008)
7. de Berg, M., van Kreveld, M., Overmars, M., Schwarzkopf, O.: Computational Geometry: Algorithms and Applications. Springer-Verlag (2000)
8. de Rezende, P.J., Lee, D.: Point set pattern matching in d-dimensions. Algorithmica **13**(4), 387–404 (1995)
9. Downey, R.G., Fellows, M.R.: Parameterized complexity. Springer (1999)
10. Goodrich, M.T., Mitchell, J.S., Orletsky, M.W.: Approximate geometric pattern matching under rigid motions. IEEE Trans. PAMI **21**(4), 371–379 (1999)
11. Kabsch, W.: A solution for the best rotation to relate two sets of vectors. Acta Crystallographica **A32**(5), 922–923 (1976)
12. Mäkinen, V., Ukkonen, E.: Point pattern matching. In: Kao, M. (ed.) Encyclopedia of Algorithms, pp. 657–660. Springer (2008)
13. Nowozin, S., Tsuda, K.: Frequent subgraph retrieval in geometric graph databases. In: 8th IEEE Int'l Conf. on Data Mining, pp. 953–958 (2008)
14. Pinsky, M., Karlin, S.: An introduction to stochastic modeling. Academic Press (2010)
15. Schwartz, J.T., Sharir, M.: Identification of partially obscured objects in two and three dimensions by matching noisy characteristic curves. The Int'l J. of Robotics Res. **6**(2), 29–44 (1987)
16. Shibuya, T.: Geometric suffix tree: Indexing protein 3-d structures. Journal of the ACM **57**(3), 15 (2010)
17. Tam, G.K., et al.: Registration of 3d point clouds and meshes: a survey from rigid to nonrigid. IEEE Trans. Vis. Comput. Graphics **19**(7), 1199–1217 (2013)

Face Image Retrieval Revisited

Jan Sedmidubsky[(⊠)], Vladimir Mic, and Pavel Zezula

Masaryk University, Botanicka 68a, 602 00 Brno, Czech Republic
xsedmid@fi.muni.cz

Abstract. The objective of face retrieval is to efficiently search an image database with detected faces and identify such faces that belong to the same person as a query face. Unlike most related papers, we concentrate on both retrieval effectiveness and efficiency. High retrieval effectiveness is achieved by proposing a new fusion approach which integrates existing state-of-the-art detection as well as matching methods. We further significantly improve a retrieval quality by employing the concept of multi-face queries along with optional relevance feedback. To be able to efficiently process queries on databases with millions of faces, we apply a specialized indexing algorithm. The proposed solutions are compared against four existing open-source and commercial technologies and experimentally evaluated on the standardized FERET dataset and on a real-life dataset of more than one million face images. The retrieval results demonstrate a significant gain in effectiveness and two-orders of magnitude more efficient query processing, with respect to a single technology executed sequentially.

1 Introduction

Face recognition (FR) is a problem of verifying or identifying a face appearing in a given image. FR has become a popular biometric technology because it does not require expensive capturing devices and does not force users to carry or remember something to verify their identity, while enables to surreptitiously reveal wanted persons from surveillance cameras. The current situation in FR technology can be fairly summarized by the following quotation: *"Claims that face recognition is a solved problem are overly bold and optimistic. On the contrary, claims that face recognition in real-world scenarios is next to impossible are simply too pessimistic"* [7]. FR algorithms can achieve an impressive true accept rate of up to 99 % in controlled conditions (frontal face of cooperative users and controlled indoor illumination) [7]. However, there are still many challenges for uncontrolled environments, such as partial occlusions, difference in age of query and database faces, large pose variations, and ambient illumination. When a query face is captured in an unconstrained environment, true accept rate falls below 60 % [12]. These challenges make it more difficult to exploit automated face recognition for forensic applications using low-quality query images obtained from mobile devices or surveillance cameras.

P. Zezula—Supported by the national project No. VG20122015073.

G. Amato et al. (Eds.): SISAP 2015, LNCS 9371, pp. 204–216, 2015.
DOI: 10.1007/978-3-319-25087-8_19

We focus on the problem of face recognition from the *retrieval* perspective. Having specified a query face image, the objective is to efficiently search for database faces belonging to the same person as the query face, based on *similarity* of face features. Similarity between features of two faces is calculated by a matching method, so-called *matcher*. In general, there are two main issues when applying a given matcher on large datasets: *effectiveness* and *efficiency*. The former one refers to the quality of retrieval, which is generally decreasing with an increasing number of persons and face images in a dataset. The latter one represents the retrieval speed, which has to be taken into account especially when the database starts to contain millions of faces.

In this paper, we propose a complex solution for effective and efficient face retrieval, taking the face detection preprocessing step into account. More specifically, the main contributions of this paper are:

- proposal of fusion approaches to integrate multiple face detection methods as well as matchers for more effective face detection and retrieval purposes;
- exploitation of multi-face queries along with relevance feedback for more effective retrieval in comparison to single-face queries;
- proposal of application of an indexing solution for efficient retrieval in databases containing more than 1 million faces, which is more than two orders of magnitude larger than the size of the standardized FERET dataset;
- experimental comparison of existing face technologies with our proposals on datasets of different image quality.

2   Face Detection

Face detection aims at localizing bounding boxes of faces within an image. This task is necessary as the first stage for all face retrieval methods. Most face detection methods utilize the idea of a different-size sliding window [14,24] to decide whether the given image area represents a face or not. Since analysis of all possible windows is very time consuming, some methods prune the search space [8,22] to further reduce the number of windows. Each window is usually analyzed by localizing low-level features such as edges [19], skin color [11], or skin texture. To decide whether the given image area represents a face, the positions of localized features can be geometrically compared. Other approaches construct 2D/3D face models [8] or deformable part models [9,25] and fit them into the given image area. Such models can be constructed artificially or as a result of a learning process applied to training images [29]. There are also fusion detection methods combining different spatial information. Degtyarev et al. [4] take outputs of several face detectors to make a final decision. Other fusion methods combine sliding-window detectors with local-region ones [26] or still images with depth images [3]. In this paper, we propose a simple spatial-based fusion method which is able to integrate arbitrary detectors returning bounding boxes as the face detection result, in contrast to related approaches that need to compare the specific face features.

Fig. 1. Bounding boxes of faces determined by: (a) three independent methods of different color ($m = 3$) and (b) our approach requiring matching of at least two of them ($t_d = 2$).

Our *fusion* approach consists of two consecutive steps. In the first step, we utilize several face detection methods to localize bounding boxes of faces independently. All the boxes are then clustered by merging boxes whose areas overlap at least from 20 % into the same cluster. In the second step, we analyze each cluster to determine how many detection methods have marked the same image area as a face. If most methods mark the area as a face, there is a high probability that it is. Formally, having m independent detection methods, then each cluster should contain from 1 to m bounding boxes. We define user parameter $t_d \in \mathbb{N}\,(1 \leq t_d \leq m)$ as the lowest number of bounding boxes within a single cluster so that the image area determined by this cluster is considered as a face. If the cluster contains fewer than t_d bounding boxes, the area is considered as a non-face. By changing the value of t_d, a user can control the trade-off between precision and recall.

We evaluate the quality of the open-source **OpenCV** [1] technology and three commercial **NeuroTech** [2], **PittPatt** [3] and **Luxand** [4] technologies against the proposed fusion approach. While the OpenCV technology is based on Haar feature-based cascade classifiers, the commercial technologies do not provide any information about their implementation. Experiments are evaluated on

1. the standardized **FERET** dataset containing $11,338$ face images captured in a controlled environment with the average face width of 261 pixels,
2. our own **CELEBS** dataset with $1,026,029$ images captured in uncontrolled environments (like the image in Figure 1), and

---

[1] http://opencv.org
[2] www.neurotechnology.com/verilook.html
[3] www.pittpatt.com
[4] www.luxand.com

Table 1. Effectiveness results of face detection approaches.

	CELEBS-mini		FERET	
	recall	precision	recall	precision
OpenCV	55 %	89 %	57 %	93 %
Luxand	63 %	82 %	51 %	90 %
NeuroTech	25 %	39 %	86 %	78 %
PittPatt	78 %	90 %	84 %	77 %
Fusion ($t_d = 1$)	83 %	57 %	91 %	63 %
Fusion ($t_d = 2$)	66 %	95 %	74 %	94 %
Fusion ($t_d = 3$)	52 %	98 %	58 %	100 %
Fusion ($t_d = 4$)	18 %	99 %	49 %	100 %

3. the **CELEBS-mini** dataset as a subset of the previous dataset with $1,261$ representative face images and the average face width of only 35 pixels.

From the dataset quality point of view, FERET is considered as a *high-quality* dataset with sufficiently large face resolution, while CELEBS and CELEBS-mini constitute examples of *low-quality* datasets. To evaluate the quality, we consider only the FERET and CELEBS-mini datasets that provide ground truth in form of bounding boxes of face images. The quality is calculated by standard measures of recall and precision:

- *recall* – a ratio between the number of correctly detected faces and the number of ground-truth faces;
- *precision* – a ratio between the number of correctly detected faces and the number of detected faces.

As correctly detected faces are treated faces whose areas overlap with ground-truth face areas at least for 20 %.

Recall and precision measures are depicted in Table 1 for (1) each of four testing technologies and (2) four cases of the proposed "Fusion" approach that aggregates their results by changing parameter $t_d : 1 \leq t_d \leq 4$. The results show that there is no clear winner among four existing technologies. This confirms our intuition that individual technologies focus on the specific detection problems. For example, NeuroTech is very good for detection of high-resolution faces taken in controlled conditions, especially with indoor illumination and slightly rotated faces, but falls down on low-quality face images. On the other hand, the proposed fusion with setting $t_d = 1$, which unionizes resulting bounding boxes of individual technologies, achieves a non-trivially higher recall (about 5 %) compared to the recall of the best testing technology, disregarding the dataset quality. While setting $t_d = 1$ aims at maximizing recall, setting $t_d = 4$ maximizes precision by intersecting resulting boxes. The results show almost the 100 % precision even when $t_d = 3$, with no regards to the quality of faces. If the objective is to achieve a high precision and keep recall at a reasonable value, a suitable setting is $t_d = 2$ which keeps about the 95 % precision and 70 % recall.

In summary, the proposed fusion approach allows us to integrate any number of face detection technologies that can be treated as black boxes. It also enables to control the trade-off between recall and precision by changing parameter t_d. To detect faces in large datasets, we recommend to prefer a high precision by setting the t_d parameter to a half of the number of integrated detection technologies. In later experiments, we process the one-million CELEBS image dataset by fusing OpenCV, Luxand and NeuroTech technologies with setting $t_d = 2$. As a consequence, we can also utilize multiple face features – extracted by the integrated technologies – to improve face matching effectiveness.

3   Face Retrieval

Face retrieval requires a preprocessing step to extract characteristic features of faces detected within an image dataset and store them into a database. In the consecutive retrieval step, the features of a query face and database faces are compared by a given matching method (so-called *matcher*) to retrieve a set of the most similar database faces. The retrieved set of faces can be optionally analyzed to recognize (classify) the identity of the query face.

The most important component is a matcher that determines similarity of two faces. Existing matchers are usually designed to compute geometric properties and relationships between significant local features, such as eyes, nose and mouth [1,27]. In contrast to local features, holistic-based matchers [10,23] describe an entire face globally. They often transform face images by the Principal Component Analysis and match them through underlying statistical regularities [13] or linear combination of class-specific galleries [17]. Face matching can be further improved by transforming face images into different color spaces [28], considering information of common scene, clothing and gender [2,31], or exploiting supplemental characteristics like scars, moles and freckles [20]. Besides research papers, there are a lot of open-source and commercial solutions provided by, e.g., Luxand, KeyLemon, Betaface, NeuroTechnology, and Cognitec.

Despite face matchers, there are fusion approaches which aggregate existing biometric matching methods [6,15,16]. For example, Nanni et al. [16] fuse different fingerprint matchers to achieve more effective recognition. In the following sections, we (1) propose a general fusion approach to aggregate face matchers, (2) introduce multi-face queries along with relevance feedback to further increase the retrieval quality, and (3) speed-up the retrieval process by organizing face features in a metric-based search structure. Our main goal is to achieve higher efficiency and effectiveness compared to individual integrated matchers which can be treated as black boxes.

3.1   Fusion of Multiple Matching Methods

Most existing matchers deal with the specific problem, such as rotated faces, face resolution and ambient illumination, which makes them dependent on dataset properties. To be possibly independent of the specific dataset, we propose a transformation-based fusion method to integrate appropriate face matchers together. The proposed method can integrate such matchers that provide a functionality for computing the *distance* (i.e., similarity) between two faces. Having available m matchers, m independent distances for the same pair of faces are obtained. These distances are then *normalized* (transformed) and finally aggregated to determine the matching result.

The most crucial step is normalization of distances. Since each matcher can return distances within a completely different range, we normalize such distances into interval $[0, 1]$. A simple way is to divide the distance by the maximum distance defined independently for each corresponding matcher. However, if distributions of distances returned by individual matchers on the same sample of faces do not correlate, such normalization cannot be used – normalized distances of different matchers would not be meaningfully comparable. This is the reason we transform an original distance into a *probability* that given two faces belong to the same person.

To transform distances into probabilities, we need to know what distances return individual matchers for pairs of faces belonging to the same persons as well as to different people. In particular, for each matcher we take a training sample of l faces whose identities are provided in advance. Then, we apply the given matcher to compute the distance for each pair of faces, i.e., $l^\triangle = l \cdot (l+1)/2$ distances. We sort these distances and divide them uniformly into $l^\triangle/100$ distinct *buckets*. The i-th bucket ($i \in [1, l^\triangle/100]$) then contains exactly 100 distances (with eventual exception of the last bucket) from the following interval:

$$[d_{(i-1) \cdot 100}, d_{i \cdot 100}),$$

where $d_0 < d_1 < \ldots < d_{l^\triangle - 1}$ represent the sorted distances. Each bucket is further processed to calculate its *recognition probability* as the ratio between the number of face pairs connecting the same person and the number of all pairs in the bucket. We assume that this probability should reflect general probability of successful recognition for the given distance range. This probability generally decreases from the first bucket towards the others as the distances gradually increase but monotonicity is not required. Figure 2 visualizes a dependence curve between the distance and recognition probability for three different matchers.

As recognition probabilities are computed, we simply normalize an input distance by localizing the i-th bucket corresponding to the input distance and returning the bucket's probability as the normalized distance. In this way, we apply m matchers to independently compute the distance of the given pair of faces and then normalize these $d_1, \ldots, d_m$ distances according to the normalization process into probabilities $p_1, \ldots, p_m$. We recommend to aggregate these probabilities by selecting the highest one, i.e., $max\{p_1, \ldots, p_m\}$ – it represents the maximum probability that the given pair of faces belongs to the same person, over probabilities returned by integrated matchers. The matching result is finally determined as a reverse value of the highest probability, i.e, the resulting distance is defined as $1 - max\{p_1, \ldots, p_m\}$.

We present retrieval effectiveness of the proposed fusion method in comparison with **Luxand** and **NeuroTech** commercial technologies and the Advanced Face Descriptor [21] defined within the **MPEG-7** standardization. These technologies are compared by evaluating a testing set of 220 and 3,058 k-nearest-neighbors (kNN) queries defined for the CELEBS-mini and FERET dataset, respectively. Each query is evaluated 400 times with different values of k to express trade-offs between precision and recall:

- *recall* – a ratio between the number of retrieved faces corresponding to the query person and the number of dataset faces belonging to the query person;
- *precision* – a ratio between the number of retrieved faces corresponding to the query person and number k.

Figure 3 illustrates trade-offs between recall and precision on the CELEBS-mini and FERET dataset. Our approach significantly outperforms other technologies on the CELEBS-mini dataset. The biggest gap can be observed with respect to NeuroTech which is not capable of achieving recall higher than 10 % on CELEBS-mini. On the other hand, it achieves slightly better recall about 4 % on FERET by fixing precision at 70 %. It is caused by its intended usage to recognize faces of high resolution taken in controlled conditions. Focusing on MPEG-7 and Luxand technologies, they achieve clearly worse results on both the datasets. The most important fact is that the proposed fusion approach achieves top results with no respect to the dataset quality, which is not true for individual technologies.

We also evaluate how different aggregation policies influence retrieval effectiveness. Figure 4 illustrates the results by aggregating probabilities on the basis of the minimum, maximum, average and median function. As expected, the minimum and maximum functions provide the worst and best results, respectively. The average function also achieves the top results since it preserves a similar ordering of faces as the maximum function on both the datasets.

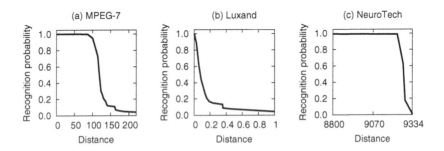

Fig. 2. Recognition probabilities computed on the CELEBS-mini dataset for three matching methods: (a) MPEG-7 [21], (b) Luxand, and (c) NeuroTech.

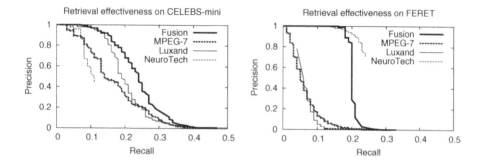

Fig. 3. Retrieval effectiveness measured on low-quality CELEBS-mini and high-quality FERET datasets for three existing matching technologies and the fusion approach.

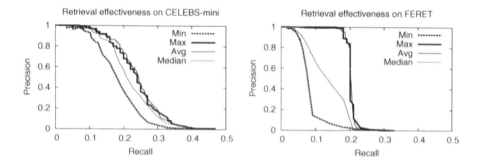

Fig. 4. Effectiveness of the fusion approach using different aggregation policies.

3.2   Multi-face Queries and Relevance Feedback

To further improve retrieval effectiveness, we propose to utilize the concept of *multi-face queries* along with *relevance feedback*. Multi-face queries allow us to specify several examples of query faces within evaluation of a single query [23]. Similarity between the set of query faces and a given database face is computed as the highest similarity between the database face and each query face. We assume that the more different examples of query faces of the same person are provided, the higher effectiveness of the retrieved set of the most similar faces is. Our assumption should be especially true when individual query faces mutu- ally differ, e.g., they represent frontal and rotated faces, faces of low and high resolution, and faces in different illumination conditions. We also incorporate the concept of *relevance feedback* for evaluation of multi-face queries. A typical usage of relevance feedback starts after evaluation of a single-face query. At this moment, a user is asked to mark retrieved true positives – faces belonging to the same person as the query face. The true positives are then exploited as the query faces for another query (the second search iteration). In this way, the user can iteratively increase the number of query faces and thus improve retrieval effectiveness.

We experimentally verify how consecutive search iterations improve retrieval effectiveness. The first iteration is initialized by evaluating the same testing set of single-face kNN queries as in the previous section. For the 80 % precision, the first iteration achieves recall of 17 % while the second and third one further increases recall up to 53 % and 66 % on the CELEBS-mini dataset, respectively. On the FERET dataset, recall of 20 % in the first iteration increases up to 46 % and 82 % in the second and third iteration, respectively. Fourth and next iterations achieve only slightly better recall than third iterations disregarding the dataset. Note that the 100 % recall is not achievable in any number of iterations if (1) faces corresponding to all true positives are not detected in the dataset or (2) the number of true positives is higher than 50 since relevance feedback is provided only the 50 most similar faces. In summary, the most important observation is that recall can dramatically increase when a user participates in marking true-positive faces in two search iterations.

3.3 Efficient Query Processing

Face retrieval technologies usually evaluate queries in a sequential way by calculating similarity between a query face and each database face. The sequential evaluation is only sufficient for databases containing a limited number of faces. Our fusion detection method (with setting $t_d = 2$) localizes $1,118,316$ faces in the CELEBS dataset. The characteristic features extracted from these faces by MPEG-7 [21], Luxand and NeuroTech approaches take about 0.8 GB, 39.1 GB and 34.3 GB, respectively, i.e., 74 GB of disk space in total. To evaluate a single-face query sequentially, such huge amount of data has to be read from disk into main memory and the proposed aggregated distance has to be computed between the query face and each database face. This operation takes about 25 minutes on the server with 8 cores at 2 GHz, 16 GB RAM and harddisks in the RAID 5 configuration – reading all the features takes 23.2 minutes while computing aggregated distances 1.7 minutes. This process could be speed-up by applying the threshold algorithm [5] in case the features of individual matchers could be efficiently retrieved from harddisk. However, we treat Luxand and NeuroTech as black-box face matchers and thus non-efficiently indexable. To evaluate queries in a couple of seconds and without a significant loss in retrieval effectiveness, we apply an approximate indexing solution.

The idea of indexing is to (1) select an appropriate kind of face features according to which the index structure is built, (2) utilize the index structure to efficiently retrieve a reasonably large *candidate set* of faces, (3) read all the features of the candidate faces from a harddisk (I/O costs), (4) *re-rank* the loaded candidate faces according to the proposed fusion method, and (5) select the k most similar re-ranked faces as the kNN query result. To construct the index structure, we need to select appropriate face features along with a distance function according to which the faces can be indexed. We have chosen the

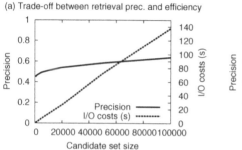
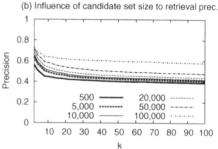

Fig. 5. Comparison of retrieval precision and efficiency for different candidate set sizes.

Advanced Face Descriptor [21] features defined within the **MPEG-7** standardization. These features are represented in form of 174-dimensional vectors of integers and can be compared by a weighted Euclidean distance function. Since the weighted Euclidean distance satisfies the postulates of metric space [30], i.e., the reflexivity, symmetry and triangle inequality, we can employ any metric-based index structure for indexing and searching MPEG-7 features. In particular, we exploit the recent M-Index structure [18] which requires about 20 minutes in the preprocessing step to index MPEG-7 features of all $1,118,316$ faces. The whole index occupies 1.5 GB and thus easily fits into main memory.

The M-Index structure is then utilized to efficiently obtain a candidate set of the most similar faces according to the MPEG-7 features. The size of the candidate set has to be significantly larger than k so that it possibly includes the desired k faces that would be retrieved sequentially by the fusion method. On the other hand, the candidate set size has to be significantly smaller with respect to the database size. Figure 5 depicts the influence of the size of candidate set to search efficiency as well as effectiveness. Specifically, Figure 5a shows the average time (in seconds) needed to read and re-rank the candidate set of different sizes for a single kNN query ($k = 10$) – times are averaged across the testing set of 220 queries defined for the CELEBS-mini dataset. Note that only I/O costs are considered since they constitute the main search bottleneck – retrieving candidate sets of different sizes using M-Index takes more or less the same portion of time of about 0.8 s. We can observe that I/O costs grow linearly while retrieval precision only logarithmically with respect to the candidate set size. Since the ground truth is not provided on the CELEBS dataset, retrieval precision represents the fraction of faces retrieved by the indexing approach and faces obtained in a sequential way on the basis of the fusion method. Figure 5b further illustrates that retrieval precision decreases only slightly for the particular sizes of candidate sets with respect to increasing answer size k.

Based on these results, we recommend to use a smaller size of candidate set to obtain the answer in a couple of seconds. In our scenario with more than one million faces and k fixed to 100, a reasonable size of the candidate set has been experimentally determined as $1,000$. This setting enables to evaluate queries in 2 seconds which is about 750-times more efficient than evaluating integrated technologies (i.e., MPEG-7, Luxand and NeuroTech) sequentially. Although such setting decreases retrieval precision to 46 % on average, the search quality can be then improved significantly by applying user's relevance feedback.

4   Conclusions

We focus on efficient and effective face retrieval. As a preprocessing step, we propose the fusion detection method to integrate any number of existing face detectors. This method allows us to control a trade-off between recall and precision and achieves stable results disregarding the dataset quality, in contrast to individual OpenCV, Luxand, NeuroTech and PittPatt detectors. To improve retrieval effectiveness, we propose the fusion method which normalizes similarities of integrated matchers into a common domain and selects the most promising value as a matching result. This method provides higher-quality and more stable results than integrated MPEG-7, Luxand and NeuroTech matchers evaluated on both CELEBS-mini and FERET datasets. To get near to real-time retrieval, we organize face features within an index structure. By indexing one-million CELEBS dataset according to MPEG-7 features and re-ranking the retrieved set of $1,000$ candidate faces, we are able to evaluate a single query within 2 s, which is about 750-times more efficient than the sequential scan. Although this setting decreases retrieval precision to 46 %, the search quality can be further enhanced significantly by applying relevance feedback.

In the future, we plan to integrate more face matching technologies to improve the retrieval quality. We would also like to evaluate different ways of re-ranking of candidate sets, e.g., according to variants of threshold algorithms [5].

References

1. Chan, C.H., Tahir, M.A., Kittler, J., Pietikäinen, M.: Multiscale local phase quantization for robust component-based face recognition using kernel fusion of multiple descriptors. IEEE Trans. on Pattern Analysis and Mach. Int., 1164–1177 (2013)
2. Chen, B.C., Chen, Y.Y., Kuo, Y.H., Hsu, W.H.: Scalable face image retrieval using attribute-enhanced sparse codewords. IEEE Transactions on Multimedia 15(5), 1163–1173 (2013)
3. Choi, W., Pantofaru, C., Savarese, S.: Detecting and tracking people using an rgb-d camera via multiple detector fusion. In: International Conference on Computer Vision Workshops, pp. 1076–1083 (2011)
4. Degtyarev, N., Seredin, O.: A geometric approach to face detector combining. In: Sansone, C., Kittler, J., Roli, F. (eds.) MCS 2011. LNCS, vol. 6713, pp. 299–308. Springer, Heidelberg (2011)
5. Fagin, R., Lotem, A., Naor, M.: Optimal aggregation algorithms for middleware. Journal of Computer and System Sciences 66(4), 614–656 (2003)
6. Górecki, T.: Sequential combining in discriminant analysis. Journal of Applied Statistics, 398–408 (2015)

7. Hua, G., Yang, M.H., Learned-Miller, E., Ma, Y., Turk, M., Kriegman, D.J., Huang, T.S.: Introduction to the special section on real-world face recognition. IEEE Trans. on Pattern Analysis and Machine Int. **33**(10), 1921–1924 (2011)

8. Huang, Y., Liu, Q., Metaxas, D.N.: A component-based framework for generalized face alignment. IEEE Trans. on Systems, Man, and Cybernetics, 287–298 (2011)

9. Cech, J., Franc, V., Matas, J.: A 3d approach to facial landmarks: detection, refinement, and tracking. In: Int. Conf. on Pattern Recognition (ICPR 2014), p. 6 (2014)

10. Jafri, R., Arabnia, H.R.: A survey of face recognition techniques. Journal of Information Processing Systems, 41–68 (2009)

11. Kakumanu, P., Makrogiannis, S., Bourbakis, N.: A survey of skin-color modeling and detection methods. Pattern Recognition, 1106–1122 (2007)

12. Klontz, J.C., Klare, B.F., Klum, S., Jain, A.K., Burge, M.J.: Open source biometric recognition. In: BTAS 2013, pp. 1–8 (2013)

13. Lai, J.H., Yuen, P.C., Feng, G.C.: Face recognition using holistic fourier invariant features. Pattern Recognition, 95–109 (2001)

14. Lampert, C.H., Blaschko, M.B., Hofmann, T.: Beyond sliding windows: object localization by efficient subwindow search. In: International Conference on Computer Vision and Pattern Recognition (CVPR 2008), pp. 1–8 (2008)

15. Nanni, L., Lumini, A., Brahnam, S.: Likelihood ratio based features for a trained biometric score fusion. Expert Systems with Applications, 58–63 (2011)

16. Nanni, L., Lumini, A., Ferrara, M., Cappelli, R.: Combining biometric matchers by means of machine learning and statistical approaches. Neurocomputing, 526–535 (2015)

17. Naseem, I., Togneri, R., Bennamoun, M.: Linear regression for face recognition. IEEE Trans. on Pattern Analysis and Machine Int., 2106–2112 (2010)

18. Novak, D., Batko, M., Zezula, P.: Metric Index: An Efficient and Scalable Solution for Precise and Approximate Similarity Search. Inf. Sys. **36**(4), 721–733 (2011)

19. Segundo, M.P., Silva, L., Bellon, O.R.P., Queirolo, C.C.: Automatic face segmentation and facial landmark detection in range images. IEEE Transactions on Systems, Man, and Cybernetics, 1319–1330 (2010)

20. Park, U., Jain, A.K.: Face matching and retrieval using soft biometrics. IEEE Transactions on Information Forensics and Security, 406–415 (2010)

21. Sikora, T.: The mpeg-7 visual standard for content description-an overview. IEEE Transactions on Circuits and Systems for Video Technology **11**(6), 696–702 (2001)

22. Subburaman, V.B., Marcel, S.: Alternative search techniques for face detection using location estimation and binary features. Computer Vision and Image Understanding, 551–570 (2013)

23. Tan, X., Chen, S., Zhou, Z.H., Zhang, F.: Face recognition from a single image per person: A survey. Pattern Recognition, 1725–1745 (2006)

24. Tsao, W.K., Lee, A.J.T., Liu, Y.H., Chang, T.W., Lin, H.H.: A data mining approach to face detection. Pattern Recognition, 1039–1049 (2010)

25. Uřičář, M., Franc, V., Hlaváč, V.: Detector of facial landmarks learned by the structured output SVM. In: Int. Conf. on Computer Vision, Imaging and Computer Graphics Theory and Applications. vol. 1, pp. 547–556. SciTePress (2012)

26. Wang, X., Han, T., Yan, S.: An hog-lbp human detector with partial occlusion handling. In: 12th International Conference on Computer Vision, pp. 32–39 (2009)

27. Wu, Z., Ke, Q., Sun, J., Shum, H.Y.: Scalable face image retrieval with identity-based quantization and multireference reranking. In: Int. Conf. on Computer Vision and Pattern Recognition (CVPR 2010), pp. 3469–3476. IEEE (2010)

28. Yang, J., Liu, C., Zhang, L.: Color space normalization: Enhancing the discriminating power of color spaces for face recognition. Pattern Recognition, 1454–1466 (2010)
29. Yang, M.H., Kriegman, D., Ahuja, N.: Detecting faces in images: A survey. IEEE Transactions on Pattern Analysis and Machine Intelligence, 34–58 (2002)
30. Zezula, P., Amato, G., Dohnal, V., Batko, M.: Similarity Search: The Metric Space Approach. Advances in Database Systems, vol. 32. Springer (2006)
31. Zhang, L., Kalashnikov, D.V., Mehrotra, S.: A unified framework for context assisted face clustering. In: Int. Conf. on Multimedia Retrieval, pp. 9–16. ACM (2013)

Semiautomatic Learning of 3D Objects from Video Streams

Fabio Carrara, Fabrizio Falchi, and Claudio Gennaro[✉]

ISTI-CNR, Via G. Moruzzi 1, 56124 Pisa, Italy
{fabio.carrara,fabrizio.falchi,claudio.gennaro}@isti.cnr.it

Abstract. Object detection and recognition are classical problems in computer vision, but are still challenging without a priori knowledge of objects and with a limited user interaction. In this work, a semiautomatic system for visual object learning from video stream is presented. The system detects movable foreground objects relying on FAST interest points. Once a view of an object has been segmented, the system relies on ORB features to create its descriptor, store it and compare it with descriptors of previously seen views. To this end, a visual similarity function based on geometry consistency of the local features is used. The system groups together similar views of the same object into clusters relying on the transitivity of similarity among them. Each cluster identifies a 3D object and the system learn to autonomously recognize a particular view assessing its cluster membership. When ambiguities arise, the user is asked to validate the membership assignments. Experiments have demonstrated the ability of the system to group together unlabeled views, reducing the labeling work of the user.

1 Introduction

In this work, a user assisted clustering system for online visual object recognition is presented. Our approach enables a single smart camera to learn and recognize objects exploiting change detection in the scene: given the evolution of the scene during time, the system incrementally builds a knowledge that can be exploited for the subsequent recognitions of the object when reappear on the scene. The user is queried when ambiguities cannot be automatically resolved.

Object detection is carried out by a local feature based background subtraction method [1] which distinguishes the foreground local features of the image from the background ones and segments new objects in the scene relying on FAST interest points. Each detected object, together with its extracted ORB local features, is maintained in a local database forming the knowledge base for object recognition. All the views of detected objects are incrementally organized in clusters based on the similarity among them. A similarity function between two object views is defined relying on local features matching and geometry constraints on their positions. The main goal of the system is to maintain gathered views in clusters where each cluster contains only views of the same 3D object, even if it has been observed under different poses or illuminations

© Springer International Publishing Switzerland 2015
G. Amato et al. (Eds.): SISAP 2015, LNCS 9371, pp. 217–228, 2015.
DOI: 10.1007/978-3-319-25087-8_20

(see Figure 1). Clusters can be labeled anytime by the user and object recognition is performed assessing the membership of a view to a particular cluster.

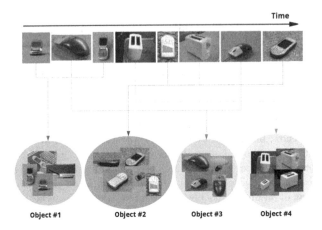

Fig. 1. Visualization of the system goal: online clustering of detected objects as recognition task.

The system has not been designed for a particular smart camera platform in mind, but it has been tested on the Raspberry Pi platform equipped with a Pi Camera module. Experiments have been made using the Stanford 3D Objects [1] [9] public available dataset in order to evaluate the ability to build a knowledge base for object recognition.

The rest of the paper is organized as follows: Section 2 describes the main features of some studied object recognition methods. Section 3 presents our method, describing the similarity function we have defined between detected objects. Section 4 describes the strategy used for similar object clustering. Section 5 reports the experiments performed and the metrics used to evaluate our method. Conclusive remarks are addressed at the end of this paper.

2   Related Work

Many solutions have been proposed to the problem of 3d object model learning for recognition task, starting from different 2d views of the object of interest.

Murase and Nayar [7] model each object as a manifold in the eigenspace obtained compressing the training image set for that object. Given an unknown input image, it is projected on the eigenspace and labeled relying on the manifold it lies on. Moreover, the exact point of the projection gives a pose estimation of the input object. However, only batch training is possible and the training set must be composed by a large number of normalized images with different poses and illuminantion and uncluttered background.

[1] http://cvgl.stanford.edu/resources.html

More recent studies address object modelling relying on local features of training images.

Weber et al. [5,10] developed a method to learn object class models from unlabeled and unsegmented cluttered scenes. The authors combine appereance and shape in the object model using the constellation model. In this model, objects are represented as flexible constellations of rigid parts (features). Most robust parts are automatically identified applying a clustering algorithm to parts detected from the training set. An expectation-maximization algorithm is applied to tune the parameters of a joint probability density function (pdf) on the shape of the constellation and the output of part detectors. An enhanced version of this method using Bayesian parameter estimation is proposed by Fei-Fei et al. [4] for the purpose of scale-invariant object categorization capable of both batch and incremental training. However, this approach performs poorly with few training images and is more suitable to the modeling of classes of objects rather than individual 3D objects. Moreover it does not cope with multi-pose 3D object recognition.

The work in this paper follows the approach of Lowe [6], who addresses the problem of view clustering for 3D object recognition using a training set of images with uncluttered background. Each view is described by its SIFT local features and adjacent views are clustered together relying on feature matching and similarity transformation error. The presented method relies on a different geometric consistency check based on homographies which is capable of relating views under different perspectives with low false positive rates.

3   Object Extraction and Matching

A specialized local feature based background subtraction method [1] has been implemented to segment stable foreground objects from a video stream relying on their FAST keypoints. A 2-level background model is created and updated using temporal statistics on the positions of the keypoints. The first level is trained to segment keypoints in background and foreground, while the second level segments the foreground keypoints in moving or stationary keypoints. Stationary foreground keypoints are used to extract views of stable foreground objects from the video removing the part of the image containing keypoints coming from the cluttered background.

The system ciclycally a) updates the background model untils it is steady, b) waits for stable new objects to be detected in the scene, c) extracts the view of the detected object, d) compares it to already collected views and e) organizes cluster of views.

Each extracted view o_i is described by a) K_i, the set of the positions of its local features (keypoints) and b) D_i, the set of their extracted ORB descriptors. ORB is a rotation invariant version of the BRIEF binary descriptor based on binary tests between pixels of the smoothed image patch [8]. It is suitable for realtime applications since it is faster than both SURF and SIFT but it has similar matching performance and is even less affected by image noise [2].

3.1 Observations Matching

A similarity function $S : (o_1, o_2) \rightarrow [0, 1]$ is defined on a pair of object views (o_1, o_2), representing the quality of the visual match between them. The similarity value among two views is computed in steps shown in Figure 2 and described below.

(a)   (b)   (c)

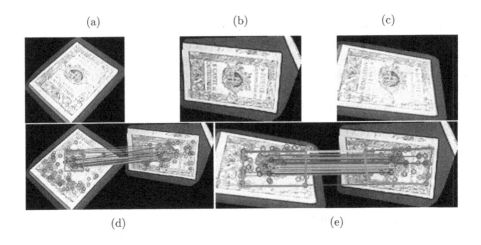

(d)   (e)

Fig. 2. Example of similarity computation among two different views (a) and (b) of the same 3D object. The homography relating the matching keypoints is shown in (d). The view (a) is transformed using the found homography in (c) and all matching steps are reapplied: the second homography found (e) confirms the match. The computed similarity value is 0.48.

Feature Matching. Let K_1, K_2 be the sets of keypoints of the compared views and D_1, D_2 their sets of corresponding descriptors.

A preliminary list L of descriptor matches is created finding for each descriptor in D_1 its nearest neighbor in D_2 using the bruteforce method. Distances between descriptors are computed using the method suggested by the authors of the descriptor. In case of ORB, Hamming distance between the binary representation of the descriptors is used.

Matches in L are then filtered keeping only the ones having distance between descriptors below T_m. We chose $T_m = 64$ as suggested by Rublee et al. [8] despite not being the most stringent value to filter bad matches, but we preferred high recall of matches rather than high precision at this step.

If there are less than 4 matches left in the list, the following steps cannot be applied and the similarity value is set to 0.

RANSAC Filtering of Matches. Bad matches in L are filtered out checking whether the points that match are geometrically consistent.

Two images of the same planar surface in space are related by a *homography* [3]. A homography is a invertible transformation represented by a 3×3 real matrix that maps the 2D coordinates of points in a image plane into the 2D coordinates in another plane.

$$H = \begin{pmatrix} h_{11} & h_{12} & h_{13} \\ h_{21} & h_{22} & h_{23} \\ h_{31} & h_{32} & 1 \end{pmatrix}$$

Let K_1^*, K_2^* be the sets of keypoints corresponding to the descriptors belonging to L. In order to find the homography that relates correctly the most of the points in K_1^* and K_2^*, RANSAC is applied [3]. RANSAC is an non-deterministic algorithm to estimate parameters of a mathematical model from a set of observed data which contains outliers. RANSAC algorithms iteratevly executes the following steps:

1. takes 4 matches (couples of points) at random from K_1^* and K_2^*,
2. computes the homography H relating those points,
3. counts the number of other matches that are correctly related by H (inliers).

After a certain number of iterations, the matrix H which gave the maximum number of inliers is returned.

Using the homography found by the RANSAC algorithm (Figure 2d), we can further filter the matches in L, keeping only the inliers of the perspective transformation.

Quasi-degenerate and flipping homographies can be detected analizing the homography matrix. Three checks are done:

- flipping homographies can be discarded checking if $det(H) < 0$.
- very skewed or prospective homographies can be discarded if $det(H)$ is too small or too big: given a parameter N, H is discarded if $det(H) > N$ or $det(H) < \frac{1}{N}$.
- homographies transforming the matching keypoints bounding box in a concave polygon can be filtered out with a convexity check.

In those cases, it is very unlikely that the views under analysis are really related by this perspective transformation, therefore the system assumes there is no similarity between them and returns a similarity value of 0.

Second Stage RANSAC. Some views may pass the homography matrix check even if the perspective transform described by H is very unlikely to be observed. In order to filter out false positives homography matrices, the image of the first view o_1 is transformed in $\acute{o}_1$ using the homography to be validated (Figure 2c) and the similarity computation steps are repeated considering the views $\acute{o}_1$ and o_2. Features are re-detected and re-extracted from $\acute{o}_1$, matched with o_2 and a second RANSAC is executed to estimate a new homography $\hat{H}$ describing the prospective transformation among $\hat{o}_1$ and o_2 4. If the original views o_1 and o_2 were really different views of the same object, $\hat{H}$ should be very near to the identity transformation (Figure 2e), otherwise the similarity between o_1 and o_2 is set to 0.

Similarity Output. After the system found a good homography relating the views, the ratios $\hat{r}_1, r_2$ among the number of inliers and the total number of detected features are computed for each view:

$$\hat{r}_1 = \frac{I}{|\hat{K}_1|}, \qquad r_2 = \frac{I}{|K_2|}$$

where I are the number of inliers of the homography estimated between views $\hat{o}_1$ and o_2, $|\hat{K}_1|$ and $|K_2|$ are respectively the number of detected keypoints in $\hat{o}_1$ and in o_2. The similarity value among original views under analysis $S(o_1, o_2)$ is defined as the harmonic mean between $\hat{r}_1$ and r_2 (Figure 3):

$$S(o_1, o_2) = \frac{2\hat{r}_1 r_2}{\hat{r}_1 + r_2}$$

	A	B	C	D
A	1	0	0	0.63
B	0	1	0.57	0
C	0	0.62	1	0
D	0.68	0	0	1

(a)   (b)   (c)   (d)   (e)

Fig. 3. Values of similarity among object views (a-d) reported in table (e).

4   Online Object Clustering

Everytime a new view of an object is gathered from the video stream, the system a) assigns it to a cluster and b) maintains clusters of views that potentially represent the same 3D object (Figure 4).

Each cluster is identified by a label assigned to views. The system puts a new view in a cluster relying on the similarity it has with other already clustered views, following an agglomerative clustering approach. The new view can bring informations useful to cluster reorganization: for example, let c_1 and c_2 be two clusters of views representing the same 3D object viewed from two different poses. An intermediate view of the 3D object could suggest the system to merge c_1 and c_2 in a unique cluster (see Figure 5).

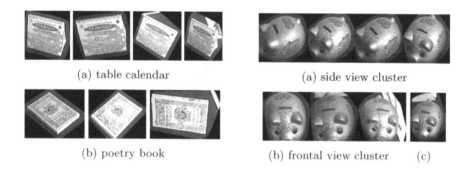

(a) table calendar

(a) side view cluster

(b) poetry book

(b) frontal view cluster   (c)

Fig. 4. Example of two object view clusters (a) and (b).

Fig. 5. Example of *cluster merging*: The new view (c) is similar to both clusters and can lead to a cluster merge.

Given a new view $\hat{o}$, a list L_s of similar views is generated scanning the local database. For each object view o_i the similarity value $s_i = S(\hat{o}, o_i)$ is computed and if it is above a similarity threshold T_s, o_i is inserted in L_s.

When trying to label $\hat{o}$, the following scenarios can occur:

1. $\hat{o}$ does not match with any views, hence a new cluster is created and a new label is assigned to $\hat{o}$.
2. $\hat{o}$ matches with one or more views all belonging to the same cluster, hence the system assigns the corresponding cluster label to $\hat{o}$.
3. $\hat{o}$ matches with more than two views beloging to different clusters. Many actions may be taken by the system in this situation:

   (a) the clusters containing the views similar to the new one are merged together in a unique bigger cluster to which the new view will belong (Figure 5).
   (b) the new view is inserted into only one among the candidates clusters.
   (c) a new cluster is created containing only the new view.

Up to now, the system does not decide automatically in the third scenario and asks the user which action should be taken. Interaction between multiple cameras and similarity values between views and clusters may be exploited to take the correct action automatically, but are not discussed in this paper and are left to future work.

In the case a new view is incorrectly put in a new cluster instead of being grouped with the other views representing the same object, the agglomerative cluster algorithm can eventually build a unique cluster if intermediate views of the same object will be collected by the system.

5   Experiments

The presented system autonomously groups object views into clusters without knowing their labels, but cannot recognize them before the user labels at least

Fig. 6. Object learning and recognition in a test video sequence: objects are added, moved and removed from the scene. The system segments objects from the background and incrementally creates clusters of similar object views. An object is recognized assessing the membership of its current view to a pre-existent cluster.

some of them, hence the system cannot be compared with traditional trained classifiers. Instead the ability of the system to build good and easy to label clusters is measured.

To do so, the publicly available *3D Objects* dataset [9] has been used. This dataset is composed by images of 10 object categories. For each category, 9-10 objects are present and for each object, several images are reported in which the specific object is shown in different poses. Each image comes with a foreground mask which denotes exactly in which part of the image the object is located. Images are taken from 8 different angles using 3 different scales and 3 different heigths for the camera, leading to around 5500 labeled images of 100 specific objects (see Figure 7).

Let $O = \{(o_1, l_1), (o_2, l_2), \ldots\}$ the set of labeled views. The entire dataset O is randomly shuffled and splitted in training set O_{train} (90%) and testing set O_{test} (10%): training views are presented to the system as coming from the output of the foreground extraction stage. The system builds clusters of views while they

Category	Object	Views					
cellphone	cellphone_1				...		
	...	...					
	cellphone_9					...	
mouse	mouse_1				...		
	...	...					
toaster	toaster_1				...		
	...	...					

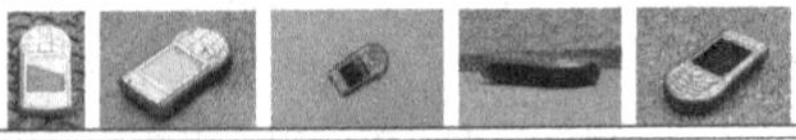

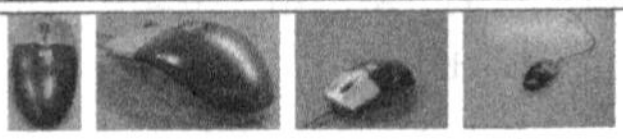

Fig. 7. Excerpt from the Stanford "3D Objects" dataset: only some views of some objects of some classes are reported.

Algorithm 1. Clustering algorithm simulating user interaction used for evaluation tests

---

    for all $(o_i, l_i) \in O_{train}$ **do**

        find the set O_s of views similar to o_i, $O_s = \{o_j \in Database : S(o_i, o_j) > T_s\}$

        if $|O_s| = 0$ **then**

            put o_i in a new cluster

        else if $|O_s| = 1$ **or** (all views in O_s belong to the same cluster) **then**

            put o_i in the cluster of the similar views

        else   ▷ simulate user interaction

            find set C_s of all clusters to which the similar views belong

            for all $c \in C_s$ **do**

                find the majority groundtruth label of c (the label appearing the most in the cluster)

            end for

            create a new cluster merging together all clusters having its majority label equal to l_i (the label of o_i)

            put o_i in the newly created cluster

        end if

    end for

---

are processed. In the case a supervised clustering is needed, the test code uses the groundtruth labels of involved views to simulate user interaction applying Algorithm 1.

Once the clusters are built, they must be labeled to produce a labeled training set. Since the user usually does not want to waste time in cleaning clusters or labeling singular objects, the test code simulates a labeling technique based on *major voting*: an entire cluster is labeled with the label of the most frequent object present in it.

The training set thus labeled is used for training a k-NN classifier. The *cluster* k-NN classifier finds the k most similar views (the ones with the higher value of similarity S) and assigns a score for each label of those views. The winning label is assigned to the processed test view. Another k-NN classifier is trained using the training set with correct labels and another labeling of the test set is generated in the same way.

Test set labelings are evaluated extracting precision, recall and F-score for each 3D object and then aggregating them using macro- and micro-averaging techniques defined as follows:

	micro-avgd	macro-avgd
precision	$p_{micro} = \dfrac{\sum_{i=1}^{n} TP_i}{\sum_{i=1}^{n}(TP_i + FP_i)}$	$p_{macro} = \dfrac{\sum_{i=1}^{n} p_i}{n}$
recall	$r_{micro} = \dfrac{\sum_{i=1}^{n} TP_i}{\sum_{i=1}^{n} P_i}$	$r_{macro} = \dfrac{\sum_{i=1}^{n} r_i}{n}$
f-score	$F_{micro} = \dfrac{2 p_{micro} r_{micro}}{p_{micro} + r_{micro}}$	$F_{macro} = \dfrac{2 p_{macro} r_{macro}}{p_{macro} + r_{macro}}$

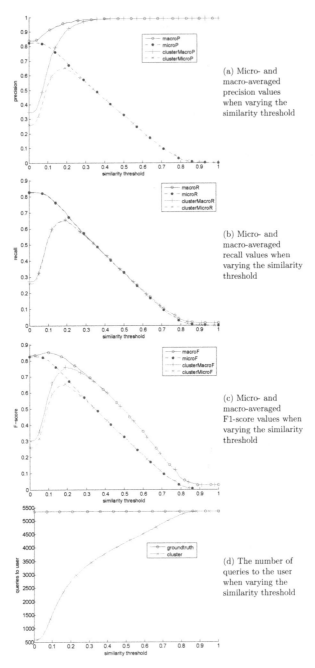

(a) Micro- and macro-averaged precision values when varying the similarity threshold

(b) Micro- and macro-averaged recall values when varying the similarity threshold

(c) Micro- and macro-averaged F1-score values when varying the similarity threshold

(d) The number of queries to the user when varying the similarity threshold

Fig. 8. Comparison of the performance of the recognition task, solved by a k-NN classifier trained with the groundtruth training set (blue lines with circle markers) and by a k-NN classifier with training set made by cluster labeling (red lines with cross markers). Solid and dashed lines indicate respectively micro- and macro-averaged metrics.

where n is the number of object classes, TP_i the true positives, FP_i the false positives, P_i the total number of views, p_i the precision, r_i the recall and F_i the F-score of the object i.

Macro-averaged metrics tends to give the same weight to each class, while micro-averages metrics takes into account possible biases introduced by each class and gives a more accurate global performance index. Another measured metric is the number of interactions the system must have with the user in order to label the training set: the *groundtruth* k-NN classifier needs the user to label each training view individually, which corresponds to a number of query to the user equal to the number of views in the training set. The *cluster* k-NN classifier needs to interact with the user a) when a cluster merging can not be resolved automatically during the online clustering and b) when a cluster has to be labeled.

6   Conclusions

In Figure 8, the performance of the two classifiers for various similarity thresholds T_s are reported.

It can be seen that for T_s around 0.2, the *cluster* k-NN classifier has almost the same performance of the *groundtruth* k-NN classifier, having only around half the interactions with the user.

However, performance degradation of the *cluster* k-NN classifier is due to the fact that we simulated a unique user interaction after the training phase which used *major voting* paradigm to label all clusters at once. Since the system is incrementally building richer and richer clusters, this is not the best way to interact with the user asking for labels: user interaction may be proactively requested only when big homogeneous clusters are involved, maximizing the amount of information collected. Moreover, smarter techniques than major voting may be implemented to simulate a more precise user labeling session. In the performed tests, many singleton or small clusters are present at the end of the training phase, raising the number of queries to the user needed to label the entire training set.

References

1. Carrara, F., Amato, G., Falchi, F., Gennaro, C.: Efficient foreground-background segmentation using local features for object detection. In: Proceedings of the International Conference on Distributed Smart Cameras, ICDSC 2015, September 08–11, 2015, Seville, Spain (submitted for publication). http://puma.isti.cnr.it/rmydownload.php?filename=cnr.isti/cnr.isti/2015-TR-012/2015-TR-012.pdf
2. De Beugher, S., Brône, G., Goedemé, T.: Automatic analysis ofin-the-wild mobile eye-tracking experiments using object, face and persondetection. In: Proceedings of the International Conference on Computer Vision Theory and Applications (VISIGRAPP 2014), vol. 1, pp. 625–633 (2014)
3. Dubrofsky, E.: Homography estimation. Ph.D. thesis, University of British Columbia (2009)

4. Fei-Fei, L., Fergus, R., Perona, P.: Learning generative visual models from few training examples: an incremental bayesian approach tested on 101 object categories. Computer Vision and Image Understanding **106**(1), 59–70 (2007)

5. Fergus, R., Perona, P., Zisserman, A.: Object class recognition by unsupervised scale-invariant learning. In: 2003 IEEE Computer Society Conference on Computer Vision and Pattern Recognition, 2003. Proceedings, vol. 2, pp. II–264. IEEE (2003)

6. Lowe, D.G.: Local feature view clustering for 3d object recognition. In: Proceedings of the 2001 IEEE Computer Society Conference on Computer Vision and Pattern Recognition, CVPR 2001, vol. 1, pp. I–682. IEEE (2001)

7. Murase, H., Nayar, S.K.: Visual learning and recognition of 3-d objects from appearance. International Journal of Computer Vision **14**(1), 5–24 (1995)

8. Rublee, E., Rabaud, V., Konolige, K., Bradski, G.: Orb: an efficient alternative to sift or surf. In: 2011 IEEE International Conference on Computer Vision (ICCV), pp. 2564–2571. IEEE (2011)

9. Savarese, S., Li, F.F.: 3D generic object categorization, localization and pose estimation. In: ICCV, pp. 1–8 (2007)

10. Weber, M., Welling, M., Perona, P.: Unsupervised learning of models for recognition. Springer (2000)

Banknote Recognition as a CBIR Problem

Joan Sosa-García[✉] and Francesca Odone

Dipartimento di Informatica, Bioingegneria, Robotica E Ingegneria Dei Sistemi,
Università degli Studi di Genova, Genova, Italy
joan.sosa.garcia@dibris.unige.it, francesca.odone@unige.it

Abstract. Automatic banknote recognition is an important aid for
visually impaired users, which may provide a complementary evidence
to tactile perception. In this paper we propose a framework for banknote recognition based on a traditional Content-Based Image Retrieval
pipeline: given a test image, we first extract SURF features, then adopt
a Bag of Features representation, finally we associate the image with the
banknote amount which ranked best according to a similarity measure
of choice. Compared with previous works in the literature, our method is
simple, computationally efficient, and does not require a banknote detection stage. In order to validate effectiveness and robustness of the proposed approach, we have collected several datasets of Euro banknotes on
a variety of conditions including partial occlusion, cluttered background,
and also rotation, viewpoint, and illumination changes. We report a
comparative analysis on different image descriptors and similarity measures and show that the proposed scheme achieves high recognition rates
also on rather challenging circumstances. In particular, Bag of Features
associated with L2 distance appears to be the best combination for the
problem at hand, and performances do not degrade if a dimensionality
reduction step is applied.

Keywords: Banknote recognition · Computer vision for visually
impaired · Content-based image retrieval · Bag of features representations

1 Introduction

Automatic banknote recognition is wide-spread in vending machines, banking
system, supermarkets, currency exchange services. In all those applications, the
processing occurs in very controlled conditions. In recent years a very different
application scenario arose: banknote recognition as a tool for elderly and visually impaired citizens. An automatic system that can assist people with severe
vision impairment to independently recognize banknotes is supposed to be highly
portable, or even wearable, to perform the recognition in a wide variety of conditions, including cluttered backgrounds, changing illumination, and different
viewpoints. Moreover, a recognition system should be able to recognize banknotes from each side, direction and rotation. At the same time, this application

© Springer International Publishing Switzerland 2015
G. Amato et al. (Eds.): SISAP 2015, LNCS 9371, pp. 229–236, 2015.
DOI: 10.1007/978-3-319-25087-8_21

offers some peculiarities: first, banknotes are normally held by the user or, in any case, we can assume they are close to the user. Then, assuming the user acquires an image by means of a smartphone or a wearable camera, the bill will occupy the majority of the image. Also, the reference class of users we are considering, will not guarantee images of good quality or carefully chosen framing. Thus we must consider the problem of occlusions or partial information.

Banknote recognition has been addressed in several ways, although in general size, colour [13] and texture [6] or a combination of both [4] are employed. In most cases the methods are based on rather complex procedures which include detection and recognition steps[1, 7, 11, 13, 20–23] — most methods assume the whole bill is included in the image and few works take into account illumination variations. Usually RGB images are used, although other sensor information has been employed, such as a combination of infra-red (IR) and visible images which seems to be rather appropriate with relatively new currencies such as euro [1].

Local features are adopted in [17] and in [5]. The latter proposes a component-based framework for banknote recognition which is effective in collecting more class-specific information and robust in dealing with partial occlusion and viewpoint changes, but the computational cost of the feature matching procedure is significant. To control the computational cost of the recognition process and attenuate the effect of noise, often they are associated with bag of words representations [2, 9, 10, 14, 16, 18]. In this work we adopt this strategy to address banknote recognition.

In this paper, we propose a new banknote recognition scheme based on a classical CBIR pipeline. Given a test image, we first extract SURF features, which we chose for their computational cost, then adopt a Bag of Features representation, finally we associate the image with the banknote amount which ranked best according to an appropriate similarity measure following a kNN procedure. Compared with previous works in the literature, our method is simpler and computationally more efficient. Besides, it is worth noticing that does not require a banknote detection stage – i.e. the whole image is represented in a single vector. At the same time it is very effective even in the case of blur, partial occlusions, clutter, as we will show in our experimental analysis.

The remainder of this paper is organized as follows: Section 2 describes the proposed banknote recognition system. Section 3 reports an exhaustive experimental analysis of the proposed system on the introduced datasets, while Section 4 is left to a final discussion.

2   Banknotes Retrieval

In this section we describe our banknote recognition pipeline based on image retrieval.

The Gallery Dataset. Our gallery is composed of 112 images, 16 per class, where the classes are EUR 5 (first and second series), 10 (first and second series), 20, 50 and 100. The dataset includes images of both front and reverse sides and

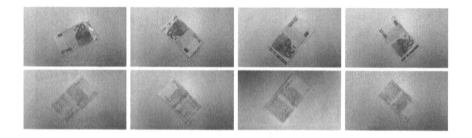

Fig. 1. EUR 20 samples from the gallery dataset.

different orientations (8 equally spaced angles between 0° and 315°), which have been acquired on a uniform background and manually annotated (see Fig. 1).

Banknote Recognition as a Retrieval Problem. A typical CBIR pipeline includes an offline phase, which consists in the preparation of the gallery, and an online recognition phase. In the *offline* phase, the gallery images are represented as feature bags, as follows. First, we extract SURF features (chosen for their computational efficiency) on each image. Then, we compute a *visual dictionary* by K-means. Further, we associate a global image descriptor, based on the dictionary, with each gallery image. Our reference approach is a classical Bag of Features (BoF) [3,14,16,18], although we will show how, in some cases, the recently proposed Mean of Bag of Features per Quadrant (MBoFQ [19]) is quite effective. The global BoF descriptor corresponds to a *vector quantization* of the image features with respect to the visual dictionary. The feature vector $v = (v_1, ..., v_D)$ is finally normalized by the so-called *power-law* normalization, which have been shown very effective in image retrieval (see[15]): $v_i = \mid v_i \mid^{\beta} \times sign(v_i)$, with $0 \leq \beta < 1$ a fixed constant. The updated vector v is then L2-normalized. We fixed $\beta = 0.1$ in all experiments.

In the *online* phase, a new query image is first represented coherently with the gallery (in the same manner), then it is compared with all the gallery images via a kNN procedure. Different similarity measures can be adopted, in the experiments section we will compare different classical choices. Then gallery images are ranked according to their similarity, and a voting procedure is applied to the first k positions of the ranking list.

3   Method Assessment

Test Set. For a quantitative performance evaluation, we acquired a test-set of 370 query images, with about 36 to 90 images for each class of banknote. Each image has a resolution of 665×1182 pixels. The images (acquired by three different users with smart phones cameras) cover a wide variety of conditions, such as partial occlusion, blur, rotations, changes of illumination and viewpoints (see Fig. 2). Differently from other literature datasets [23], typically collected by scanners or

Fig. 2. Samples of query images taken under different conditions. Row 1: blurry images; row 2: clutter; row 3: occlusions; row 4: viewpoint changes; row 5: rotations; row 6: illumination changes.

in constrained conditions, our dataset represents a better approximation of real world scenarios, covering a wider variety of conditions. In particular, we consider situations which may arise in a practical usage, with a banknote lying on a flat surface in front of the user, or being held by the user.

Different Image Descriptors. We first evaluate the use of SURF features in combination with different image descriptors in the context of banknote recognition. Our analysis compares **BoF** [3], **Fisher** [15], **VLAD** [8], **Spatial Pyramid (SPM)** [12], **MBoFQ** [19]. We choose different vocabulary sizes in order to obtain descriptors of similar dimensionality, considering a trade off between smaller representations and performances. In all experiments, the vector size of Fisher, VLAD, SPM and MBoFQ is equal to 20000 while BoF allows us to obtain comparable performances with 10000. Table 1 reports the retrieval performances on the query dataset obtained with $k = 12$ and the similarity evaluated by the euclidean distance. The BoF achieves an overall recognition rate of 96.49%. A similar performance is obtained by MBoFQ but with larger vectors. Notice that,

Table 1. Recognition rates (%) with different image descriptors.

Descriptor	Denomination class (EUR)					Average
	5	10	20	50	100	
BoF	97.53	91.82	98.90	**98.08**	100	**96.49**
MBoFQ	96.30	95.45	**100**	90.38	100	**96.49**
VLAD	91.36	**96.36**	95.60	86.54	94.44	93.51
FISHER	**98.77**	90.91	98.90	92.31	97.22	95.41
SPM	83.95	67.27	93.41	82.69	94.44	82.16

Table 2. Recognition rates (%) for different values of k (Euclidean Distance).

Descriptor	top$-k$ positions				
	12	7	5	3	1
BoF	**96.49**	**97.57**	**96.22**	**96.49**	**96.49**
MBoFQ	**96.49**	95.14	95.14	95.14	94.32
VLAD	93.51	94.86	93.78	94.05	94.59
FISHER	95.41	96.49	95.95	95.95	95.14
SPM	82.16	80.54	80.54	80.27	87.03

previous methods [20, 23] obtained results never above 95% on different but comparably difficult datasets.

Different Values of k. Table 2 shows the recognition rate for different values of k. The highest recognition rates are obtained with the BoF method for all values of k. It can be observed that the performance of MBoFQ vectors decreases when smaller number of k-positions is considered. Note that the performance of VLAD, FISHER and SPM descriptors increases for some values of k with respect the initial value $k = 12$, but in general their behavior is unstable.

Different Similarity Measures. Table 3 shows the results obtained by using different similarity measures in the ranking phase on the query dataset. By comparing the results of Table 3, the recognition rate is more accurate by using Euclidean Distance for all image descriptors considered.

The Effect of Image Size. Figure 3 shows the performance of the descriptors at different image size. The initial query dataset is resized at several image resolutions, starting from 90% resolution of the initial images to a 40% image resolution. Each of the resolution reductions of the initial dataset produces a new query dataset and the system is evaluated on each individual dataset. It can be observed in Figure 3 that there is a significant increase in the performance up to 98.1% of BoF descriptors at 70% image resolution. By observing the results shown in Figure 3, the recognition rates of the BoF and MBoFQ methods remain

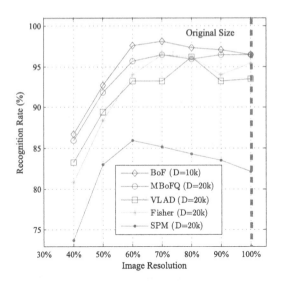

Fig. 3. Recognition rates (%) at several image resolutions.

Table 3. Recognition rates (%) via different image descriptors and similarity measures.

Descriptor	Similarity Measure				
	L1	L2	$Sign - \mathcal{X}^2$	$\mathcal{X}^2$	Hist. Inter. (dis)
BoF	71.89	**96.49**	76.22	72.43	**95.68**
MBoFQ	81.08	**96.49**	**84.05**	84.05	93.24
VLAD	84.05	93.51	21.89	25.95	31.62
FISHER	**84.86**	95.41	22.16	16.49	24.59
SPM	64.59	82.16	58.10	65.95	71.35

above 95% from 90% to 60% image resolution. Where the latter corresponds to an image size equal to 400×710, similar to the resolution of webcams.

Dimensionality Reduction. Today, in image retrieval, it is common practice to include a dimensionality reduction step over the final feature vector. This process helps reducing the size of the descriptor, *improving* retrieval performances, but as an additional benefit controls data redundancy. Table 4 compares different descriptors after a PCA and whitening procedure [8] on banknote recognition. In this experiment, the reference and query dataset are represented following the procedure described in Section 2 and then each image vector is reduced applying PCA. The ranking of the reference dataset given a query image is obtained by comparing the reduced vectors (128-D) of the query and reference images. Given the results of Table 4, it can be seen that the dimensionality reduction does not affect much the recognition performance. Besides, there is a slight increase in the performance for MBoFQ, VLAD and SPM vectors. High-speed processing is performed by using low-dimensionality vectors and therefore the computational

Table 4. Comparison of image descriptors of low dimensionality (128-D).

Descriptor	Initial Size	Recognition Rate (%)	Reduced Size	Recognition Rate (%)
BoF	10000	**96.49**		95.14
MBoFQ		**96.49**		**96.76**
VLAD	20000	93.51	128	94.05
FISHER		95.41		95.14
SPM		82.16		86.49

costs decrease. This speed-up allows to enlarge the reference dataset as well as incorporate new banknotes from other countries.

4   Discussion

In this paper we proposed a banknote recognition method based on a simple, although effective, CBIR pipeline. We evaluated our method on the Euro currency on a very challenging query dataset that covers a wide variety of conditions, such as partial occlusion, blur, rotation, changes of illumination, scale and viewpoints. Our experimental analysis compared different image descriptors; here we found out that a standard BoF has a high recognition performance (about 98%) on the proposed datasets and is robust to handle scale changes, partial occlusions (e.g. user's hands). Good performance is also achieved by other image descriptors, such as MBoFQ and Fisher vectors. Our reference application domain is the development of aids for visually impaired people. Currently we are testing the generality of the method to different scenarios, by means of a demo application.

We conclude by observing our approach can be easily extended to other currencies, simply by acquiring an appropriate image gallery. For its simplicity and low time-space computational cost, the proposed system can be easily incorporated into wearable devices to assist visually impaired people to automatically recognize banknotes in their daily activity.

References

1. Aoba, M., Kikuchi, T., Takefuji, Y.: Euro banknote recognition system using a three-layered perceptron and rbf networks. Transactions on Mathematical Modeling and Its Applications **44**, 99–109 (2003)
2. Chum, O., Philbin, J., Sivic, J., Isard, M., Zisserman, A.: Total recall: automatic query expansion with a generative feature model for object retrieval. In: ICCV, pp. 1–8 (2007)
3. Csurka, G., Dance, C., Fan, L., Willamowski, J., Bray, C.: Visual categorization with bags of keypoints. In: SLCV, ECCV 2004, vol. 1, p. 22 (2004)
4. García-Lamont, F., Cervantes, J., López, A.: Recognition of mexican banknotes via their color and texture features. Expert Systems with Applications **39**(10), 9651–9660 (2012)
5. Hasanuzzaman, F.M., Yang, X., Tian, Y.: Robust and effective component-based banknote recognition for the blind. IEEE Transactions on Systems, Man, and Cybernetics, Part C: Applications and Reviews **42**(6), 1021–1030 (2012)

6. Hassanpour, H., Farahabadi, P.M.: Using hidden markov models for paper currency recognition. Expert Systems with Applications **36**(6), 10105–10111 (2009)

7. Jahangir, N., Chowdhury, A.R.: Bangladeshi banknote recognition by neural network with axis symmetrical masks. In: 10th International Conference on Computer and Information Technology, ICCIT 2007, pp. 1–5. IEEE (2007)

8. Jégou, H., Chum, O.: Negative evidences and co-occurences in image retrieval: the benefit of PCA and whitening. In: Fitzgibbon, A., Lazebnik, S., Perona, P., Sato, Y., Schmid, C. (eds.) ECCV 2012, Part II. LNCS, vol. 7573, pp. 774–787. Springer, Heidelberg (2012)

9. Jégou, H., Douze, M., Schmid, C.: On the burstiness of visual elements. In: CVPR, pp. 1169–1176 (2009)

10. Jégou, H., Harzallah, H., Schmid, C.: A contextual dissimilarity measure for accurate and efficient image search. In: CVPR, pp. 1–8. IEEE (2007)

11. Kosaka, T., Omatu, S., Fujinaka, T.: Bill classification by using the lvq method. In: 2001 IEEE International Conference on Systems, Man, and Cybernetics, vol. 3, pp. 1430–1435. IEEE (2001)

12. Lazebnik, S., Schmid, C., Ponce, J.: Beyond bags of features: spatial pyramid matching for recognizing natural scene categories. In: CVPR, vol. 2, pp. 2169–2178. IEEE (2006)

13. Lee, J.K., Jeon, S.G., Kim, I.H.: Distinctive point extraction and recognition algorithm for various kinds of euro banknotes. International Journal of Control Automation and Systems **2**, 201–206 (2004)

14. Nister, D., Stewenius, H.: Scalable recognition with a vocabulary tree. In: CVPR, vol. 2, pp. 2161–2168. IEEE (2006)

15. Perronnin, F., Liu, Y., Sánchez, J., Poirier, H.: Large-scale image retrieval with compressed fisher vectors. In: CVPR, pp. 3384–3391. IEEE (2010)

16. Philbin, J., Chum, O., Isard, M., Sivic, J., Zisserman, A.: Object retrieval with large vocabularies and fast spatial matching. In: CVPR, pp. 1–8. IEEE (2007)

17. Reiff, T., Sincak, P.: Multi-agent sophisticated system for intelligent technologies. In: IEEE International Conference on Computational Cybernetics, ICCC 2008, pp. 37–40. IEEE (2008)

18. Sivic, J., Zisserman, A.: Video google: a text retrieval approach to object matching in videos. In: ICCV, pp. 1470–1477. IEEE (2003)

19. Sosa-García, J., Odone, F.: Mean bof per quadrant - simple and effective way to embed spatial information in bag of features. In: VISAPP (2015)

20. Takeda, F., Nishikage, T.: A proposal of structure method for multicurrency simultaneous recognition using neural networks. Transactions of the Japan Society of Mechanical Engineers **66**(648) (2000)

21. Takeda, F., Nishikage, T.: Multiple kinds of paper currency recognition using neural network and application for euro currency. In: Proceedings of the IEEE-INNS-ENNS International Joint Conference on Neural Networks, IJCNN 2000, vol. 2, pp. 143–147. IEEE (2000)

22. Takeda, F., Omatu, S.: High speed paper currency recognition by neural networks. IEEE Transactions on Neural Networks **6**(1), 73–77 (1995)

23. Takeda, F., Sakoobunthu, L., Satou, H.: Thai banknote recognition using neuralnetwork and continues learning by dsp unit. In: Palade, V., Howlett, R.J., Jain, L. (eds.) KES 2003. LNCS, vol. 2773, pp. 1169–1177. Springer, Heidelberg (2003)

Efficient Image Search with Neural Net Features

David Novak[1]([✉]), Jan Cech[2], and Pavel Zezula[1]

[1] Masaryk University, Brno, Czech Republic
{david.novak,zezula}@fi.muni.cz
[2] Czech Technical University, Prague, Czech Republic
cechj@cmp.felk.cvut.cz

Abstract. We present an efficiency evaluation of similarity search techniques applied on visual features from deep neural networks. Our test collection consists of 20 million 4096-dimensional descriptors (320 GB of data). We test approximate k-NN search using several techniques, specifically FLANN library (a popular in-memory implementation of k-d tree forest), M-Index (that uses recursive Voronoi partitioning of a metric space), and PPP-Codes, which work with memory codes of metric objects and use disk storage for candidate refinement. Our evaluation shows that as long as the data fit in main memory, the FLANN and the M-Index have practically the same ratio between precision and response time. The PPP-Codes identify candidate sets ten times smaller then the other techniques and the response times are around 500 ms for the whole 20M dataset stored on the disk. The visual search with this index is available as an online demo application. The collection of 20M descriptors is provided as a public dataset to academic community.

1 Introduction: Content-Based Image Retrieval

The content-based image retrieval (CBIR) is an area that naturally requires similarity techniques to match the image data. A successful CBIR system must stand on two pillars: *effective* image processing to achieve high quality of the retrieval, and *efficient* search to make the system work in real time and on a large scale. Specifically, the image processing typically leads to certain features (descriptors, stimuli) that capture the application-driven characteristics of the image data; the actual searching process is then realized in space of these features.

Currently, the state-of-the-art image recognition approach is based on deep convolutional neural networks (details follow). The objective of this paper is to study how current similarity techniques can manage visual features obtained by this approach. These features are 4096-dimensional real vectors, which makes it a challenge to efficiently search in large collections. These vectors can be compared using various distance functions, and thus we focus mainly on metric-based search techniques. We analyze the space properties of the visual features extracted from a 20 million image collection and we evaluate the search efficiency of different approaches.

This work was supported by Czech Research Foundation project P103/12/G084.

The Deep Convolutional Neural Networks (DCNN) have a long history in computer vision and machine learning. LeCun et al. [6] introduced an architecture with many hidden layers where weights of lower-level layers are shared as convolution filters. Two decades later, a similar architecture was applied to a large scale visual recognition problem by Krizhevsky et al. [5] and won the ILSVRC 2012 challenge by a large margin. The network was trained to recognize 1k selected ImageNet categories from more than 1M training images. The network consisting of 5 convolutional and 3 fully connected layers takes raw size-normalized images as an input. The network having about 60M parameters represents a very flexible classifier with a great class capacity.

After the great success of [5], researches started to consider a possibility to re-use the representation power of the Krizhevsky's network to solve other recognition problems, i.e. to adapt the classifier to recognize classes the network was not trained for, by using a much smaller dataset [3,12]. The last network layer that outputs the class scores is in fact a linear classifier taking a non-linear representation of an image (the previous layer response). The response of the last hidden layer of Krizhevsky's network is coined the DeCAF feature in [3]. The feature is demonstrated to hold a great representation power and ability to generalize to other recognition tasks by training a simple linear classifier. Moreover, a semantic information is carried implicitly which is shown by tendency of the DeCAF features to cluster semantically similar images of categories on which the network was never explicitly trained.

The DCNN methods were also studied in the context of CBIR [13], using the deep features and a suitable metric. A good generalization to various datasets is shown, but attempts to learn a metric are reported to have a minor effect.

2   Similarity Indexing and Searching

In this section, we describe selected techniques for similarity-based indexing.

FLANN. FLANN is a popular library for performing fast approximate nearest neighbour search developed by Muja and Lowe [7]. It is often used in various computer vision problems where a large dataset is involved and it is incorporated into OpenCV. FLANN contains two main algorithms: (1) a forest of randomized k-d trees, and (2) a hierarchical k-means tree. Additionally, FLANN includes a method for an automatic selection of the most suitable algorithm and its parameters given the target dataset. The choice depends on the nature of the set of features where the search is performed; on its size, on the dimensionality, on the structure of the data, and of course on the desired precision of the approximate nearest neighbour search. The auto-tuning algorithm optimizes a score that consists of a weighted combination of the search time, tree build time and memory overhead. The weights are optionally set by a user. The optimization can run on a representative fraction of the dataset only, which further speeds the auto-tuning up and makes the approximate search system easy to set up.

Metric-Based Approaches. Further, we focus on the *metric access methods* (MAMs), that model the data as a metric space $(\mathcal{D}, \delta)$, where $\mathcal{D}$ is the domain

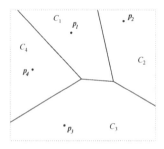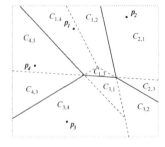

Fig. 1. Voronoi partitioning (left) and of second level Voronoi partitioning (right).

and δ is a metric (distance) function. Specifically, we describe a class of MAMs based on recursive Voronoi partitioning of the space as depicted in Figure 1 for pivots $p_1, \ldots, p_4$. The left part shows standard partitioning which generates four Voronoi cells $C_1, \ldots, C_4$; on the right, each of these cells is partitioned using the rest of the pivots. Cell $C_{i,j}$ then contains objects for which p_i is the closest pivot and p_j is the second closest. This principle can be used recursively l-times and it is often formalized as *permutations of pivots*: objects from Voronoi cell $C_{i_1,\ldots,i_l}$ can be "mapped" onto a vector $\langle i_1, \ldots, i_l \rangle$, which is an l-prefix of a certain permutation of the pivot indexes [8].

This principle has been successfully applied by several MAMs, for instance by a structure called M-Index [8]. This index builds a dynamic trie-like structure over the recursive Voronoi diagram, so that only the *overfilled* cells are partitioned to another level. Given a k-nearest neighbor query k-NN(q), the M-Index forms a *candidate set* of indexed objects by accessing data objects x from the "most promising" Voronoi cells; these candidate objects are refined by evaluation of $\delta(q, x)$ and the best k objects are returned. The Voronoi cell data can be stored either in memory or on the disk in continuous chunks. This approach can be further improved by combining several independent Voronoi partitions [10] in a similar way as in case of randomized k-d forest.

The same space partitioning is used also in a recent technique called PPP-Codes [11]. This MAM defines a mapping of the metric objects onto small codes composed of the pivot permutation prefixes from several pivot spaces. These codes are kept in memory; given a k-NN query, the PPP-Codes search algorithm combines candidate sets from the independent pivot spaces into a small but very accurate candidate set. Only objects from this candidate set are retrieved from the disk and refined. As these objects are read one-by-one (via their identifiers), this approach assumes an efficient key-value store, ideally kept on an SSD disk.

3 Efficiency Evaluation

According to the state of the art in computer vision, we use the DeCAF$_7$ feature produced by the last hidden layer of the neural network model provided by the Caffe project[1] [4], which was trained according to [5]. This 4096-dimensional float

---

[1] http://caffe.berkeleyvision.org

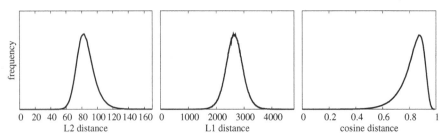

Fig. 2. Distance histograms of the data space with different metric functions; corresponding intrinsic dimensionality values are: L_2: 26.8, L_1: 36.0, cosine distance: 46.9.

vector was extracted from a collection called Profiset [1] consisting of 20 million images provided for research purposes by a microstock photography company. This set of 20M features is public for research purposes at http://disa.fi.muni.cz/profiset/.

In the beginning of this evaluation, we we provide analysis of the feature space properties. Figure 2 shows histogram of distances calculated on a sample of 1M images with three metrics: L_2, L_1 and cosine distance; the figure caption shows also respective values of *intrinsic dimensionality* calculated as $\mu^2/(2 \cdot \sigma^2)$, where μ and σ^2 are the mean and variance of the distance histogram [2].

The core of this section is evaluation of k-NN processing efficiency using L_2. Denoting A the approximate k-NN search answer and A^P the precise NN answer, the answer quality is measured by $recall(A) = precision(A) = \frac{|A \cap A^P|}{K} \cdot 100\%$. The key performance indicator is the wall-clock time of the query processing. All results were averaged over 1,000 queries from the outside of the dataset. We use several subsets of the collection of sizes from 100K to 20M. The evaluation was realized on a 12-core Intel Xeon @ 2.0 GHz machine with 60 GB of main memory, and SSD disk with transfer rate about 270 MB/s with random accesses.

3.1 In-memory Indexes

First, the in-memory FLANN and M-Index were tested on subsets up to 3M objects (48 GB in main memory). The FLANN auto-tuning procedure (running on a 100K sample) chose the randomized k-d tree forest with 32 trees; parameter "number of accessed leaves" varied for different required values of 1-NN recall. The M-Index was configured to use four Voronoi trees [10], each with 512 pivots and the parameter "number of accessed objects" was altered.

Plots in Figure 3 show dependence between the single-thread search times and k-NN recall for FLANN and M-Index on 1M dataset (various values of k). We can see that the results are very similar; this is quite surprising since the partitioning principles and the implementation platforms (C++ vs. Java) of the two indexes differ significantly. FLANN is able to return some results within milliseconds while the M-Index has a minimum response time about 20 ms. This is caused especially by initial calculation of distances between the query object and the set of 4×512 pivots. On the other hand, the recall values grow faster for M-Index, especially for higher values of k.

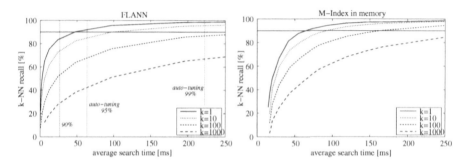

Fig. 3. Recall of k-NN queries vs. search time (single thread) for FLANN and main memory M-Index on 1M data collection.

Figure 4 shows the same type of dependence for 10-NN and collection sizes varying between 100K and 3M. We can see that both indexes scale quite well; the M-Index has again "slower start" but it outruns FLANN in the end.

3.2 Disk-Oriented Indexes

Finally, we analyze how disk-oriented indexes perform on collections up to 20M. First, we test the disk version of M-Index with the same configuration as in the memory case (four indexes, each with 512 pivots, 1M dataset). Since we use four independent space partitioning (in a similar way as locality-sensitive hashing approaches do), the physical data is now replicated *four times*. The left graph in Figure 5 shows the k-NN recall with respect to percentage of data accessed by the index; these results are independent of the memory/data implementation. The right graph compares the search time for various implementations: memory vs. disk and single- vs. multi-thread query evaluation. We can see that the disk variant is feasible with multi-treading. For all disk-oriented experiments, the disk caches were dropped before running every 1000-query batch.

Further, we focus on efficiency of the PPP-Codes index, which has been designed for larger collections of voluminous data objects [11]. In our case, it

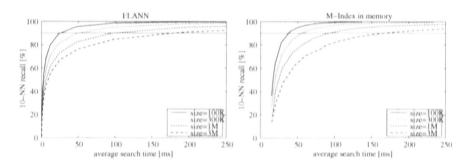

Fig. 4. Recall of 10-NN operations vs. search time (single thread) for FLANN and main memory M-Index on collections of different sizes.

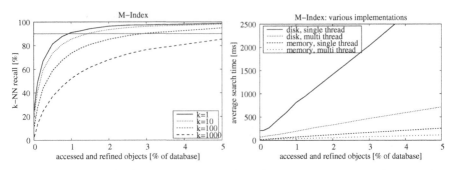

Fig. 5. k-NN recall (left) and search times (right) for different settings (memory/disk and single-/multi-thread processing) for variable perc. of accessed data (1M collection).

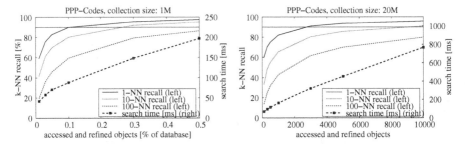

Fig. 6. k-NN recall and search times with respect to accessed objects: 1M and 20M.

uses the same four sets of 512 pivots as the M-Index, it keeps a memory structure (about 1 GB for the 20M collection) and the actual data objects are compressed on the disk (124 GB for the 20M collection).

For comparison with the results above, left graph in Figure 6 shows values of k-NN recall (left vertical axis) and of search times (right axis) with respect to percentage of accessed objects of 1M collection. We can see that PPP-Codes access one order of magnitude fewer objects than M-Index and the search times are about 1/3 of the M-Index with multi-thread processing on the disk. The search time improvement is not proportional to candidate set reduction since the PPP-Codes have more demanding in-memory processing phase and the candidate set objects on the disk are accessed one-by-one [11].

The right graph presents the results on the 20M collection. In this case, the horizontal axis shows the absolute number of accessed objects (out of 20M) and we can see that high recall values are achieved for response times around 500 ms. In practice, lower response times are achieved by not dropping the disk caches.

4   Conclusions

The fusion of the deep neural networks and similarity-based indexing has many good applications in the area of content-based image retrieval. The high dimensionality and bulkiness of the visual features from the neural networks calls for analysis of actual search efficiency of current indexing techniques on large datasets of this data. We have introduced a test collection with 20M 4096-dimensional image features and tested the k-NN search efficiency of selected indexing techniques. The results indicate that if the data fit into main memory, the metric-based structure M-Index [8] is as efficient as the FLANN [7] library. With the disk version of M-Index, the search would not stay real-time for large datasets because the index accesses over 1 % of the data to produce good results.

The PPP-Codes index [11] better fits this type of datasets as it can achieve fine results accessing around 0.02 % out of the 20M dataset; the search times are around 500 ms. There is an online demonstration application available at http://disa.fi.muni.cz/demos/profiset-decaf/ which presents k-NN visual search on this 20M dataset with the PPP-Codes index [9]. The collection of the 20M descriptors is publicly available at http://disa.fi.muni.cz/profiset/.

References

1. Budikova, P., Batko, M., Zezula, P.: Evaluation platform for content-based image retrieval systems. In: Gradmann, S., Borri, F., Meghini, C., Schuldt, H. (eds.) TPDL 2011. LNCS, vol. 6966, pp. 130–142. Springer, Heidelberg (2011)
2. Chávez, E., Navarro, G.: Measuring the dimensionality of general metric spaces. Technical report, Department of Computer Science, University of Chile (2000)
3. Donahue, J., Jia, Y., Vinyals, O., Hoffman, J., Zhang, N., Tzeng, E., Darrell, T.: DeCAF: a deep convolutional activation feature for generic visual recognition. In: International Conference in Machine Learning (ICML), pp. 647–655 (2014)
4. Jia, Y., Shelhamer, E., Donahue, J., Karayev, S., Long, J., Girshick, R., Guadarrama, S., Darrell, T.: Caffe: convolutional architecture for fast feature embedding. In: International Conference on Multimedia (2014)
5. Krizhevsky, A., Sutskever, I., Hinton, G.E.: Imagenet classification with deep convolutional neural networks. Advances In Neural Information Processing Systems **25**, 1097–1105 (2012)
6. LeCun, Y., Boser, B., Denker, J., Henderson, D., Howard, R., Hubbard, W., Jackel, L.: Backpropagation applied to handwritten zip code recognition. Neural Computation **1**(4) (1989)
7. Muja, M., Lowe, D.G.: Scalable Nearest Neighbour Algorithms for High Dimensional Data. IEEE Trans. on PAMI **36**(11), 2227–2240 (2014)
8. Novak, D., Batko, M., Zezula, P.: Metric Index: An Efficient and Scalable Solution for Precise and Approximate Similarity Search. Information Systems **36**(4), 721–733 (2011)
9. Novak, D., Batko, M., Zezula, P.: Large-scale image retrieval using neural net descriptors. In: Proceedings of SIGIR 2015 (to appear, 2015)
10. Novak, D., Zezula, P.: Performance Study of Independent Anchor Spaces for Similarity Searching. The Computer Journal **57**(11), 1741–1755 (2014)
11. Novak, D., Zezula, P.: Rank aggregation of candidate sets for efficient similarity search. In: Decker, H., Lhotská, L., Link, S., Spies, M., Wagner, R.R. (eds.) DEXA 2014, Part II. LNCS, vol. 8645, pp. 42–58. Springer, Heidelberg (2014)
12. Oquab, M., Bottou, L., Laptev, I., Sivic, J.: Learning and transferring mid-level image representations using convolutional neural networks. In: CVPR (2014)
13. Wan, J., Wang, D., Hoi, S., Wu, P., Zhu, J., Zhang, Y., Li, J.: Deep learning for content-based image retrieval: a comprehensive study. In: Proc. of 22nd ACM International Conference on Multimedia (2014)

Textual Similarity for Word Sequences

Fumito Konaka and Takao Miura[✉]

Department of Advanced Sciences, HOSEI University,
3-7-2 KajinoCho, Koganei, Tokyo 184–8584, Japan
fumito.konaka.2t@stu.hosei.ac.jp, miurat@hosei.ac.jp

Abstract. In this work, we introduce new kinds of sentence similarity, called *Euclid similarity* and *Levenshtein similarity*, to capture both word sequences and semantic aspects. This is especially useful for Semantic Textual Similarity (STS) so that we could retrieve SNS texts, short sentences or something including collocations. We show the usefulness of our approach by some experimental results.

Keywords: Euclid similarity · Levenshtein similarity · Semantic textual similarity

1 Introduction

Nowadays there exist a variety of documents spread over internet. One of the typical examples is Social Networking Service (SNS), which is provided through some platform to build community or social relations among people sharing interests, activities, backgrounds or real-life connections in terms of messages, tweets or documents (*Wiki*). A social network service is provided using some mechanisms such as Blog and Twitter, some profiles and social links. These messages are characteristic because they consist of short texts chained many followers (or *retweets*), contain a few duplicate but special buzzwords (i.e., lol, :)) and ignore grammatical rules very often.

Information retrieval for these kinds of information have been widely discussed. Among others, a model of *Bag-of-Words* is common in text information processing. That is, every document is described as a vector over words with an assumption of Distributed Semantic Model (DSM) [5] which means words in similar contexts carry similar semantics. However, there exist serious deficiencies for SNS texts since sentences in SNS are generally short and sparse so that word sequences may carry characteristic semantics. For example, in two statements "I hope to marry her" and "I hope to divorce her", we have same words except one, thus the statements have completely different semantics.

In this work, we introduce new kinds of similarity, *Euclid similarity* and *Levenshtein similarity* between sentences to provide a new approach towards information retrieval of sentences including BLOG/Twitter. The basic idea comes from *semantic* distance. Euclid similarity allows us to obtain better similarity based on multiple word expression (as n-gram and collocation) considered

© Springer International Publishing Switzerland 2015
G. Amato et al. (Eds.): SISAP 2015, LNCS 9371, pp. 244–249, 2015.
DOI: 10.1007/978-3-319-25087-8_23

as a unit. They happen to co-occur often closely positioned. We also introduce semantic similarity into Levenshtein distance and the new difference of two words reflect the similarity. This provides us with independence of sentence length for similarity. In fact, we may have same context but much differnce of size[1] and we see the approach works better than dependent ones.

The rest of the paper is organized as follows. In section 2 we describe semantic similarity for sentences as well as some related works. Section 3 contains a framework of our approach including extended Levenshtein distance. Section 4 contains some experimental results. In section 5 we conclude this work.

2   Semantic Similarity

Semantic Textual Similarity (STS) provides us with new kinds of text retrieval. In fact, it allows us to capture semantic structure directly by means of word/phrase sense and the interrelationship same as grammatical structure instead of word distribution (Bag-Of-Words, BOW) model. For instance, *a small elephant looks at a big ant* is completely different in size from *a big elephant looks at a small ant*, but the two are same from the viewpoint of BOW model.

Here we like to focus attention on sequences as a new feature. The typical issue is *collocation*, which is a sequence of words or terms that co-occur more often than would be expected by chance (*WIKI*). Note that an *idiom* means a phrase carrying semantics different from the constituents[2], and that *collocation* means an expression of several words which likely happens more often[3].

On the contrary, we have different expression to describe identical situation. For example, two sentences "His lecture came across well." and "His lecture resonated well." talk about identical fact since come across means resonate though different length. We should examine words and collocation enough for powerful retrieval.

There have been several work of the similarity proposed so far. Tubaki et al. discuss a fundamental model based on word-vector space and sentence structures[5]. In fact, they examine model how to learn word description using sentence structure optimization and decomposition, and propose semantic similarity defined by kernel functions.

Islam et al. has proposed another similarity putting attention on word strings and word similarity[2]. By this approach *miss-spell* aspects could be involved.

Feng et al. has discussed similarity between sentences using *similarity of word sequences*[1]. The approach provides us with some improvement caused by short sentences, although they ignore collocation aspects. However, let us note that these approach show the results depending on the length (the number of words) heavily. Also no discussion is found about collocation.

---

[1] Some texts of different size talk about same content many times: "Please don't let this get you down", "Keep fighting and never give up", "Be strong". Love from Britain, Dec.12, 2014 in FaceBook message for Julia Lipnitskaya.

[2] "*I eat eyeball*" means "*I am scolded*" but not have any food.

[3] We may say "*something like that*" more likely as a custom.

3   Semantic Distance

In this section, we discuss how we should think about semantic distance between sentences putting attention on *sequence* of words and introduce 2 kinds of similarity, *Euclid similarity* and *Levenshtein similarity* to do that. Here we consider similarity as a certain value in $0 \sim 1$ and the smaller value means less similar.

3.1   Euclid Similarity

Assume word similarity $Sim_E(w_i, \bar{w}_i)$ for any two words w_i and $\bar{w}_i$ and we like to extend the definition for capturing sentence similarity between S_1, S_2. If both S_1, S_2 contain the same number (say, k) of words where each i-th word is w_i and $\bar{w}_i$ respectively. Then we define the similarity $Sim_E(S_1, S_2)$, called *Euclid Similarity*, as follows: $Sim_E(S_1, S_2) = \dfrac{\sqrt{\sum_{i=1}^{k} Sim(w_i, \bar{w}_i)^2}}{k}, w_i \in S_1, \bar{w}_i \in S_2.$

Next, let us define similarity of any two sentences using Euclid similarity. Note we assume the same number of words in two sentences in the definition of Euclid similarity. In a sentence S, we call consecutive n words in S by a *shingle*. To introduce the similarity, first we decompose the two sentences into the same number (k) of shingles, and then we give the similarity between the two shingles.

In English, it is well-known that any collocation (n-gram) may carry its own meaning with the length at most $n = 4$. We decompose S into the sequence of shingles of $n = 1, .., 4$ so that we have $s/4 \sim s$ shingles. To decompose two sentences S_1 and S_2 into k shingles, we obtain possible range of the decomposition and select the common possibility suitable for both S_1 and S_2.

We obtain two ranges I_1, I_2 for S_1, S_2 respectively and calculate a new range I_0 as $I_1 \cap I_2$ We say the similarity is 0 if no possibility is found (i.e., $I_0 = \phi$).

For each k in I_0, we decomopose S_1, S_2 into k shingles $(w_1, .., w_k)$ and $(w'_1, .., w'_k)$. Let $w, \bar{w}$ be two shingles and define the similarity $Sim_E(w, \bar{w})$ of the two shingles. Any shingle may or may not contain collocations which we can be see by examining dictionary. If the case, we put the constituent words together into one so that we still consider the shingle as a sequence of words. When we have several collocations in the shingle, we *make* copies of shingle containing different collocations to obtain similarity alternatively. Now the similarity $Sim_E(w, \bar{w})$ is defined as follows: $Sim_E(w, \bar{w}) = \max\limits_{i,j} Path(a_i, b_j)$. Here a_i, b_j mean each word/collocation in w and $\bar{w}$ respectively, none of the two is *stopword* or something like that[4]. $Path(a_i, b_j)$ is calculated using WordNet[5][4].

3.2   Levenshtein Similarity

Here let us introduce another kind of sentence similarity *Levenshtein similarity*, denoted by $Sim_L(S_1, S_2)$. Compared to Euclid similarity, we examine all the pairs of words appeared in S_1 and S_2 while keeping word sequences.

---

[4] In our approach, we extract only nouns and verbs so that, for instance, pronouns are removed.

[5] http://wordnet.princeton.edu/

For sentences S_1, S_2, we define *Levenshtein similarity* between S_1 and S_2, denoted by $Sim_L(S_1, S_2)$, as follows:

Levenshtein Similarity Calculate $Sim_L(S_1, S_2)$
(1) $M_{i,0} = i (0 \le i \le m), M_{0,j} = j (0 \le j \le n)$
(2) $M_{i,j} = \min($
$\qquad M_{i-1,j-1} + (1 - Path(w_i, \bar{w}_j)),$
$\qquad M_{i,j-1} + (1 - Path(w_i, \bar{w}_j)),$
$\qquad M_{i-1,j} + (1 - Path(w_i, \bar{w}_j)))$
(3) $Sim_L(S_1, S_2) = 1 - M_{m,n} / \max(m, n)$

In the definition we examine WordNet many times for whole sentence. This means our Levenshtein definition captures semantic similarity instead of character matching. Let us note we examine all the word pairs looking at WordNet.

3.3   Using Semantic Distances

Clearly it is hard to decide how well we obtain sentence similarity, because the result depends on contexts, domains and language nature. In this work, we model the situation by a parameter λ as well as the two similarities where $0 \le \lambda \le 1.0$:
$Sim(S_1, S_2) = \lambda \times Sim_E(S_1, S_2) + (1 - \lambda) \times Sim_L(S_1, S_2)$.

The parameter λ tells us how well sequences give similarity, it is impossible to estimate λ automatically. In the following section, we show some experimental results of our model.

4   Experiments

In this experiment, we examine 30 pairs of nouns among 65 pairs discussed in [3] and referred in [1], [2], [3]. Then we have interpreted these nouns with the *first* interpretation in the Collins Cobuild dictionary [6] and applied TreeTagger[7] for morphological processing in advance.

We examine extended semantic distance. We construct Levenshtein similarity distinguishing nouns from verbs by giving weight 0.5 for Euclid similarity, 0.3 for Levenshtein similarity on nouns and 0.2 for Levenshtein similarity on verbs. These values have been devised through preliminary experiments.

To evaluate the experiment, we examine precision of ranking proposed by [3] with two baseline results [1], [2].

Let us illustrate all the results of Precisions and Ranking in tables 1 and 3 respectively. In table 1, TopRank means the number of pairs ranked highly. A table 1 shows that, compared to our result, Islam approach works 10 percent worse and Feng et al. approach equally.

In table 3, we see that our approach contains a pair of "coast&forest" in top 10 rank which is 21th in Answer, and that our approach doesn't contain any pair in top 10 rank which is 21th or below in Answer.

---

[6] http://www.collinsdictionary.com/dictionary/english-cobuild-learners
[7] http://www.cis.uni-muenchen.de/~schmid/tools/TreeTagger/

Table 1. Precision

TopRank	Ours	Islam	Feng
1	1	0	1
5	0.8	0.6	0.8
10	0.6	0.8	0.6
15	0.6	0.8	0.67
20	0.7	0.85	0.75
25	0.88	0.96	0.88
30	1	1	1
Average	0.8	0.72	0.81

Table 2. Unified Explanatory Notes

coast & shore		
Position	"coast"	"shore"
1	the	the, shore, or the
2	coast	shore, of, a, sea
3	is	lake, or, wide, river
4	an, area	is, the, land, along
5	of, land	the, edge, of, it
6	that, is	some one, who, is_on, shore
7	next, to	is_on, the, land, rather
8	the, sea	than, on, a ship
coast & forest		
Position	"coast"	"forest"
1	the	a, forest, is
2	coast, is, an, area	a, large, area
3	of, land, that, is	where, trees
4	next, to, the, sea	grow, close, together

Table 3. Ranking

Rank	Answer	Ours	Islam	Feng
1	midday&noon	midday&noon	cock&rooster	midday&noon
2	cock&rooster	cock&rooster	midday&noon	cock&rooster
3	cemetery&graveyard	serf&slave	gem&jewel	cemetery&graveyard
4	gem&jewel	forest&woodland	boy&lad	gem&jewel
5	forest&woodland	gem&jewel	automobile&car	boy&lad
6	coast&shore	cemetery&graveyard	implement&tool	cord&string
7	implement&tool	coast&forest	cemetery&graveyard	serf&slave
8	boy&lad	boy&rooster	cord&string	automobile&car
9	automobile&car	journey&voyage	coast&shore	grin&smile
10	cushion&pillow	automobile&car	serf&slave	boy&rooster
11	grin&smile	hill&woodland	journey&voyage	boy&sage
12	serf&slave	boy&lad	magician&wizard	magician&wizard
13	cord&string	magician&oracle	forest&graveyard	journey&voyage
14	autograph&signature	grin&smile	grin&smile	asylum&fruit
15	journey&voyage	magician&wizard	furnace&stove	magician&oracle
16	magician&wizard	boy&sage	cushion&pillow	coast&shore
17	furnace&stove	autograph&signature	hill&woodland	autograph&signature
18	hill&mound	forest&graveyard	glass&tumbler	cushion&pillow
19	oracle&sage	coast&shore	coast&forest	furnace&stove
20	hill&woodland	cord&string	forest&woodland	autograph&shore
21	glass&tumbler	implement&tool	magician&oracle	glass&tumbler
22	coast&forest	autograph&shore	autograph&signature	forest&woodland
23	magician&oracle	glass&tumbler	boy&rooster	implement&tool
24	boy&rooster	asylum&fruit	boy&sage	forest&graveyard
25	forest&graveyard	furnace&stove	hill&mound	hill&woodland
26	boy&sage	cushion&pillow	bird&woodland	oracle&sage
27	cord&smile	oracle&sage	autograph&shore	hill&mound
28	asylum&fruit	bird&woodland	oracle&sage	bird&woodland
29	bird&woodland	hill&mound	asylum&fruit	cord&smile
30	autograph&shore	cord&smile	cord&smile	coast&forest

To examine our approach and the baselines, let us discuss the differences. In table 3, our approach says "coast & shore" and "coast & forest" are ranked as 19th and 7th respectively which are 6th and 22th in Answer.

Explanatory notes (in the Collins) of the words "coast", "shore" and "forest" are *The coast is an area of land that is next to the sea."*, *"The shores or the shore of a sea, lake, or wide river is the land along the edge of it. Someone who is on shore is on the land rather than on a ship."* and *"A forest is a large area where trees grow close together."* respectively.

First we see big difference of the notes length of "coast & shore". Table 2 contains the unified result of the notes. We examine "shore" and obtain collocation, in fact, we have Paths values in similarities. $Path(coast, shore) = 0.5$, $Path(area, land) = 0.08$, $Path(sea, ship) = 0.08$. Note that 1st, 3rd, 5th, 6th and 7th have similarity 0.

As for "coast & forest", we get almost same lengh of the explanatory notes (of "coast" and "forest"). Again a table 2 contains the unified result of the notes. Also Path values show $Path(area, area) = 1$ and $Path(land, tree) = 0.17$. The 1st and 4th words contain similarity 0.

Similarly Path values see that "land & tree" are more similar rather than "area &land" and "sea & ship". That's why "coast & shore" is ranked lower and "coast & forest" higher. Though the result heavily depends on the notes (in the Collins), we might expect our approach generally collects possible notation fluctuations about focused terms in a sentence.

5   Conclusion

In this work, we introduced two kinds of similarity, Euclid similarity and Levenshtein similarity to model sequence and semantics of words for the purpose of STS and short text retrieval. Then we introduced semantic similarity between sentences.

Our experimental result shows that the precision results are generally nice results, say 10 percent better than Islam[2] in top 10 pairs, for example.

References

1. Feng, J., Zhou, Y.M., Martin, T.: Sentence similarity based on relevance. In: Proceedings of IPMU, vol. 8, p. 833 (2008)
2. Islam, A., Inkpen, D.: Semantic text similarity using corpus-based word similarity and string similarity. ACM Transactions on Knowledge Discovery from Data (TKDD) **2**(2), 10 (2008)
3. Li, Y., McLean, D., Bandar, Z.A., O'shea, J.D., Crockett, K.: Sentence similarity based on semantic nets and corpus statistics. IEEE Transactions on Knowledge and Data Engineering **18**(8), 1138–1150 (2006)
4. Pedersen, T., Patwardhan, S., and Michelizzi, J.: WordNet: similarity: measuring the relatedness of concepts. In: Demonstration Papers at HLT-NAACL 2004, pp. 38–41. Association for Computational Linguistics, May 2004
5. Tsubaki, M., Duh, K., Shimbo, M. and Matsumoto, Y.: Modeling and learning semantic co-compositionality through prototype projections and neural networks. In: Conf. on Empirical Methods in Natural Language Processing (EMNLP) (2013)

Motion Images: An Effective Representation of Motion Capture Data for Similarity Search

Petr Elias, Jan Sedmidubsky$^{(\boxtimes)}$, and Pavel Zezula

Masaryk University, Botanicka 68a, 602 00 Brno, Czech Republic
xsedmid@fi.muni.cz

Abstract. The rapid development of motion capturing technologies has caused a massive usage of human motion data in a variety of fields, such as computer animation, gaming industry, medicine, sports and security. These technologies produce large volumes of complex spatio-temporal data which need to be effectively compared on the basis of similarity. In contrast to a traditional way of extracting numerical features, we propose a new idea to transform complex motion data into RGB images and compare them by content-based image retrieval methods. We see these images not only as human-understandable visualization of motion characteristics (e.g., speed, duration and movement repetitions), but also as descriptive features for their ability to preserve key aspects of performed motions. To demonstrate the usability of this idea, we evaluate a preliminary experiment that classifies 1,034 motions into 14 categories with the 87.4 % precision.

1 Introduction and Related Work

Motion capturing devices, such as Microsoft Kinect, ASUS Xtion, OptiTrack or Vicon, can interactively digitize human movement into complex spatio-temporal *motion capture (mocap) data*. The mocap data describe human movements by simplified stick figures of human skeletons. The stick figure consists of bones that are connected by joints. The positions of joints are estimated for each video frame in form of 3D coordinates. The captured motion data have a great potential to be automatically processed, e.g., in sports to compare performance of athletes, in law-enforcement to identify special-interest persons or detect suspicious events, in health care to determine the success of rehabilitative treatments or detect disorders in musculoskeletal system, or in computer animation to synthesize and generate realistic human motions for production of high-quality games or movies.

In any case, these potential applications require to efficiently and effectively analyze, segment, search and classify complex spatio-temporal motion data on the basis of *similarity*. Determining similarity of such data is a nontrivial task which can even require an expertise when evaluated by humans, e.g., referees and trainers in sports or physiotherapists and doctors in medicine. To determine similarity of motions automatically and as much effectively as possible, various kinds of motion *features* are usually extracted and compared on the basis of distance-based functions or machine-learning approaches.

© Springer International Publishing Switzerland 2015
G. Amato et al. (Eds.): SISAP 2015, LNCS 9371, pp. 250–255, 2015.
DOI: 10.1007/978-3-319-25087-8_24

Motion features can be extracted in form of joint angles (rotations) [13], distances between joints [12,17], relative joint velocity and acceleration [14], joint trajectories [3], absolute and relative 3D joint coordinates [16] or covariance matrices of subsequent frames [6,15]. Depending on the application, various features can be also combined [1,2] or only a key subset of descriptive joints [10] can be considered. The extracted features are then usually represented as sequences of high-dimensional vectors which can be compared by (1) distance-based functions like the dynamic time warping and its variants [10] and (2) machine-learning approaches such as the support vector machines [1,4,6], neural networks [3] or hidden Markov models [7].

A large diversity of motion features and comparison methods supports the fact that there is no global-winner similarity model which would be usable for a wide range of applications, e.g., for recognizing person-related aspects such as gender [5] as well as for recognizing specific movement actions [10] like running or fighting. In this paper, we propose a completely different idea which effectively visualizes mocap data into color images and compares them by standard content-based image retrieval (CBIR) methods. In this way we transform a hard problem of computing similarity of spatio-temporal motion data into a well-known problem of similarity of images. A related idea is proposed by Milovanovic et al. [8] who visualize mocap data in a similar fashion but for very specific application of gait recognition. We rather focus on a broader scope of movements ranging from simple gestures to complex performances. The main difference is that we make the body position and orientation relative for each skeleton configuration to efficiently utilize the used color space. The relative position and body orientation allow us to visualize the same movements by the same colors disregarding their placement within a motion. The power of the proposed concept is experimentally verified on the action-recognition application.

2   Motion Image: Motion Capture Data as an Image

We propose an idea to transform mocap data of a single motion into a color image. The transformation objective is to represent different motions, such as walking, jumping or kicking, as visually distinctive images. The whole transformation idea is mainly motivated by the following facts:

– Visualization of mocap data provides humans with better understanding of motion characteristics, such as speed, duration and movement repetitions;
– Comparison of images based on their visual similarity is a known concept nowadays – many CBIR methods, e.g., methods based on MPEG-7, SIFT, SURF and DeCAF features, might be employed;
– Image representation is more robust as it preserves key aspects of original mocap data, in contrast to traditional high-dimensional numerical features such as joint angle rotations, velocities and distances between joints.

Regardless of motion capturing device or data storing format, we define every *motion* m as a sequence $(p_1, \ldots, p_n)$ of *poses* p_i ($i \in [1, n]$) where n equals to the

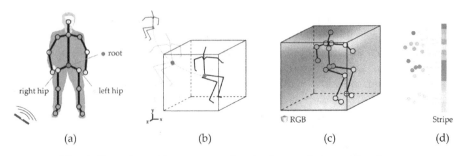

Fig. 1. Transformation process of a single pose into a stripe image.

motion length (i.e., the number of captured frames). Each pose p_i is a vector $(c_1, \ldots, c_l)$ of 3-dimensional real-world coordinates $c_i \in \mathbb{R}^3$ ($i \in [1,l]$) of the specific body *joints*, where l denotes the number of these joints. Fig. 1a illustrates a simplified body model with only $l = 17$ joints.

We firstly show how to transform a single pose p_i into a "stripe" image and then present transformation of the whole motion m by concatenating stripe images of individual poses $p_1, \ldots, p_n$. A pose is transformed into the stripe image by positioning the human skeleton into a 3-dimensional cube, converting the positioned skeleton into RGB color space, and visualizing the converted skeleton as the stripe image. The whole process is described in the following four steps.

1. **Centering** – the skeleton is centered within a 3-dimensional cube $[0..255]^3$ of a base of 256 centimeters. This size is considerably high enough to envelop the human skeleton performing an arbitrary movement. The skeleton is centered by positioning the *root* joint (c_1) in the middle of such cube, i.e., at the position $[128, 128, 128]$.
2. **Rotation** – after centering, the skeleton is rotated within the cube around y-axis in such a manner that the subject faces the positive x-axis. The angle of rotation is determined by positioning the left and right hip joints parallel to the z-axis. Fig. 1b graphically illustrates the process of centering and rotating the skeleton within the 3-dimensional cube.
3. **RGB conversion** – the centered and rotated skeleton is converted into RGB color space $[0..255]^3$ which perfectly suits for this purpose. The RGB space is defined by mixing red, green and blue color, each of them with domain of exactly 256 values. Thus we can easily obtain particular color for every skeleton joint based on its location within the cube – see Fig. 1c.
4. **Visualization** – the converted skeleton within the RGB cube is visualized by the *stripe image* of $l \times 1$ pixels, where $l = 31$ corresponds to the number of all joints. Pixel $[i, 1]$ ($i \in [1,l]$) represents the location color of i-th joint in the RGB cube. The example of the stripe image is shown in Fig. 1d.

The advantage of stripe images is that they preserve enough information to be easily reverted to relative coordinates of the original mocap data with error approximately up to 1 cm. Moreover, the skeleton rotation ensures the construction of the

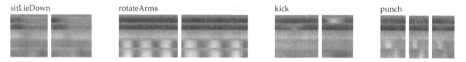

Fig. 2. 9 motion images belonging to 4 different action categories. For example, three repetitions of the "rotateArms" movement can be easily recognized by humans.

same stripe images for the same poses performed in different directions with respect to real-world coordinates.

To transform a whole motion into a single image, we simply concatenate stripe images of its individual poses in the same order. In particular, motion $m = (p_1, \ldots, p_n)$ is processed to construct n stripe images which are gradually concatenated into a *motion image* of $l \times n$ pixels. Rows of the resulting motion image represent how the position of individual joints changes over time and columns constitute the skeleton configuration at a given time. Figure 2 shows examples of generated motion images representing four different actions.

3 Similarity of Motion Images and Its Evaluation

Once original mocap data are encoded within a single image, we might transform a challenging problem of computing similarity of spatio-temporal motion data into a problem of similarity of images. To determine similarity of motion images, each image has to be firstly preprocessed in order to extract its descriptive content-based features, e.g., MPEG-7, SIFT, SURF and DeCAF. The extracted image features can be then compared by distance functions to determine their similarity. In this work, we choose standard 282-dimensional MPEG-7 visual features (scalable color, color structure, color layout, edge histogram and homogeneous texture) that are compared by a weighted aggregation distance function combining the Manhattan and Euclidean distances. The more details about the MPEG-7 features and the comparison function are available in [11].

We demonstrate the usability of the proposed concept of motion images on the action-recognition application. The purpose of such application is to determine what kind of movement activity the person performs within a query motion. We evaluate this scenario on the HDM05 [9] dataset that provides an action-level ground truth. This ground truth classifies 1,034 motion actions into 14 categories – see Fig. 3. All the 1,034 actions have been transformed into motion images from which MPEG-7 visual features have been extracted. The experiment has been evaluated by taking the extracted features of each motion as a query and searching for its nearest neighbor. If the nearest neighbor is assigned the same category as the query motion, the query is considered to be successful. The precision of the experiment is measured as a ratio between the number of successful queries and the number of all (1,034) queries.

As a preliminary result we achieve the precision of 87.4 % by classifying all the 1,034 motions. Such high classification rate contributes to the fact that the

Action (# of samples)	id	1	2	3	4	5	6	7	8	9	10	11	12	13	14
cartwheel (6)	1	100													
grabDepR (105)	2		89				4			3					5
kick (49)	3		4	65	6				4	4	4		2		10
punch (48)	4			2	98										
rotateArms (46)	5					98								2	
sitLieDown (43)	6						91	9							
standUp (43)	7		5				16	70	2						7
throwR (23)	8		9	4	4				9	61		4			9
jump (25)	9									84	12				4
hopOneLeg (18)	10		6						6	6	83				
neutral (75)	11											68	1		31
tpose (198)	12												100		
exercise (19)	13			5		5				5	5			79	
turn (336)	14												11		89

Fig. 3. Confusion matrix denoting the classification precision (in %) of 1,034 motions.

proposed concept of motion images constitutes a promising idea for determining similarity of mocap data. On the other hand, the biggest problem can be observed with the "neutral" category where 31 % of actions are misclassified into the "turn" category. This is caused by rotating each motion pose into a fixed direction within the RGB cube, thus the "turn" actions look similar to neutral movements. Also 16 % of actions in the "standUp" category is misclassified into the "sitLieDown" category as both categories produce very similar images with colors just in reverse order. The incorrectly classified cases bring new challenges to utilize enhanced RGB transformations and advanced image retrieval methods.

4   Conclusions and Future Research Directions

We present a new concept of visualization and similarity for mocap data. Contrary to state-of-the art approaches that extract specific high-dimensional numerical features, we transform mocap data into image representations by loosing original information only about the absolute skeleton position and orientation. By comparing transformed images on the basis of only MPEG-7 visual features, we correctly classify 87.4 % motions (out of 1,034 samples) into 14 categories. Motion images can also serve as human-understandable visualization from which interesting characteristics, such as movement repetitions, can be easily observed.

We see a great potential for further improvement in using sophisticated image features or domain-specific features trained by reinforcement learning methods. We intend to inspect how various motion normalizations, different visualizations, and various perceptually uniform color spaces affect the quality of similarity. We also plan to compare effectiveness of the proposed solution against state-of-the-art approaches for mocap data classification.

Acknowledgments. This research was supported by GBP103/12/G084.

References

1. Baumann, J., Wessel, R., Krüger, B., Weber, A.: Action graph: a versatile data structure for action recognition. In: International Conference on Computer Graphics Theory and Applications (GRAPP 2014). SCITEPRESS (2014)
2. Chen, X., Koskela, M.: Classification of RGB-D and motion capture sequences using extreme learning machine. In: Kämäräinen, J.-K., Koskela, M. (eds.) SCIA 2013. LNCS, vol. 7944, pp. 640–651. Springer, Heidelberg (2013)
3. Cho, K., Chen, X.: Classifying and visualizing motion capture sequences using deep neural networks. CoRR abs/1306.3874 (2013)
4. Evangelidis, G., Singh, G., Horaud, R.: Skeletal quads: human action recognition using joint quadruples. In: 22nd International Conference on Pattern Recognition (ICPR 2014), pp. 4513–4518 (2014)
5. Hu, M., Wang, Y., Zhang, Z., Wang, Y.: Combining spatial and temporal information for gait based gender classification. In: International Conference on Pattern Recognition (ICPR 2010), pp. 3679–3682. IEEE Computer Society (2010)
6. Hussein, M.E., Torki, M., Gowayyed, M.a., El-Saban, M.: Human action recognition using a temporal hierarchy of covariance descriptors on 3D joint locations. In: Joint Conference on Artificial Intelligence (IJCAI 2013), pp. 2466–2472 (2013)
7. Liang, Y., Lu, W., Liang, W., Wang, Y.: Action recognition using local joints structure and histograms of 3D joints. In: Computational Intelligence and Security (CIS), pp. 185–188 (2014)
8. Milovanovic, M., Minovic, M., Starcevic, D.: Walking in colors: Human gait recognition using kinect and CBIR. IEEE MultiMedia **20**(4), 28–36 (2013)
9. Müller, M., Röder, T., Clausen, M., Eberhardt, B., Krüger, B., Weber, A.: Documentation Mocap Database HDM05. Tech. Rep. CG-2007-2, Universität Bonn (2007)
10. Müller, M., Baak, A., Seidel, H.P.: Efficient and robust annotation of motion capture data. In: ACM SIGGRAPH/Eurographics Symposium on Computer Animation (SCA 2009), p. 10. ACM Press (2009)
11. Salembier, P., Sikora, T.: Introduction to MPEG-7: Multimedia Content Description Interface. John Wiley & Sons Inc., New York (2002)
12. Sedmidubsky, J., Valcik, J., Balazia, M., Zezula, P.: Gait recognition based on normalized walk cycles. In: Bebis, G., Boyle, R., Parvin, B., Koracin, D., Fowlkes, C., Wang, S., Choi, M.-H., Mantler, S., Schulze, J., Acevedo, D., Mueller, K., Papka, M. (eds.) ISVC 2012, Part II. LNCS, vol. 7432, pp. 11–20. Springer, Heidelberg (2012)
13. Sedmidubsky, J., Valcik, J., Zezula, P.: A key-pose similarity algorithm for motion data retrieval. In: Blanc-Talon, J., Kasinski, A., Philips, W., Popescu, D., Scheunders, P. (eds.) ACIVS 2013. LNCS, vol. 8192, pp. 669–681. Springer, Heidelberg (2013)
14. Thanh, T.T., Chen, F., Kotani, K., Le, B.: Automatic extraction of semantic action features. In: Signal-Image Technology & Internet-Based Systems (SITIS 2013), pp. 148–155. IEEE Computer Society (2013)
15. Vieira, A., Lewiner, T., Schwartz, W., Campos, M.: Distance matrices as invariant features for classifying mocap data. In: 21st International Conference on Pattern Recognition (ICPR 2012), pp. 2934–2937 (2012)
16. Wang, J., Liu, Z., Wu, Y., Yuan, J.: Mining actionlet ensemble for action recognition with depth cameras. In: International Conference on Computer Vision and Pattern Recognition (CVPR 2012), pp. 1290–1297. IEEE Computer Society (2012)
17. Zhao, X., Li, X., Pang, C., Zhu, X., Sheng, Q.Z.: Online human gesture recognition from motion data streams. In: 21st International Conference on Multimedia (MM 2013), pp. 23–32. ACM (2013)

Implementation and Engineering Solutions

Brute-Force k-Nearest Neighbors Search on the GPU

Shengren Li and Nina Amenta[(✉)]

University of California, Davis, USA
amenta@cs.ucdavis.edu

Abstract. We present a brute-force approach for finding k-nearest neighbors on the GPU for many queries in parallel. Our program takes advantage of recent advances in fundamental GPU computing primitives. We modify a matrix multiplication subroutine in MAGMA library [6] to calculate the squared Euclidean distances between queries and references. The nearest neighbors selection is accomplished by a truncated merge sort built on top of sorting and merging functions in the Modern GPU library [3]. Compared to state-of-the-art approaches, our program is faster and it handles larger inputs. For instance, we can find 1000 nearest neighbors among 1 million 64-dimensional reference points at a rate of about 435 queries per second.

1 Introduction

Many important operations in data science involve finding nearest neighbors for each element in a query set Q from a fixed set R of high-dimensional reference points. The k-nearest neighbors problem takes sets Q and R as input, and a constant k, and returns the k nearest neighbors (kNNs) in R for every $q \in Q$. In this paper we consider the high-dimensional version of this problem and we give a state-of-the-art implementation of a brute-force GPU algorithm.

High-dimensional data may be structured data with many variables, but it also arises as long feature vectors derived from unstructured data such as text, images, video, time-series or shapes. Finding nearest-neighbors is the first step in using kernel and non-parametric regression to interpolate functions over the data [8,26]. When learning classifiers, a nearest-neighbor algorithm [17] is often the most accurate predictor in practice, especially in well-designed feature spaces [13,15]. An important topic of continuing research is using the nearest-neighbor algorithm with a distance function chosen specifically to improve the classification accuracy on a particular reference data set R. One successful group of algorithms in this area [21,27,47–49] chooses the distance function locally for each query q, based on a large set of nearest neighbors in R. Our brute-force algorithm would be particularly good in this situation, since it easily handles large values of k as well as large R.

High-dimensional nearest-neighbor search suffers from the "curse of dimensionality" [14]. This makes it impossible to construct index data structures of reasonable size on R that can answer a nearest-neighbor query exactly in time

© Springer International Publishing Switzerland 2015
G. Amato et al. (Eds.): SISAP 2015, LNCS 9371, pp. 259–270, 2015.
DOI: 10.1007/978-3-319-25087-8_25

sub-linear in $n = |R|$, not only in the worst case but also in many reasonable definitions of average case. Sub-linear solutions even to the approximate version of the problem are surprisingly difficult, and only in recent years have algorithms, most based on Locality Sensitive Hashing [19], provided provable worst-case sub-linear query times using polynomial-sized index structures. Thus brute-force approaches remain an important part of the solution space.

The GPU, with its massive SIMD parallelism, is well-suited to brute-force approaches, providing exact worst-case results at the rate of a couple of milliseconds per query for moderately-sized problems (e.g., a few million reference points). As GPU speed and memory size continues to increase — AMD recently released a 32GB GPU — the problem sizes appropriate for the GPU increase as well. Large problems will always have to be handled from disk [28], but even there, hybrid CPU-GPU implementations [32,46] rely on the GPU to solve large subproblems by brute-force.

The efficiency of brute-force GPU implementations can themselves vary greatly, particularly with respect to the optimization of data movement through the memory hierarchy. Using better libraries for common operations such as sorting and matrix multiplication can easily improve performance by an order of magnitude over naive implementations. Our brute-force implementation makes heavy use of recent highly optimized CUDA libraries.

Any brute-force implementation consists of two steps. First, we compute a matrix $d^2(Q, R)$ giving the squared distance of each $q \in Q$ to each $r \in R$. To implement this step, we modify the inner loop of a well-optimized open source matrix multiplication kernel, the SGEMM kernel in MAGMA library [6]. In the second step, for each query, we search its row of the matrix to find the k smallest squared distances. There is considerable variation in how this step can be carried out in brute-force implementations. Our implementation uses the *merge-path* function from the Modern GPU library [3], which has proved to be very useful in other contexts, to implement a truncated merge sort, in which only the k smallest items move forward from one level of merging to the next.

Together, these two steps form a CUDA program, that, to the best of our knowledge, is currently the fastest kNN implementation on the GPU. Our code scales linearly in $m = |Q|$, $n = |R|$, the dimension d, and k, and unlike other codes, it handles large values of k (up to $k = 3000$). We compare our implementation to the two recent published algorithms for which code is available, cuknns [1] and kNN CUDA [2], and to an implementation with the segmented sort function in the Modern GPU library.

2   Related Work

There are many GPU approaches to brute-force kNN, applying different strategies for the two major components of the algorithm.

Squared Distance Matrix: The two main existing approaches to computing the squared distance matrix are to implement it directly with a custom kernel

[9,23,24,29,30,33,35–37], or to derive the distances from an already well-optimized matrix multiplication routine [10,18,24,31,45]. Custom direct implementations are typically optimized by *tiling*, which divides the distance matrix into equal-sized submatrices (or tiles) and then assigns a thread block to each. The tile size is set so that a group of query and reference points can be accommodated in the fast shared memory and reused by threads within the same block.

The matrix multiplication approach to computing the squared distance matrix $d^2(Q, R)$ is based on the equation

$$d^2(Q, R) = N_Q + N_R - 2Q^T R, \tag{1}$$

where the elements of the ith row of N_Q are $\|Q_i\|^2$, and the elements of the jth column of N_R are $\|R_j\|^2$. These can be computed using custom CUDA kernels [18,24] or Thrust library [7] primitives [31]. A matrix multiplication routine from a highly optimized library, e.g., cuBLAS [4] calculates the more expensive third term, and the speed of the highly optimized library routine compensates for the additional arithmetic operations.

Selecting Nearest Neighbors: The approaches for selecting nearest neighbors are more diverse. Kuang and Zhao [33] simply sort all the distances to each query using GPU radix sort; this relies on the speed of modern sorting libraries. Dashti et al. [18] use radix sort as well, but on the entire matrix. The candidate distances are first sorted all together and then stably sorted by query index to separate the results for each query. Kato and Hosino [29,30] build a max-heap for each query and parallel threads push new candidates to the heap using atomic operations. Beliakov and Li [12] calculate the kth smallest distance to each query directly using a GPU selection algorithm [11] based on Kelley's cutting plane method, a convex optimization technique.

Many approaches divide the distances to each query into blocks. Liang et al. [35–37] find the local kNN within each block by testing each distance against all the others in parallel; a single thread per query then merges the lists. Arefin et al. [9] maintain an unsorted array of size k for each query and a pointer to the largest element in the array. A single thread maintains this structure at each level with a linear scan.

Several other approaches use a parallel reduction pattern, that is, a hierarchical pattern of comparisons. Barrientos et al. [10] create multiple heaps for each query and then merge the heaps at each level. Miranda et al. [39] choose the kNN at each level using quickselect. Komarov et al. [31] also use quickselect, implemented with the CUDA warp vote function __ballot(), bit count function __popc() and bit shift operations.

Truncated sort was introduced by Sismanis et al. [45]. Elements are discarded from the sort when it is clear that they cannot belong to the smallest k. They describe several algorithms, and show that their truncated bitonic sort has outstanding performance on the GPU. Garcia et al. [23,24] use a truncated insertion sort.

Besides brute-force approaches, some of the asymptotically more efficient approximate kNN algorithms have been implemented on the GPU. Pan et al. [42–44] and Lukac et al. [38] construct variants of Locality Sensitive Hashing. The running times of these methods are competitive with existing brute-force implementations, but they return approximate results; like the brute-force approach, the main bottleneck is the selection of the kNNs from a large set of candidates from R, so our techniques may be useful in implementing these approaches as well. There are also heuristic techniques that use various kinds of filtering to try and avoid computing the entire squared distance matrix $d^2(Q, R)$ [16,20,46].

3   Implementation

Let $m = |Q|$ be the size of the list Q of query points and let $n = |R|$ be the number of reference points. In the input, the query and reference lists are organized as $d \times m$ and $d \times n$ matrices, where d is the dimension. These matrices are stored as row-major 1D arrays, so that, for each dimension i, the ith components of all the points are contiguous; this facilitates coalesced access to global memory.

In the squared distance matrix $d^2(Q, R)$, we represent each of the distances as a 64-bit integer, as follows. The high 32 bits contain the floating point distance between the reference and the query, and the low 32 bits contain the integer index of the reference point. When merging lists of kNNs for a particular query, this composite representation allows us to swap the positions of two candidate distances by swapping two 64-bit integers instead of swapping the distances and indices separately.

3.1   Computing the Squared Distance Matrix

We leverage the efficiency of GPU matrix multiplication, which is a very well-studied operation, to compute $d^2(Q, R)$. Listings 1.1 and 1.2 compare the computation of the squared Euclidean distance matrix and matrix multiplication. The only difference between them is in the innermost loop.

Listing 1.1. Squared Euclidean distances

```
for i = 0 to m-1
  for j = 0 to n-1
    distance[i,j] = 0
    for k = 0 to d-1
      diff = Q[k,i] - R[k,j]
      distance[i,j] += diff * diff
```

Listing 1.2. Dot products

```
for i = 0 to m-1
  for j = 0 to n-1
    product[i,j] = 0
    for k = 0 to d-1
      product[i,j] += Q[k,i] * R[k,j]
```

Our computation of $d^2(Q, R)$ is a modification of a very efficient CUDA matrix multiplication kernel [6,22,34,40], replacing the internal loop with the squared Euclidean distance computation and then combining the resulting squared distance with the index of the reference point $r \in R$ to generate the 64-bit candidate representation described above.

This distance computation inherits a number of optimizations from the matrix multiplication kernel. The most important is tiling. The squared distance matrix $d^2(Q, R)$ is divided into tiles of size $m_{blk} \times n_{blk}$. The input for computing a tile is a $d \times m_{blk}$ stripe of Q and a $d \times n_{blk}$ stripe of R. Each tile is processed by a block of threads. These data chunks are loaded into shared memory in a coalesced fashion and reused by threads within the same thread block. The tile size is tuned to the Fermi architecture.

Since the introduction of the Fermi architecture, accessing data in registers is much faster than accessing data from shared memory. To take advantage of this, one more level of tiling is employed at the thread level. Each thread computes a $m_{thd} \times n_{thd}$ matrix with stride m_{blk}/m_{thd} and n_{blk}/n_{thd}. For each dimension, n_{thd} values are loaded from shared memory to registers and reused to compute all $m_{thd} \times n_{thd}$ partial results.

The kernel also uses loop unrolling and double buffering [34]. Loop unrolling replaces a loop with a single block of straight-line code. Not only is the cost of looping eliminated, but also more instruction level parallelism can be obtained by the compiler. Double buffering takes advantage of the Fermi GPU's dual-issue architecture. It overlaps the arithmetic operations of the current iteration with the memory operations of the following iteration.

Other brute-force kNN search implementations [10,18,24,31,45] take advantage of fast GPU matrix multiplication, but they use it as a subroutine as described in Section 2. Clearly, our approach saves both memory and computation time. The only drawback is that we can only use it with open source matrix multiplication codes. Fortunately, MAGMA [6] is competitive with proprietary matrix multiplication kernels (see Section 4).

3.2   Selecting Nearest Neighbors

Overview: A naive approach to finding the k-nearest neighbors for each query would sort the n candidates by distance and then return the first k. Following Sismanis et al. [45], we use a truncated sorting algorithm instead, which discards candidates as it becomes clear that they cannot belong to the top k.

The truncated merge sort is designed to use the GPU shared memory efficiently. In the first stage, we divide the n candidates of a query into chunks of size at least k that fit into shared memory, and sort each chunk in parallel with a block of threads. In the second stage, we iteratively merge pairs of sorted chunks and discard the larger half of each pair, so that the number of sorted chunks in play decreases by a factor of two at each iteration (notice that this property lets the k-nearest neighbors of one query be found in $O(n)$ work). The second stage stops when only one chunk is left, which contains the k-nearest neighbors. In both the sorting and the merging stages, the operations for different queries are executed in parallel, the sorts and merges on the different chunks of each query are executed in parallel, and the chunk-level sorts and merges are themselves parallel operations.

Merge Path: We use the Merge Path algorithm [3,25,41] for both sorting and merging. In this section we briefly describe Merge Path and why it is so efficient.

Let a and b be the two input arrays, sorted from smallest to largest; for simplicity assume all elements are unique. Let $s(i, j)$ denote the set consisting of the first i elements from a and the first j elements from b: $a[0] \ldots a[i-1]$ and $b[0] \ldots b[j-1]$. Finally, define the list S_p of possible choices of $s(i,j)$ such that $i+j = p$, ordered by i. For instance, if $a = [2, 5, 11, 13]$ and $b = [3, 8, 12, 17]$, we get $S_3 = [[3, 8, 12], [2, 3, 8], [2, 3, 5], [2, 5, 11]]$. The correct first p elements of the output has to be one of the items of S_p ($[2, 3, 5]$ in the example); call this s_p.

Now consider mapping the function $f(i, j) = b[j-1] < a[i]$ over S_p, where we define $b[-1] = -\infty$. In the example, we get $[12 < 2 = \text{False}, 8 < 5 = \text{False}, 3 < 11 = \text{True}, -\infty < 13 = \text{True}]$. And in fact, $f(i, j)$ is always False to the left of s_p and True at and to the right of s_p [41]. So we can find s_p by binary search on the Boolean array $f(S_p)$, computing only the items of $f(S_p)$ that we need to evaluate. Once we know s_p, we can break the problem of merging a and b into two independent parts, one merging the first p output elements and the other merging the rest.

In fact, we break the problem into several independent parts. Assuming that there are r processors, we evenly divide the output array c into non-overlapping segments of size $l = \frac{|a|+|b|}{r}$. Processor x finds s_{lx} and then generates the output between positions lx and $l(x+1) - 1$. Each processor works independently of the others.

Merge Path works well on the GPU because it divides the work into roughly balanced subtasks. Choosing the size of the subtasks is the key tuning parameter; choosing r too large increases the number of subproblems and allows the binary search to dominate, while choosing r too small allows the sequential merges to dominate and fails to create enough work for all the processors.

Using MGPU: Modern GPU (MGPU) [3] is a library of high-performance CUDA primitives, including Merge Path, that takes advantage of parallelism at both the kernel and thread block levels. We demonstrate that the MGPU primitives, particularly Merge Path, lead to a very efficient nearest-neighbors selection algorithm.

In the first (sorting) stage of the selection algorithm, each thread block loads a chunk of nearest neighbor candidates into shared memory and then calls `mgpu::CTAMergesort` to sort them. In `mgpu::CTAMergesort`, each thread first sorts a small number of candidates in registers. Next, the sorted arrays are merged (still in shared memory) using a parallel reduction pattern. Each merge operation is done with Merge Path. In the first step of the reduction, two threads work on merging each pair of arrays. As the array length doubles, so does the number of cooperating threads per array, so that each sequential merge operation ends up handling the same number of items (determined by the parameter r, above).

At each iteration of the second (merging) stage, each thread block loads two of the sorted chunks into shared memory and then uses the Merge Path algorithm. Just the smaller half of the output c will be stored back to global memory, so we only need to assign threads to construct the first half of the output array.

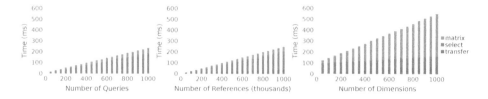

Fig. 1. Total running time, with the proportions of computing squared distance matrix (denoted by *matrix*), selecting nearest neighbors (denoted by *select* and with $k = 1000$) and transferring data (denoted by *transfer*) as we vary the number of queries, references and dimensions, respectively.

The chunk sizes we use depend on the choice of k, when k is large; for $k < 500$, we use the chunk size for $k = 500$ since making it smaller does not improve the running time.

4   Results

Our experimental environment employs CUDA Toolkit 6.5 [5] and a GeForce GTX 460 graphics card, which uses the Fermi architecture and has 1023 MB global memory.

Since our implementation is brute-force, the distribution of input data does not influence the performance of our program, so we use test data composed of uniformly distributed random numbers between -1 and 1.

The size of input is determined by the number of queries (m), references (n) and dimensions (d). We generated three test datasets to demonstrate the influence of each of these factors on the running time:

- $m \in [50, 1000]$, $n = 100000$ and $d = 64$.
- $m = 90$, $n \in [50000, 1000000]$ and $d = 64$.
- $m = 500$, $n = 100000$ and $d \in [50, 1000]$.

Evaluation and Analysis: Figure 1 shows the running time of our program and each of its three major components, computing squared distance matrix, selecting nearest neighbors and transferring data between CPU and GPU, on the test data. The running time is indeed linear in each of the factors (m, n, d), although the overall running time is $O(mnd)$. In any fixed dimension, the running time for the nearest neighbors selection step increases more quickly as the input size grows, and eventually dominates the time required for the matrix multiplication. The number of dimensions is irrelevant to the performance of the nearest neighbors selection.

Because the optimal chunk size in the nearest neighbors selection phase is achieved at $k = 500$, choosing k smaller than that does not improve the running time by much. The running time increases linearly with k for $k > 500$, however.

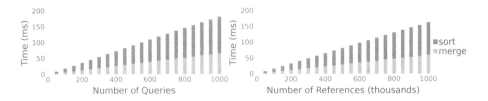

Fig. 2. Total running time of selecting nearest neighbors with $k = 1000$ and the proportions of sorting and merging candidate chunks (denoted by *sort* and *merge*, respectively) as we vary the number of queries and references, respectively.

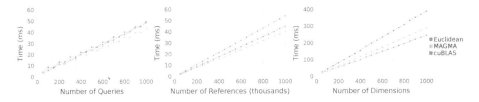

Fig. 3. Running time comparison of our squared distance matrix computation kernel (denoted by *Euclidean*) and the SGEMM subroutine in MAGMA [6] and cuBLAS [4] as we vary the number of queries, references and dimensions, respectively.

Next, we take a closer look at selecting nearest neighbors and its two kernels, sorting and merging candidate chunks, in Figure 2. We observe that the running time of the sorting step increases more quickly with input size.

Comparisons: To evaluate the performance of the kernel that computes our squared distance matrix $d^2(Q, R)$, we compare its performance to the two SGEMM (single precision general matrix-matrix multiply) implementations in MAGMA [6] and cuBLAS [4] (see Figure 3). Recall that both custom squared distance kernels and squared distance computations that use matrix multiplication as a subroutine are less efficient than the heavily optimized matrix multiplication subroutines.

Our implementation is modified from the SGEMM subroutine in MAGMA, but it is only marginally slower. The proprietary SGEMM matrix multiply implementation in cuBLAS performs better than MAGMA as the number of dimensions increases. In principle, any efficient matrix multiplication kernel can be modified to compute squared distances; we could not use cuBLAS only because it is not open source.

Finally, we evaluate the running time of our Merge Path nearest neighbors selection step with that of two recent nearest neighbor algorithms for which code is available. These are truncated insertion sort [2,24] and truncated bitonic sort [1,45]. We also compare against segmented sort applied to all n candidates for each query, as implemented in the Modern GPU library [3]. These comparisons

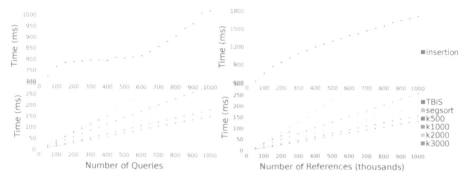

Fig. 4. Running time comparison of our nearest neighbors selection component for different k (denoted by *k500*, *k1000*, *k2000* and *k3000*, respectively), truncated insertion sort [2] (denoted by *insertion* and with $k = 100$), truncated bitonic sort [1] (denoted by *TBiS* and with $k = 500$) and segmented sort in the Modern GPU library [3] (denoted by *segsort* and with $k = n$) as we vary the number of queries and references, respectively.

are shown in Figure 4. The running time of the truncated insertion sort is shown in a separate graph because it is significantly slower, even for $k = 100$.

Our kernels are configured to find 500, 1000, 2000 and 3000 nearest neighbors per query, respectively. 3000 is the maximum size of candidate chunk that our kernels can handle (limited by the size of shared memory). In both graphs, the truncated bitonic sort works well up to a certain point, after which it stops producing correct results. Our program is twice as fast at $k = 500$, and only when we reduce k to 16 does TBiS become faster than our program with $k = 500$. Segmented sort [3] is robust at large input sizes, but it is slower and requires much more memory.

5   Conclusions

Finding ways to use highly-optimized GPU library functions is an effective way to achieve both speed and robustness in this important application. Our algorithm advances the state of the art for all but the smallest values of k. It is unique in its ability to handle large values of k, and large input datasets. The performance of our algorithm for very small values of k is limited mainly by the performance of the selection step. Possibly this could be improved by allowing one thread block to perform multiple truncated merge sorts in parallel. The drawback of this approach would be that it complicates the kernel.

Approximate kNN search approaches where nearest-neighbor candidates are filtered so that not all squared distances need to be computed could benefit from using our truncated merge sort to select the true nearest neighbors from the candidates. This is true for Locality Sensitive Hashing as well as for heuristic approaches.

Acknowledgments. We are grateful for NSF grant IIS-0964357, which supported this work.

References

1. cuknns: GPU accelerated k-nearest neighbor library (2012). http://autogpu.ee. auth.gr/doku.php?id=cuknns:gpu_accelerated_k-nearest_neighbor_library
2. kNN CUDA (2013). http://vincentfpgarcia.github.io/kNN-CUDA/
3. Modern GPU (2013). http://nvlabs.github.io/moderngpu/
4. cuBLAS in CUDA toolkit 6.5. (2014). https://developer.nvidia.com/cuBLAS
5. CUDA toolkit 6.5. (2014). https://developer.nvidia.com/cuda-toolkit-65
6. MAGMA 1.6.1. (2015). http://icl.cs.utk.edu/magma/
7. Thrust (2015). https://developer.nvidia.com/Thrust
8. Altman, N.S.: An introduction to kernel and nearest-neighbor nonparametric regression. The American Statistician **46**(3), 175–185 (1992)
9. Arefin, A.S., Riveros, C., Berretta, R., Moscato, P.: GPU-FS-kNN: A software tool for fast and scalable kNN computation using GPUs. PLOS ONE **7**(8), e44000 (2012)
10. Barrientos, R.J., Gómez, J.I., Tenllado, C., Matias, M.P., Marin, M.: kNN query processing in metric spaces using GPUs. In: Jeannot, E., Namyst, R., Roman, J. (eds.) Euro-Par 2011, Part I. LNCS, vol. 6852, pp. 380–392. Springer, Heidelberg (2011)
11. Beliakov, G., Johnstone, M., Nahavandi, S.: Computing of high breakdown regression estimators without sorting on graphics processing units. Computing **94**(5), 433–447 (2012)
12. Beliakov, G., Li, G.: Improving the speed and stability of the k-nearest neighbors method. Pattern Recognition Letters **33**(10), 1296–1301 (2012)
13. Belongie, S., Malik, J., Puzicha, J.: Shape matching and object recognition using shape contexts. IEEE Transactions on Pattern Analysis and Machine Intelligence **24**(4), 509–522 (2002)
14. Beyer, K., Goldstein, J., Ramakrishnan, R., Shaft, U.: When is nearest neighbor meaningful? In: Beeri, C., Bruneman, P. (eds.) ICDT 1999. LNCS, vol. 1540, pp. 217–235. Springer, Heidelberg (1998)
15. Boiman, O., Shechtman, E., Irani, M.: In defense of nearest-neighbor based image classification. In: IEEE Conference on Computer Vision and Pattern Recognition, CVPR 2008, pp. 1–8. IEEE, June 2008
16. Cayton, L.: Accelerating nearest neighbor search on manycore systems. In: 2012 IEEE 26th International Parallel and Distributed Processing Symposium, pp. 402–413. IEEE, May 2012
17. Cover, T.M., Hart, P.E.: Nearest neighbor pattern classification. IEEE Transactions on Information Theory **13**(1), 21–27 (1967)
18. Dashti, A., Komarov, I., D'Souza, R.M.: Efficient computation of k-nearest neighbour graphs for large high-dimensional data sets on GPU clusters. PLOS ONE **8**(9), e74113 (2013)
19. Datar, M., Immorlica, N., Indyk, P., Mirrokni, V.S.: Locality-sensitive hashing scheme based on p-stable distributions. In: Proceedings of the Twentieth Annual Symposium on Computational Geometry, SCG 2004, pp. 253–262. ACM (2004)
20. Diehl, P., Schweitzer, M.A.: Efficient neighbor search for particle methods on GPUs. In: Meshfree Methods for Partial Differential Equations VII, Lecture Notes in Computational Science and Engineering, vol. 100, pp. 81–95. Springer (2015)
21. Domeniconi, C., Peng, J., Gunopulos, D.: Locally adaptive metric nearest-neighbor classification. IEEE Transactions on Pattern Analysis and Machine Intelligence **24**(9), 1281–1285 (2002)

22. Dongarra, J., Gates, M., Haidar, A., Kurzak, J., Luszczek, P., Tomov, S., Yamazaki, I.: Accelerating numerical dense linear algebra calculations with GPUs. In: Numerical Computations with GPUs, chapter 1, pp. 3–28. Springer International Publishing (2014)
23. Garcia, V., Debreuve, E., Barlaud, M.: Fast k nearest neighbor search using GPU. In: IEEE Computer Society Conference on Computer Vision and Pattern Recognition Workshops, CVPRW 2008, pp. 1–6. IEEE, June 2008
24. Garcia, V., Debreuve, É., Nielsen, F., Barlaud, M.: K-nearest neighbor search: fast GPU-based implementations and application to high-dimensional feature matching. In: Proceedings of 2010 IEEE 17th International Conference on Image Processing, pp. 3757–3760, September 2010
25. Green, O., McColl, R., Bader, D.A.: GPU merge path - a GPU merging algorithm. In: Proceedings of the 26th ACM International Conference on Supercomputing, ICS 2012, pp. 331–340. ACM (2012)
26. Härdle, W.: Applied nonparametric regression. Number 19 in Econometric Society Monographs. Cambridge University Press (1990)
27. Hastie, T., Tibshirani, R.: Discriminant adaptive nearest neighbor classification. IEEE Transactions on Pattern Analysis and Machine Intelligence **18**(6), 607–616 (1996)
28. Jégou, H., Douze, M., Schmid, C.: Product quantization for nearest neighbor search. IEEE Transactions on Pattern Analysis and Machine Intelligence **33**(1), 117–128 (2011)
29. Kato, K., Hosino, T.: Solving k-nearest neighbor problem on multiple graphics processors. In: Proceedings of the 2010 10th IEEE/ACM International Conference on Cluster, Cloud and Grid Computing, CCGRID 2010, pp. 769–773. IEEE Computer Society (2010)
30. Kato, K., Hosino, T.: Multi-GPU algorithm for k-nearest neighbor problem. Concurrency and Computation: Practice and Experience **24**(1), 45–53 (2012)
31. Komarov, I., Dashti, A., D'Souza, R.M.: Fast k-NNG construction with GPU-based quick multi-select. PLOS ONE **9**(5), e92409 (2014)
32. Kruliš, M., Skopal, T., Lokoč, J., Beecks, C.: Combining CPU and GPU architectures for fast similarity search. Distributed and Parallel Databases **30**(3–4), 179–207 (2012)
33. Kuang, Q, Zhao, L.: A practical GPU based KNN algorithm. In: Proceedings of the Second Symposium International Computer Science and Computational Technology (ISCSCT 2009), pp. 151–155. Citeseer, December 2009
34. Kurzak, J., Tomov, S., Dongarra, J.: Autotuning GEMM kernels for the Fermi GPU. IEEE Transactions on Parallel and Distributed Systems **23**(11), 2045–2057 (2012)
35. Liang, S., Liu, Y., Wang, C., Jian, L.: A CUDA-based parallel implementation of k-nearest neighbor algorithm. In: International Conference on Cyber-Enabled Distributed Computing and Knowledge Discovery, CyberC 2009, pp. 291–296. IEEE, October 2009
36. Liang, S., Liu, Y., Wang, C., Jian, L.: Design and evaluation of a parallel k-nearest neighbor algorithm on CUDA-enabled GPU. In: 2010 IEEE 2nd Symposium on Web Society (SWS), pp. 53–60. IEEE, August 2010
37. Liang, S., Wang, C., Liu, Y., Jian, L.: CUKNN: a parallel implementation of k-nearest neighbor onCUDA-enabled GPU. In: IEEE Youth Conference on Information, Computing and Telecommunication, YC-ICT 2009, pp. 415–418. IEEE, September 2009

38. Lukač, N., Žalik, B.: Fast approximate k-nearest neighbours search using GPGPU. In: GPU Computing and Applications, chapter 14, pp. 221–234. Springer (2015)

39. Miranda, N., Chávez, E., Piccoli, M.F., Reyes, N.: (Very) Fast (All) k-nearest neighbors in metric and non metric spaces without indexing. In: Brisaboa, N., Pedreira, O., Zezula, P. (eds.) SISAP 2013. LNCS, vol. 8199, pp. 300–311. Springer, Heidelberg (2013)

40. Nath, R., Tomov, S., Dongarra, J.: An improved magma gemm for Fermi graphics processing units. International Journal of High Performance Computing Applications **24**(4), 511–515 (2010)

41. Odeh, S., Green, O., Mwassi, Z., Shmueli, O., Birk, Y.: Merge path - parallel merging made simple. In: 2012 IEEE 26th International Parallel and Distributed Processing Symposium Workshops & Ph.D. Forum (IPDPSW), pp. 1611–1618. IEEE, May 2012

42. Pan, J., Lauterbach, C., Manocha, D.: Efficient nearest-neighbor computation for GPU-based motion planning. In: The 2010 IEEE/RSJ International Conference on Intelligent Robots and Systems (IROS), pp. 2243–2248. IEEE, October 2010

43. Pan, J., Manocha, D.: Fast GPU-based locality sensitive hashing for k-nearest neighbor computation. In: Proceedings of the 19th ACM SIGSPATIAL International Conference on Advances in Geographic Information Systems, GIS 2011, pp. 211–220. ACM, November 2011

44. Pan, J., Manocha, D.: Bi-level locality sensitive hashing for k-nearest neighbor computation. In: 2012 IEEE 28th International Conference on Data Engineering (ICDE), pp. 378–389. IEEE, April 2012

45. Sismanis, N., Pitsianis, N., Sun, X.: Parallel search of k-nearest neighbors with synchronous operations. In: 2012 IEEE Conference on High Performance Extreme Computing (HPEC), pp. 1–6. IEEE, September 2012

46. Teodoro, G., Valle, E., Mariano, N., Torres, R., Meira Jr, W., Saltz, J.H.: Approximate similarity search for online multimedia services on distributed CPU–GPU platforms. The VLDB Journal **23**(3), 427–448 (2014)

47. Vincent, P., Bengio, Y.: K-local hyperplane and convex distance nearest neighbor algorithms. In: Advances in Neural Information Processing Systems 14 (NIPS 2001), pp. 985–992. MIT Press (2002)

48. Weinberger, K.Q., Saul, L.K.: Distance metric learning for large margin nearest neighbor classification. Journal of Machine Learning Research **10**, 207–244 (2009)

49. Zhang, H., Berg, A.C., Maire, M., Malik, J.: SVM-KNN: Discriminative nearest neighbor classification for visual category recognition. In: 2006 IEEE Computer Society Conference on Computer Vision and Pattern Recognition, vol. 2, pp. 2126–2136. IEEE (2006)

Regrouping Metric-Space Search Index for Search Engine Size Adaptation

Khalil Al Ruqeishi$^{(\boxtimes)}$ and Michal Konečný

School of Engineering and Applied Science, Aston University, Birmingham, UK
{alruqeik,m.konecny}@aston.ac.uk

Abstract. This work contributes to the development of search engines that self-adapt their size in response to fluctuations in workload. Deploying a search engine in an Infrastructure as a Service (IaaS) cloud facilitates allocating or deallocating computational resources to or from the engine. In this paper, we focus on the problem of regrouping the metric-space search index when the number of virtual machines used to run the search engine is modified to reflect changes in workload. We propose an algorithm for incrementally adjusting the index to fit the varying number of virtual machines. We tested its performance using a custom-build prototype search engine deployed in the Amazon EC2 cloud, while calibrating the results to compensate for the performance fluctuations of the platform. Our experiments show that, when compared with computing the index from scratch, the incremental algorithm speeds up the index computation 2–10 times while maintaining a similar search performance.

1 Introduction

A typical search engine distributes its search index into multiple processors to achieve a sufficiently high throughput [4–8, 13, 16]. However, the workload of a search engine typically fluctuates. Therefore, it is desirable that a search engine adapts its size to avoid wasting resources when the workload is low and to avoid unacceptable delays when the workload is high. If the engine is deployed in an Infrastructure as a Service (IaaS) cloud, the cloud facilitates allocating or deallocating compute resources to or from the engine.

Such an adaptive search engine repeatedly determines the number of processors to use, appropriately regroups the search data to form a new search index, and re-deploys the data onto the processors according to the new index.

Fig. 1 illustrates the running of such an adaptive search engine obtained using our small-scale prototype deployed on Amazon EC2.

In this paper, we focus on an important part of our prototype engine, namely the mechanism for regrouping the search data for a small or larger number of processors. We propose an algorithm for this task and evaluate its effectiveness using controlled tests in the prototype engine. We observe that our algorithm speeds up this task 2–10 times when compared with computing these groups from scratch (see Fig. 6). In addition, the search performance does not deteriorate significantly when using our algorithm (see Fig. 5).

© Springer International Publishing Switzerland 2015
G. Amato et al. (Eds.): SISAP 2015, LNCS 9371, pp. 271–282, 2015.
DOI: 10.1007/978-3-319-25087-8_26

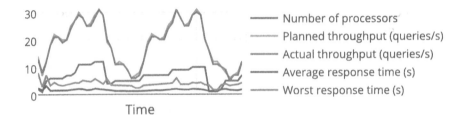

Fig. 1. Search engine updates number of processors whenever the workload changes.

The remainder of this paper is organized as follows. Section 2 recalls the background, in particular Subsection 2.1 describes the architecture of a search engine with a distributed metric space index. Section 3 reviews related work. Section 4 describes our algorithm for regrouping the search data. Section 5 presents the design and results of our experiments to validate and evaluate our algorithm. Section 6 concludes and outlines opportunities for further development.

2 Background

We build on previous research on distributing search data onto processors, in particular,

we use KmCol [7] for the initial grouping of search data. Let us recall the main components of KmCol because some of them feature in our incremental regrouping algorithm.

KmCol groups the search data in 3 steps, which leads to 3 levels of groupings. We adopt the following notation for the 4 types of object in these groupings:

- Data points: points in a metric space, representing the objects of the search
- LC-clusters: groups of nearby data points
- H-groups: groups of nearby LC-clusters, optimising for sample queries Q
- G-groups: groups of H-groups, one per processor

LC-clusters are computed using the List of Clusters (LC) algorithm [2,5–7]. LC-clusters are created to reduce the number of objects that feed into the next, more resource-intensive algorithm.

H-groups are formed from LC-clusters using K-means with the metric d_Q, derived from the set of sample queries Q, effectively optimising the engine for queries similar to Q. Due to the nature of K-means, H-groups have varying sizes.

G-groups are computed from H-groups using a procedure we call Group-Balanced, which attempts to balance their sizes.

The metric d_Q is defined using the *query-vector* model proposed in [14].

The metric d_Q makes pairs of points that are near in the natural metric seem far away from each other if they are close to many queries from Q. Conversely,

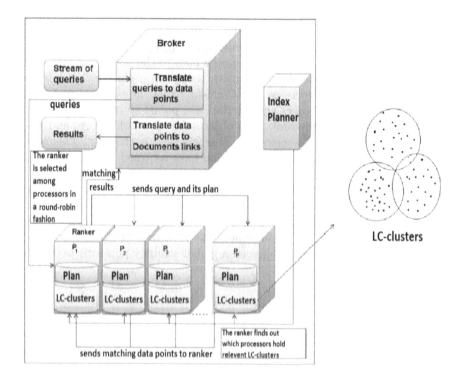

Fig. 2. Searching using distributed metric space index

the metric d_Q makes pairs of faraway points seem almost identical if they are not near any of the queries from the set Q. This means that H-groups finely separate those LC-clusters that are more likely to contain results of queries, leading to a good balance of load among processors.

2.1   Search Engine Distributed Architecture

We utilise the search engine parallel processing architecture outlined in Fig. 2. This architecture is analogous to that used in [4,5,7,13]. The Index Planner node is responsible for calculating G-groups and distributing them to the processors. It sends each processor not only its LC-clusters, but also an **index plan**, which is a map indicating for each LC-cluster on which processor it is. The index plan is used by the processor when it acts as a ranker for a query to determine which processors to contact regarding the query.

While search engines typically receive k-**nearest neighbor** (kNN) queries, i.e., "find k nearest objects to a specified object q for a small k" [7], search engines would translate such queries to **range queries** (q, r), i.e., "find all objects within distance r from q", because they are easier to distribute and process. Our engine also adopts this approach.

The ranker processor calculates the distance among the query and all of the centers across processors and formulates a *query plan*, namely the set of LC-clusters that intersect the ball of the range query (q, r).

The ranker sends the query and its query plan to the processor p_i that contains the first cluster to be visited, namely, the first LC-cluster that intersects the query ball. Then p_i processes all LC-clusters that intersect (q, r). For each such cluster, p_i compares (q, r) against the data points stored inside. The processor p_i then returns to the ranker the objects that are within (q, r) and passes the query and its plan to the next processor specified in the plan. This continues until all the processors in the query plan have returned their results to the ranker. The ranker sorts the query answers and passes the best k back to the broker as shown in Fig. 2. Each processor acts both as a processor in charge of processing a subset of LC-clusters and as a potential ranker.

Note that the architecture in Fig. 2 uses the Global Index Global Centers (GG) strategy because it uses a single node (i.e., the Index Planner) to compute the whole index. According to [6], such a global strategy performs better than local indexing strategies.

3   Related Work

According to [5], distributed metric space query processing was first studied in [12]. This work was extended in [6] for the LC-based approach, studying various forms of parallelization. As we said earlier, this study concluded that the GG strategy performs better than local indexing strategies.

An attractive feature of schemes without a global index is that they lend themselves to Peer-to-Peer (P2P) processing, which naturally supports resizing in response to load variations. For example, [11] presents a distributed metric space index as a P2P system called M-index. Unfortunately, such schemes tend to lead to a reduced search performance. Moreover, M-index is based on a pivot partitioning model, which has a high space complexity. For further related work using P2P metric space indexing see e.g. [3,9,10,15].

We note that [1,5] address the related problem of performance degradation when query load becomes unbalanced across processors. The query scheduling algorithm proposed in [5] balances the processing of queries by dynamically skewing queries towards explicit sections of the distributed index. On the other hand, [1] proposes dynamic load balancing based on a hypergraph model, but it is not concerned with multimedia search and does not use metric space index.

4   Adapting Search Engine Size

An adaptive search engine will repeatedly re-evaluate its load and, when appropriate, switch over from p active processors to a different number of active processors. Recall that the initial H-group and G-groups were computed using the Km-COL algorithm as described in Sect. 2. Each switchover comprises the following steps:

1. Determine the new number of processors p' based on recent load.
2. (Re-)compute H-groups and G-groups (i.e., the index plan) for p' processors.
3. Distribute the index plan and the relevant LC-clusters onto each processor.
4. Pause search.
5. Switch to new LC-clusters and plan, de/activating some processors.
6. Resume search.

Our main contribution is an algorithm for step 2 and experimental evidence of how different ways of implementing step 2 impact search performance after the switchover. To allow us to focus on step 2 and the resulting search performance, we perform the switchovers while the engine is inactive, omitting steps 4 and 6. We also skip step 1 as p and p' will be determined by our experiment design.

4.1   Computing H-groups and G-groups

We compute G-groups from H-groups in the same way as in the KmCol algorithm. We therefore focus on the computation of H-groups for p' processors from H-groups for p processors. We introduce the following three methods (called **transition types**):

TT-R: Compute H-groups from scratch using K-means, like KmCol.
TT-S: Reuse the H-groups from previous configuration.
TT-A: Increase the number of H-groups using Adjust-H (Algorithm 1).

Algorithm 1. Adjust-H$(d)(H, \text{new\_size})$

Tuning Parameters: d: a metric on C
Input: H: a set of H-groups partitioning C,
        new_size: the target number of H-groups
        (new_size $> |H|$)

Output: updated H with $|H| = \text{new\_size}$

1: $H_{\text{sorted}} = \text{sort\_by\_decreasing\_size}(H)$
2: **while** $\text{size}(H_{\text{sorted}}) \neq \text{new\_size}$ **loop**
3:    $\text{largest\_group} = H_{\text{sorted}}.\text{getFirst}()$
4:    $\text{new\_groups} = K\text{-means}(d)(\text{largest\_group}, 2)$ // split
5:    $H_{\text{sorted}}.\text{insert\_sorted}(\text{new\_groups})$
6:    $H_{\text{sorted}}.\text{delete}(\text{largest\_group})$
7: **end loop**
8: **return** H_{sorted}

Notice that the number of H-groups will never be decreased. This is appropriate because, as we show in Sect. 5, reducing the number of H-groups does not improve search performance.

Adjust-H takes as parameters the number new_size ($= p' \cdot w$) and the old H-groups. On line 1, it starts by arranging the H-groups in an ordered collection, with the largest group first. On lines 2–7, the number of H-groups is increased by repeatedly splitting the largest H-group into two using K-means, until there are new_size many of them. Thanks to the following observation, we do not need to study the effect of repeated TT-A on search performance:

Proposition 1 (Repeated TT-A is equivalent to a single TT-A).
For any set H and sequence $|H| < p_1 < p_2 < \ldots < p_n$, it holds:
$Adjust\text{-}H\big(\ldots Adjust\text{-}H(Adjust\text{-}H(H, p_1), p_2), \ldots, p_n\big) = Adjust\text{-}H(H, p_n)$

Proof. A repeated execution of Adjust-H results in successive executions of the loop that forms the algorithm. There are no commands to change the H-groups between the successive executions of the loop. Thus the result of the repeated loop executions is the same as running the loop only once with new_size set to the final value p_n.   □

To pursue our goal to speed up switchovers while keeping a good search performance, we will test the search performance implications of the three transition types TT-R, TT-S and TT-A.

Based on preliminary observations, we formed the following hypotheses:

H1 The time it takes to compute H-groups grows significantly with the number of these H-groups.

H2 Increasing the number of H-groups does not reduce search performance. Equivalently, when reducing p, TT-S does not lead to a worse search performance than TT-R.

H3 Computing a number of H-groups and then splitting them up using TT-A does not impair search performance when compared to computing the same number of H-groups directly using TT-R.

We provide experimental evidence supporting these hypotheses in Sect. 5.

Using these hypotheses, on the assumption that they are correct, we propose the algorithm Regroup (Algorithm 2) to decide which of the three transition types to use.

The algorithm takes as parameters the numbers $w_{\min}$ and w_{init}. TT-R uses w_{init} to compute H-groups from scratch, while $w_{\min}$ is used by TT-A to recompute H-groups. Due to hypothesis H2, the values of these tuning parameters do not significantly affect search performance. We therefore use the fairly low values $w_{\text{init}} = 2$ and $w_{\min} = 1.5$ in our experiments in order to reduce the time

---

Algorithm 2. Regroup($w_{\text{init}}, w_{\text{min}}$)($p', H, d_Q, Q$)

Tuning Parameters: $w_{\text{init}}, w_{\text{min}} \geq 1$

Input: p': new number of processors,

   H: a set of H-groups partitioning C (optional, needed if Q absent),

   d_Q: a metric on C (optional, needed if Q absent),

   Q: sample set of queries (optional, needed if H absent)

Output: G: a partition of C with $|G| = p'$,

    updated H and d_Q

1: **if** Q is provided **then**

2:   $d_Q :=$ Query-Vector-Metric(C, Q)

3:   $H := K$-means(d_Q)($p' * w_{\text{init}}, C$) // TT-R

4: **elseif** $|H| < p' * w_{\text{min}}$ **then**

5:   $H :=$ Adjust-H(d_Q)($H, p' * w_{\text{min}}$) // TT-A

6: **end** // TT-S: the if block not executed

7: $G :=$ Group-Balanced(H, p')

8: **return** G, H, d_Q

---

it takes to compute the H-groups. At the beginning, if a new Q is provided, it is necessary to update the metric d_Q and recompute the H-groups from scratch (TT-R, lines 2 and 3). If the number of H-groups is smaller than $p' * w_{\text{min}}$, the number of H-groups is increased (TT-A, line 5). If there is no change in Q and $p > p'$, then H is reused (TT-S). Finally, on line 7, new G-groups are computed from the H-groups, using Group-Balanced, an algorithm borrowed from KmCol.

5   Experimental Evidence Supporting Hypotheses

In the experiments, the three transition types are compared in terms of their effect on *search performance* and the time it takes to compute H-groups for the new number of processors. (a component of *switch-over performance*). The performance is influenced by the following parameters:

1. **Search engine size evolution** (*SE*): We consider only a single switchover at a time and write it as $p \rightarrow p'$. E.g., 5→8 encodes a single switchover from 5 to 8 processors. In our experiments, we use increasing or decreasing transitions of the sequence $2, 3, 5, 8, 12, 18$ and contrast sets of transitions sharing a similar ratio p/p' or sharing the same p'.

2. **Dataset** (*D*): A dataset represents the set of objects that needs to be searched. In our experiments, we used a randomly selected set of 1,000,000 objects from the CoPhIR Dataset[1]. Each object comprises 282 floating-point number coordinates.

---

[1] http://cophir.isti.cnr.it/

3. **Sample queries** (Q): As explained in Sect. 2, the set defines the metric d_Q which is used to partition LC-clusters into H-groups. In our experiments, we used as Q a randomly selected set of 1,000 objects from the CoPhIR Dataset.

4. **Query profile** (QP): Query profile simulates how users send queries to the search engine. It is determined by a sequence of queries and the timing when each query occurs. In our experiments, we use as queries 100,000 randomly selected objects from the CoPhIR Dataset. We fire the queries at a constant query rate. This rate is not a parameter of the experiment because it is determined automatically in the process of measuring maximum throughput as described below.

Search performance is measured using *maximum throughput* defined as follows. The current output throughput of a search engine is the rate at which answers to queries are sent to clients. This is equal to the input throughput, i.e., the rate at which the queries are arriving, except when the queries are accumulating inside the engine.

Maximum throughput is the highest output throughput achieved when flooding the input with queries. We have observed that the network stack efficiently facilitates the queuing of queries until the engine is able to accept them. Each of the search engine nodes (Fig. 2) was deployed on a separate Amazon EC2 medium virtual machine instance. In each experiment, we used the following steps to obtain sufficiently reliable throughput measurements despite significant performance fluctuations of the Amazon cloud platform:

- Conduct two speed tests: an initial and a final test. The two tests are identical. Each test comprises 4 repetitions of a fixed task based on distributed searching.
- If the speed variation within these 4 repetitions is over 5%, the cloud is not considered sufficiently stable.
- Also if the initial and final speed measurements differ by over 2%, the cloud is not considered sufficiently stable.
- The average of the speed measurements in the initial and final tests is used to calibrate the maximum throughput measurements obtained in the experiment to account for longer-term variations in the cloud performance.

When the stability tests failed repeatedly, we relaxed the thresholds and took the average of the measurements obtained from 3 repetitions of the experiment.

We observed that in many experiments, the throughput fluctuates at the beginning and then stabilises. To discount the initial instability, we run each search experiment as a sequence of blocks of 100 queries and we wait until there are four consecutive blocks with a performance variation below 30%. We discount the preceding blocks that have a higher variance.

We artificially slowed down all processors by a factor of 5 to compensate for the slow network in the EC2 cloud (around 240 MBits/s), simulating a faster network, which would be found in typical clusters (around 1 GBit/s).

The full code for our experimental search engine and the experiments described in this section is available on http://duck.aston.ac.uk/ngp.

5.1 The Number of H-groups

Experiment E1. To test hypothesis H1, we computed different numbers of H-groups and observed how the computation time grows with size while the remaining parameters are fixed.

The results (Fig. 3(a)) confirm hypothesis H1.

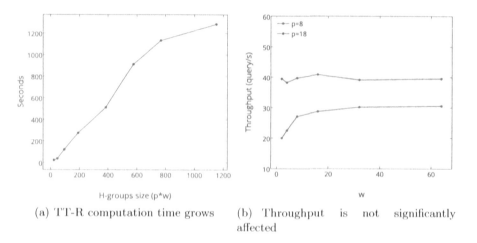

(a) TT-R computation time grows

(b) Throughput is not significantly affected

Fig. 3. Impact of increasing the number of H-groups ($= w * p$) on performance.

Experiment E2. In a similar setup as experiment E1, we checked whether the extra computation time spent creating more H-groups translates to better search performance, in contradiction to hypothesis H2. We have done this for $p = 8$ and $p = 18$ and the same values of w as for E1. The results of E2 in Fig. 3(b) show that the throughput is not significantly affected by w, confirming H2.

5.2 Search Performance of TT-S

Experiments E3 and E4. Reusing H-groups for $p' < p$ (TT-S) is much faster than recomputing H-groups (TT-R). The alternative phrasing of hypothesis H2 states that this speed up does not come at a cost to the search performance. Here we report on experiments that confirm hypothesis H2 in the alternative phrasing:

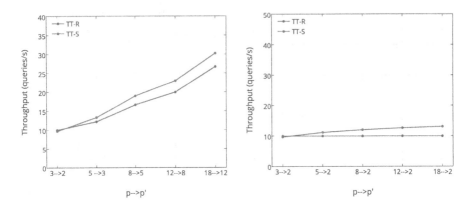

Fig. 4. TT-S and TT-R produce similar throughput, measured separately for increasing p' (E3) and increasing p/p' (E4).

The same switchover $p \rightarrow p'$ is performed using TT-R and independently using TT-S and the resulting search performance is measured.

These two experiments differ in the set of switchovers considered as follows:

- E3 varies p' and fixes the ratio p/p'.
- E4 varies the ratio p/p' and fixed p'.

The results of these experiments shown in Fig. 4 support H2: TT-S does not lead to worse search performance than TT-R when switching over to a smaller number of processors.

5.3   Comparing TT-A and TT-R

Experiments E5 and E6. In this section, we test hypothesis H3 by comparing the results of experiments that measure the search performance after computing H-groups using TT-R and TT-A. Moreover, we capture the computation time of the transitions to measure the speed-up of TT-A over TT-R.

As with E3 and E4, the experiments differ in the set of switchovers considered as follows:

- E5 varies the ratio p/p' and fixed p'.
- E6 varies p' and fixes the ratio p/p'.

The results of experiments $E5$ and $E6$ in Fig. 5 support hypothesis H3, namely they show that the maximum throughputs after TT-A is similar to, sometimes even better than the maximum throughput after TT-R. Plots in Fig. 6 show that in this context the speed-up of TT-A versus TT-R is 2–10 times.

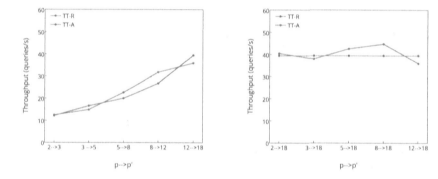

Fig. 5. TT-A and TT-R lead to a similar maximum throughput after switchovers with various p' and with various ratios.

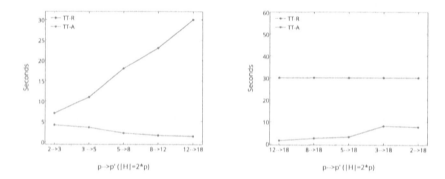

Fig. 6. TT-A is faster than TT-R in switchovers with with various p' and various ratios.

6 Conclusions

We have proposed a new algorithm for planning an incremental regrouping of a metric-space search index when a search engine is switched over to a different size. This algorithm is inspired by the results of a set of experiments we conducted. These experiments also indicate that our algorithm facilitates 2–10 times faster switchover planning and leads to a similar search performance when compared with computing the index from scratch.

In this work, we studied only the re-computation of the metric-space index when the search engine changes size. We plan to develop and study the remaining aspects of an adaptive search engine, such as determining when and how to change the engine size and re-distributing the search data among processors processors according to the newly computed search index while keeping the engine responsive.

References

1. Catalyurek, U.V., Boman, E.G., Devine, K.D., Bozdağ, D., Heaphy, R.T., Riesen, L.A.: A repartitioning hypergraph model for dynamic load balancing. Journal of Parallel and Distributed Computing **69**(8), 711–724 (2009)
2. Chávez, E., Navarro, G.: A compact space decomposition for effective metric indexing. Pattern Recognition Letters **26**(9), 1363–1376 (2005)
3. Doulkeridis, C., Vlachou, A., Kotidis, Y., Vazirgiannis, M.: Peer-to-peer similarity search in metric spaces. In: Proceedings of the 33rd International Conference on Very Large Data Bases, pp. 986–997. VLDB Endowment (2007)
4. Gil-Costa, V., Marin, M.: Approximate distributed metric-space search. In: Proceedings of the 9th Workshop On Large-Scale And Distributed Informational Retrieval, pp. 15–20. ACM (2011)
5. Gil-Costa, V., Marin, M.: Load balancing query processing in metric-space similarity search. In: 2012 12th IEEE/ACM International Symposium on Cluster, Cloud and Grid Computing (CCGrid), pp. 368–375. IEEE (2012)
6. Gil-Costa, V., Marin, M., Reyes, N.: Parallel query processing on distributed clustering indexes. Journal of Discrete Algorithms **7**(1), 3–17 (2009)
7. Marin, M., Ferrarotti, F., Gil-Costa, V.: Distributing a metric-space search index onto processors. In: 2010 39th International Conference on Parallel Processing (ICPP), pp. 433–442. IEEE (2010)
8. Marin, M., Gil-Costa, V., Bonacic, C.: A search engine index for multimedia content. In: Luque, E., Margalef, T., Benítez, D. (eds.) Euro-Par 2008. LNCS, vol. 5168, pp. 866–875. Springer, Heidelberg (2008)
9. Marin, M., Gil-Costa, V., Hernandez, C.: Dynamic P2P indexing and search based on compact clustering. In: Second International Workshop on Similarity Search and Applications, SISAP 2009, pp. 124–131. IEEE (2009)
10. Novak, D., Batko, M., Zezula, P.: Metric index: An efficient and scalable solution for precise and approximate similarity search. Information Systems **36**(4), 721–733 (2011)
11. Novak, D., Batko, M., Zezula, P.: Large-scale similarity data management with distributed metric index. Information Processing & Management **48**(5), 855–872 (2012)
12. Papadopoulos, A.N., Manolopoulos, Y.: Distributed processing of similarity queries. Distributed and Parallel Databases **9**(1), 67–92 (2001)
13. Puppin, D.: A search engine architecture based on collection selection. Ph.D. thesis, PhD thesis, Dipartimento di Informatica, Universita di Pisa, Pisa, Italy (2007)
14. Puppin, D., Silvestri, F., Laforenza, D.: Query-driven document partitioning and collection selection. In: InfoScale 2006: Proceedings of the 1st International Conference on Scalable Information Systems. ACM Press, New York (2006)
15. Yuan, Y., Wang, G., Sun, Y.: Efficient peer-to-peer similarity query processing for high-dimensional data. In: 2010 12th International Asia-Pacific Web Conference (APWEB), pp. 195–201. IEEE (2010)
16. van Zwol, R., Rüger, S., Sanderson, M., Mass, Y.: Multimedia information retrieval: new challenges in audio visual search. In: ACM SIGIR Forum, vol. 41, pp. 77–82. ACM (2007)

Improving Parallel Processing of Matrix-Based Similarity Measures on Modern GPUs

Martin Kruliš[✉], David Bednárek, and Michal Brabec

Parallel Architectures/Algorithms/Applications Research Group,
Faculty of Mathematics and Physics, Charles University in Prague,
Malostranské nám. 25, Prague, Czech Republic
{krulis,bednarek,brabec}@ksi.mff.cuni.cz

Abstract. Dynamic programming techniques are well-established and employed by various practical algorithms which are used as similarity measures, for instance the edit-distance algorithm or the dynamic time warping algorithm. These algorithms usually operate in iteration-based fashion where new values are computed from values of the previous iteration, thus they cannot be processed by simple data-parallel approaches. In this paper, we propose a way how to utilize computational power of massively parallel GPUs to compute dynamic programming algorithms effectively and efficiently. We address both the problem of computing one distance on large inputs concurrently and the problem of computing large number of distances simultaneously (e.g., when a similarity query is being resolved).

Keywords: GPU · CUDA · Dynamic programming · Edit distance · Dynamic time warping

1 Introduction

Many similarity measures are based on an algorithmic paradigm called *dynamic programming*. It is often employed when a problem (in our case similarity measure) is defined using a recursive formula that would directly lead to an algorithm with exponential time complexity. The dynamic programming approach prunes out redundant work or the recursive formula and yields a polynomial algorithm, where each subproblem is computed exactly once.

In this work, we focus on dynamic programming algorithms whose subproblems form a matrix. Each partial result in the matrix is computed from a small subset of previous results, which permits a limited degree of concurrent evaluation. Typical examples of this approach are the edit distance problem originally described by Levenshtein [4], the dynamic time warping [9], or the Smith-Waterman algorithm [14] for molecular sequence alignment.

We have selected the Wagner-Fischer dynamic programming algorithm [17] for the Levenshtein distance problem as a representative for our implementation since its computational simplicity emphasizes the communication and synchronization overhead associated with parallel computation.

© Springer International Publishing Switzerland 2015
G. Amato et al. (Eds.): SISAP 2015, LNCS 9371, pp. 283–294, 2015.
DOI: 10.1007/978-3-319-25087-8_27

The data transfers between the threads computing different elements may render a naïve parallel implementation memory-bound, although in theory, the matrix does not require any memory representation as we are interested only in its bottom-right corner. We will not employ any optimizations designed specifically for the Levenshtein distance (like the Myers' algorithm [10]), so our proposed solution is applicable for similar dynamic programming algorithms as well.

The communication overhead becomes even more important in a massively parallel environments like GPUs. In addition to the high degree of parallelism, the specific execution model and complicated memory hierarchy of the GPUs forces the programmers to deal with many technical issues which are not present in multicore CPU programming.

This paper investigates the problems of matrix-based dynamic programming on modern GPUs and proposes a solution that is suitable for their contemporary architectures. We have taken the same approach as Tomiyama et al. [15], but we have improved their solution by optimizing internal data transfers, which are one of the most serious parallelization bottlenecks in dynamic programming. We have utilized new thread-cooperative instructions that were introduced in the NVIDIA Kepler architecture [11] (CUDA compute capability 3.0 and higher). The results indicate that this optimization can improve overall performance by a factor of 1.75 with respect to an unoptimized GPU solution.

In addition, we have investigated possible parallelization approaches to a multi-distance problem – i.e., when multiple distances need to be computed. This situation is typical for similarity queries, when a distance between query and each database objects needs to be computed. Even though this may seem to be an embarrassingly data parallel problem, the nontrivial memory requirements of the algorithm impose serious performance limitations when executed on manycore GPUs.

The paper is organized as follows. Section 2 overviews work related to GPU implementations of dynamic programming algorithms. In Section 3, we revise the fundamentals of GPU architectures with emphasis on data exchange. Section 4 presents our proposed solution and its implementation details. The empirical evaluation is summarized in Section 5 and Section 6 concludes our paper.

2   Related Work

Most parallel algorithms are based on an observation of Delgado et al. [2], who studied the data dependencies in the dynamic programming matrix. Two possible ways of processing the matrix were defined in their work – *uni-directional* and *bi-directional* filling. The original idea allows limited concurrent processing, but it needs to be modified for massively parallel environment.

One of the first papers that covers the whole issue of the parallelization of Levenshtein distance on GPUs was presented by Tomiyama et al. [15]. Their approach divides the dynamic programming matrix into parallelogram blocks. Independent parallelograms are computed by separate CUDA thread blocks while each block is computed in a highly cooperative manner. The main focus of the work addressed the problem of appropriate block size and its automatical selection. On the other hand, their experiments are currently out of date, since they were performed on a GPU with compute capability 1.3 only.

Perhaps the most recent work on the topic was presented by Chacón [1]. It is based on Myers' bit-parallel algorithm [10] which uses an observation made by Ukkonen [16], that all the adjacent elements in the matrix differs at most by ± 1. The algorithm converts the data-dependency problem into a bit-carry problem, which can be computed more efficiently using bit-wise operations.

Similar approach was taken by Xu et al. [18], but their work focused on multi-query string matching, where multiple words are matched against a larger text. They have used a pattern concatenation technique, so that multiple queries can be processed by one distance computation.

A similar problem that uses dynamic programming is *dynamic time warping* (DTW) defined by Müller [9]. Since the performance issue is also quite important for DTW applications, Sart et al. discussed parallelization techniques for GPUs and FPGAs [13]. They focused mainly on a specific version of DTW algorithm, which reduces the dependencies of the dynamic programming matrix, thus allows more efficient parallelization.

Another example of a problem suitable for dynamic programming is the Smith-Waterman algorithm [14], which is used for protein sequences alignment. To our best knowledge, the first attempt to parallelize this algorithm on GPUs was made by Liu et al. [6]. Their work was performed before the release of CUDA framework, thus the computation had to be transformed into a rendering problem, which was then implemented using OpenGL and GLSL shaders [12].

Manavski et al. [8] reimplemented the Smith-Waterman algorithm using CUDA technology. Slightly different solution was presented by Ligowski et al. [5]. Their work focused on searching in the entire database of proteins. Khajeh-Saeed et al. [3] utilized the computational power of multiple GPUs to solve this problem. Perhaps the most recent version was presented again by Liu et al. [7] and it combines observations from the previous work.

Unlike these related papers, our work does not focus on a single dynamic programming algorithm. Individual algorithms may be optimized using specific observations, which are limited to a single application. We are proposing an optimization, that can be used in several dynamic programing algorithms or even adopted for similar problems where data transfers are the issue.

3   GPU Fundamentals

GPU architectures differ from CPU architectures in multiple ways. In this section, we revise the GPU architectures (especially of NVIDIA GPUs [11]) fundamentals with particular emphasis on aspects, which have great importance in the light of the studied problem.

3.1   GPU Device

A GPU card is a peripherial device connected to the host system via the PCI-Express (PCIe) bus. The GPU is quite independent, since it has its own memory and processing unit; however, it is not completely autonomous and its operations must be controlled by host. Furthermore, the GPU computational cores cannot access the host memory directly, so the input data must be transferred to the GPU over the PCIe and computed results must be transferred back.

The GPU processor comprises several *streaming multiprocessors* (SMPs), which can be roughly related to the CPU cores. The SMPs share only the main memory bus and the L2 cache of the GPU, but otherwise they are almost independent. Each SMP consists of a number of *GPU cores*[1], one or more instruction decoders and schedulers[2], L1 cache, and shared memory. Each core has its own arithmetical units for integer and float operations and a private set of registers. As the number of GPU cores is larger than the number of schedulers, a scheduler issues the same instruction into several cores which thus synchronously execute the same code within different threads. This set is called a *warp* and always constitutes 32 threads in the current NVIDIA architectures. Although a part of the threads in the warp may be masked off, the respective cores are not freed for other work. Furthermore, each SMP keeps a list of active warps whose execution is interleaved in order to keep the cores busy when some warps are waiting due to synchronization or memory access. Thus, a number of warps is effectively running in parallel in each SMP.

3.2   Thread Execution

The true machine code is not exposed to the programmer, allowing to hide hardware details and change code between generations. Thus, an additional layer of abstraction is built above the underlying hardware; in our case determined by the CUDA specification.

Code routines, called *kernels*, are fed into the GPU by the CPU. When scheduled to run with a particular data set, the kernel becomes a CUDA *grid*. Each grid is divided into CUDA *blocks*, each block consists of several *warps* and each warp consist of 32 threads. While the warp size is fixed, the sizes of blocks and grids are determined by the application.

The CUDA software/hardware stack schedules grids over entire GPUs and assigns blocks to individual SMPs. Thus, all threads in a block may access the same *shared* memory associated to the SMP, and synchronize using several primitives. On the other hand, blocks in a grid share only the global memory and their ability to synchronize is limited to atomic instructions. In most cases, synchronization between blocks as well as data exchange with the host CPU occurs only at the beginning and at the end of the grid.

---

[1] E.g., 32 cores in Fermi or 192 cores in Kepler architecture [11].
[2] E.g., one scheduler in Fermi or four schedulers in Kepler.

Threads in one warp are executed in a lockstep[3] – i.e., they are all issued the same instruction at a time. This model allows easier synchronization of work performed by the warp and it also permits warp-specific instructions to be implemented, such as the *warp shuffle* instructions which are used for direct data interchange between threads in a warp.

On the other hand, branching in the code (e.g., in an 'if' statement) is expensive under the lockstep execution as it must be converted to masking some threads off and, thus, wasting resources.

3.3 Memory Organization

Another important issue is the memory organization, which is depicted in Figure 1. We need to distinguish four types of memory:

- host memory (RAM),
- global memory (VRAM),
- shared memory,
- and private memory (GPU core registers).

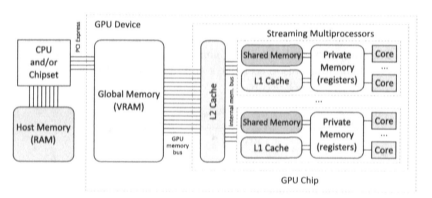

Fig. 1. Host and GPU memory organization scheme

The *host memory* is the operational memory of the computer. Input data needs to be transferred from the host memory to the global memory of the GPU and the results need to be transferred back.

The *global memory* is accessible from all GPU cores/threads, so the input data and the computed results are stored here. It is connected via global memory bus, which has high latency and high bandwidth. The bus transfers data in wide aligned blocks, so the threads in a warp are encouraged to access data which are close together.

The *shared memory* is present on each SMP and dedicated to the running thread block. It is rather small (tens of kB) but almost as fast as the GPU registers. The shared memory can play the role of a program-managed cache,

---

[3] Also called Single Instruction Multiple Threads (SIMT) model.

or it can hold intermediate results of the threads in the block. The memory is divided into banks (usually 16 or 32), so that subsequent 32-bit words are in subsequent banks. When multiple threads access the same bank (except if they read the same address), their operations are serialized.

Finally, the *private memory* belongs exclusively to a single thread and corresponds to the GPU core registers. Private memory size is very limited (tens to hundreds of words per thread), therefore it is suitable just for a few local variables.

4   Implementation

As illustrated by Figure 2, the dependencies between elementary calculations in the two-dimensional matrix allow parallel computation for all elementary tasks on any diagonal line. Thus, the computation may be done by a single sequential sweep through all the diagonals in the matrix whilst each diagonal is computed in parallel. Unfortunately, such simple approach to parallelization suffers from two deficiencies: First, the size of the diagonal (n) varies throughout the matrix which requires frequent addition and removal of computing units (threads) during the sweep. Second, regardless of the exact assignment of the physical threads to the elementary tasks, each thread processing a task must interchange information with at least one other thread when the computation advances to a subsequent diagonal. The computational cost of the elementary tasks is assumed to be small, thus the thread communication cost plays an important role as well.

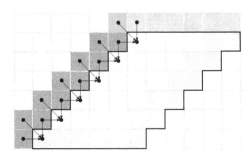

Fig. 2. A single block with emphasized input, output, and data dependencies

When more than one instance of the problem is computed simultaneously, it is also possible to compute the instances in parallel while each instance is evaluated sequentially (e.g., row-wise). However, as we will demonstrate in the evaluation, this approach suffers from the inability to fit sufficient number of instances into fast memory, because the memory has limited size.

Thus, although the multi-instance problem seems to be embarrassingly parallel, non-trivial algorithms must be investigated to overcome the memory size limitation. In general, we have the following options when assigning multiple instances of the distance problem to a GPU:

- An *instance per thread* – the trivial data-parallel solution described above.
- An *instance per warp* – requires data exchange between threads using either the shuffle instructions or shared memory
- An *instance per block* – data exchange between warps must be done using shared memory.
- An *instance per GPU* – executing a sequence of instances. As each instance is spread over several SMPs, data exchange must be done via global memory. A trivial version of this approach is the diagonal sweep described above. Nevertheless, we will describe a faster *blocked* algorithm.

Down through this list, the complexity of data exchange increases. On the other hand, the number of instances present at a time decreases; thus, each instance may occupy a larger portion of the GPU memory hierarchy. Consequently, the applicability of these methods depends on the size of the instance data and experiments are needed to determine optimal approach.

4.1 Parallelogram Blocks

Due to the nature of data-dependencies in the computation matrix, warps or blocks work best when assigned to parallelogram-like portions of the matrix, as depicted in Figure 2. The picture also shows the input values for such a block, divided into *left buffer* (red), *upper-left buffer* (blue), and *upper buffer* (green). The arrows show the dependencies for the first diagonal in the block which will be computed in parallel. The output of the parallelogram block is shown in light gray, the white fields inside the parallelogram correspond to temporary values computed and later discarded during the block evaluation.

Each block is processed in three steps: load input data from the global memory, compute the values, and write the results back to the global memory for the subsequent blocks. A block of height H and width W is computed by H threads that *synchronously* perform W iterations. Between subsequent iterations, the threads must exchange a part of the data.

If H equals to the warp size, the synchronous execution is ensured by the lockstep nature of warps and the data interchange may be implemented by the shuffle instructions. For larger H, shared memory and barriers must be involved in the data exchange and synchronization.

4.2 Using Shuffle Instructions

The shuffle instructions allow direct exchange of values between threads within a warp. To utilize this feature to its full potential, we have restricted the block size (width and height) to be divisible by the warp size, thus all the warps are fully occupied. Since the shuffle instructions work only within a warp, we additionally employ shared memory transfers for data passed between threads in different warps.

The shuffle instructions rotates the data as depicted in Figure 3a. Each thread passes the last computed result to its neighbor and simultaneously receives a value which is stored to upper variable.

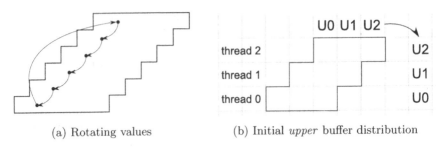

(a) Rotating values   (b) Initial *upper* buffer distribution

Fig. 3. Data transfers of the *upper* buffer values within warp

After the shuffling, the last thread (on the top) has to load its *upper* value from the shared memory. This could be done directly, however, it would waste the memory throughput as only one thread from the warp would be reading at a time.

Instead of reading from shared memory one at a time, the *upper* values are distributed among the registers of the threads and then another set of shuffle instructions is used to feed the last thread with the correct value at the right moment. Thus, the register capacity of the warp is used to hold all the *upper* values instead of the shared memory, offering smaller latency.

All *upper* values are loaded at the beginning, so that each thread loads one value to its register as shown in Figure 3b. Since the *upper* values are consumed during the sweep through the block, they can be gradually replaced by the result values computed by the first thread (in the bottom). This way, the space (registers) originally allocated for input are reused to store output values. Therefore, when the iterations are concluded, the threads write their result values back to the same global memory buffer, so they can be used as an input for another diagonal block.

4.3   Synchronization via Shared Memory

For parallelogram blocks whose height is larger than the warp size, shuffle instructions can not pass data over the warp boundaries. Therefore, the corresponding results are written into and subsequently loaded from a shared memory buffer. All the block threads must be synchronized on a barrier to ensure that all the threads have written the data before they are read. The barrier also acts as a memory fence, so the data are visible to all threads after the barrier.

4.4   Blocked Algorithm

The elements of the distance matrix are grouped into parallelogram blocks as shown in Figure 4, where the numbers denote *coarse diagonals*. Blocks that are not completely contained in the distance matrix are processed the same way as the others, which allows us to avoid checking any conditions during the block calculation. Gray area represents the input values for the left and upper blocks

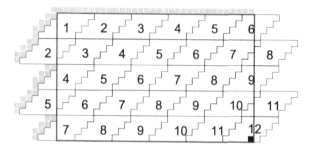

Fig. 4. Distance matrix divided into parallelogram blocks

and the black square is the final result (computed edit distance). The blocks communicate through global memory buffers which are allocated in advance, since the algorithm does not need to vary their size. The division of the matrix into the blocks follows a similar pattern as in the work of Tomiyama [15].

Similarly to the fine diagonals in Figure 2, the coarse diagonals allow parallel processing of blocks. Blocks retain the same scheme of dependencies as single elements, thus the blocks in a coarse diagonal can be processed in parallel, because they depend only on blocks from two previous coarse diagonals. The block numbers depicted in Figure 4 indicate the order in which they are computed.

Each coarse diagonal of blocks is computed by a single CUDA grid where each CUDA block corresponds to a single parallelogram block. Selecting appropriate block size (i.e., its width and height) is an important factor that affects the efficiency of the algorithm as demonstrated in the Section 5.

5   Experiments

In this section, we present experimental evaluation of two problems. In the first scenario, we used GPU to accelerate one distance computation of two large inputs and evaluate the benefits of our optimization that utilizes warp shuffling instruction. The second scenario simulates a sequential scan in a similarity search, where many distances are computed on much smaller inputs. We compare our method with a simple data-parallel approach that computes each distance in a separate thread.

In our experiments, the compared values (i.e., the string characters) had always 32-bits. This may not be typical for ANSI strings, but some encodings (e.g., UTF-32) use such chars and other algorithms often compare numerical values of that size. We present only the results where both input strings of a distance have the same size, since they show the full potential of parallel processing.

The experiments were conducted on NVIDIA Tesla K20m (Kepler architecture, compute capability 3.5) with 2496 cores and 5 GB of VRAM. Each test was repeated five times and the measurements were within 1% deviation range. Arithmetic averages of the repeated measurements are presented as the results.

5.1 Single Distance

For the single-distance test, we have selected only two string lengths (64k and 128k) due to the limited scope. Figure 5 summarizes the measured times for the two selected sizes. The compared methods are denoted as follows: *base* stands for the baseline algorithm implemented according to Tomiyama [15], *cached* stands for the algorithm that uses texture cache for the string data, and *shfl* is the optimized version that employs shuffle instructions to exchange data within the thread warp.

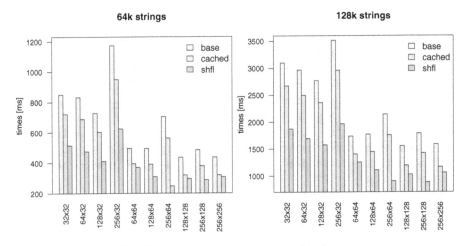

Fig. 5. Measured times for various block sizes

Presented times compare different block sizes from 32×32 to 256×256 of the individual parallelograms that are processed by different CUDA blocks. Let us emphasize, that the size is *width* × *height*, where width is the number of items processed by a thread and height is the number of thread in CUDA block. We have observed that sizes which are powers of two perform much better than other sizes (which are only multiple of warp size). A significant drop in performance was observed in cases when either the width or height of the block exceeds 256.

Both baseline version and cached version exhibit the best performance for 128×128 and 256×256 block sizes. This can be easily explained, since these sizes create the most workload for the SMPs of the GPU. On the other hand, the optimized version using shuffle functions perform better when the warps can perform more work without interruptions (i.e., without accessing the memory) even at the cost that the SMPs are slightly more underutilized. Hence the optimal blocks have width 256 and height 64 or 128 items. Proposed optimization achieved 1.75× speedup over the baseline algorithm and 1.3× speedup over the algorithm that uses texture cache for the string data.

5.2 Multiple Distances

In case of multiple distances, we used input strings of sizes from 32 to 256. The *simple* denotes the naïve parallelization of Wagner-Fischer algorithm [17], where each distance is computed in a separate thread and each thread hold exactly one row of the internal matrix in shared memory. The *shfl* denotes our proposed algorithm (that uses both texture cache and shuffling instructions).

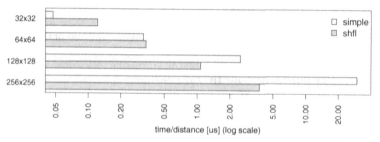

Fig. 6. Computing multiple distances concurrently

In each case, we have computed 256k distances and the results are normalized to time-per-distance values. The results indicate, that the turning point is the string length of 64 characters. Smaller strings are better processed by the simple algorithm, since it has smaller synchronization overhead. When the length grows, each thread requires more space in shared memory to hold its internal data, which inevitably leads to lower GPU core occupancy.

6 Conclusions

We have studied the problem of data dependencies in dynamic programming matrix, which are essential for parallelization of various similarity distances on GPUs. This problem cannot be easily eliminated, but we have proposed an optimization that utilizes new warp-cooperative shuffle instructions of the NVIDIA GPUs. The shuffle instructions have nontrivial impact on the overall performance, since they reduce memory data transfers and increase occupancy of the SMPs due to lowered shared memory consumption. We have also experimentally verified the importance of the new read-only texture cache introduced in the Kepler architecture.

Additionally, we have tested a multi-distance version of the algorithm where multiple distances are computed concurrently. We have established that for very small inputs, a simple data-parallel algorithm is better; however, when the length of the input strings exceed 64 the proposed algorithm gets better results.

Acknowledgments. This paper was supported by Czech Science Foundation (GAČR), projects P103-14-14292P and P103-13-08195, and by the Charles University Grant Agency (GAUK) project 122214.

References

1. Chacón, A., Marco-Sola, S., Espinosa, A., Ribeca, P., Moure, J.C.: Thread-cooperative, bit-parallel computation of Levenshtein distance on GPU. In: Proceedings of the 28th ACM International Conference on Supercomputing, pp. 103–112. ACM (2014)
2. Delgado, G., Aporntewan, C.: Data dependency reduction in dynamic programming matrix. In: 2011 Eighth International Joint Conference on Computer Science and Software Engineering (JCSSE), pp. 234–236. IEEE (2011)
3. Khajeh-Saeed, A., Poole, S., Perot, B.J.: Acceleration of the Smith-Waterman algorithm using single and multiple graphics processors. Journal of Computational Physics **229**(11), 4247–4258 (2010)
4. Levenshtein, V.I.: Binary codes capable of correcting deletions, insertions and reversals. Soviet Physics doklady **10**, 707 (1966)
5. Ligowski, L., Rudnicki, W.: An efficient implementation of Smith Waterman algorithm on GPU using CUDA, for massively parallel scanning of sequence databases. In: IEEE International Symposium on Parallel & Distributed Processing, 2009, IPDPS 2009, pp. 1–8. IEEE (2009)
6. Liu, Y., Huang, W., Johnson, J., Vaidya, S.: GPU accelerated Smith-Waterman. In: Alexandrov, V.N., van Albada, G.D., Sloot, P.M.A., Dongarra, J. (eds.) ICCS 2006. LNCS, vol. 3994, pp. 188–195. Springer, Heidelberg (2006)
7. Liu, Y., Wirawan, A., Schmidt, B.: Cudasw++ 3.0: accelerating Smith-Waterman protein database search by coupling CPU and GPU SIMD instructions. BMC Bioinformatics **14**(1), 117 (2013)
8. Manavski, S.A., Valle, G.: CUDA compatible GPU cards as efficient hardware accelerators for Smith-Waterman sequence alignment. BMC Bioinformatics **9**(Suppl 2), S10 (2008)
9. Müller, M.: Dynamic time warping. In: Information Retrieval for Music and Motion, pp. 69–84 (2007)
10. Myers, G.: A fast bit-vector algorithm for approximate string matching based on dynamic programming. Journal of the ACM (JACM) **46**(3), 395–415 (1999)
11. NVIDIA: Kepler GPU Architecture. http://www.nvidia.com/object/nvidia-kepler.html. Accessed 10 July 2015
12. Owens, J.D., Luebke, D., Govindaraju, N., Harris, M., Krüger, J., Lefohn, A.E., Purcell, T.J.: A survey of general-purpose computation on graphics hardware. In: Computer Graphics Forum, vol. 26, pp. 80–113. Wiley Online Library (2007)
13. Sart, D., Mueen, A., Najjar, W., Keogh, E., Niennattrakul, V.: Accelerating dynamic time warping subsequence search with GPUs and FPGAs. In: 2010 IEEE 10th International Conference on Data Mining (ICDM), pp. 1001–1006. IEEE (2010)
14. Smith, T.F., Waterman, M.S.: Identification of common molecular subsequences. Journal of Molecular Biology **147**(1), 195–197 (1981)
15. Tomiyama, A., Suda, R.: Automatic parameter optimization for edit distance algorithm on GPU. In: Daydé, M., Marques, O., Nakajima, K. (eds.) VECPAR 2012. LNCS, vol. 7851, pp. 420–434. Springer, Heidelberg (2013)
16. Ukkonen, E.: Finding approximate patterns in strings. Journal of Algorithms **6**(1), 132–137 (1985)
17. Wagner, R.A., Fischer, M.J.: The string-to-string correction problem. Journal of the ACM (JACM) **21**(1), 168–173 (1974)
18. Xu, K., Cui, W., Hu, Y., Guo, L.: Bit-parallel multiple approximate string matching based on GPU. Procedia Computer Science **17**, 523–529 (2013)

Time Series Subsequence Similarity Search Under Dynamic Time Warping Distance on the Intel Many-core Accelerators

Aleksandr Movchan[✉] and Mikhail Zymbler

South Ural State University, Chelyabinsk, Russia
movchanav@susu.ru

Abstract. Subsequence similarity search is one of the most important problems of time series data mining. Nowadays there is empirical evidence that Dynamic Time Warping (DTW) is the best distance metric for many applications. However in spite of sophisticated software speedup techniques DTW still computationally expensive. There are studies devoted to acceleration of the DTW computation by means of parallel hardware (e.g. computer-cluster, multi-core, FPGA and GPU). In this paper we present an approach to acceleration of the subsequence similarity search based on DTW distance using the Intel Many Integrated Core architecture. The experimental evaluation on synthetic and real data sets confirms the efficiency of the approach.

1 Introduction

Subsequence similarity search is one of the most important problems of time series data mining and appears in a wide spectrum of subject domains, e.g. climate modeling [1], economic forecasting [5], medical monitoring [6], etc. The problem assumes that a query sequence and a longer time series are given, and the task is to find a subsequence in the longer time series, which best matches with the query sequence.

Currently there is empirical evidence that the Dynamic Time Warping (DTW) [2] is the most popular similarity measure in many applications [3]. DTW is computationally expensive and there are approaches to solve this problem, e.g. lower bounding [3], computation reusing [14], data indexing [11], early abandoning [12], etc. However, DTW still costs too much and there are studies to accelerate subsequence similarity search using parallel hardware, e.g. computer-cluster [16], multi-core [15], FPGA and GPU [14,17,18].

In this paper we present a parallel algorithm for subsequence similarity search based on DTW distance adapted for use on a central processor unit (CPU) accompanied with the Intel Xeon Phi many-core coprocessor [4]. The remainder of the paper is organized as follows. Section 2 contains formal definition of the problem, briefly describes Intel Xeon Phi architecture and programming model and discusses related work. The proposed algorithm is presented in the section 3. The results of experimental evaluation of the algorithm are described in section 4. Section 5 contains summarizing comments and directions for future research.

© Springer International Publishing Switzerland 2015
G. Amato et al. (Eds.): SISAP 2015, LNCS 9371, pp. 295–306, 2015.
DOI: 10.1007/978-3-319-25087-8_28

2   Formal Definitions and Related Work

2.1   Formal Definitions

A *time series* T is an ordered sequence $t_1, t_2, \ldots, t_N$ of real data points, measured chronologically, where N is a length of the sequence.

Dynamic Time Warping (DTW) is a similarity measure between two time series X and Y, where $X = x_1, x_2, \ldots, x_N$ and $Y = y_1, y_2, \ldots, y_N$, is defined as follows.

$$DTW(X, Y) = d(N, N), \text{ where}$$

$$d(i, j) = |x_i - y_j| + min \begin{cases} d(i - 1, j) \\ d(i, j - 1) \\ d(i - 1, j - 1), \end{cases}$$

$$d(0, 0) = 0; d(i, 0) = d(0, j) = \infty; i = j = 1, 2, \ldots, N.$$

A *subsequence* T_{im} of time series T is its continuous subset starting from i-th position and consisting of m data points, i.e. $T_{im} = t_i, t_{i+1}, \ldots, t_{i+m-1}$, where $1 \leq i \leq N$ and $i + m \leq N$.

A *query* Q is a certain subsequence to be found in T. Let n is a length of the query, $n \ll N$.

Subsequence similarity search problem aims to finding a subsequence, which is the most similar to the query with respect to a given similarity measure. Let D is a similarity measure, then the resulting subsequence is $\underset{1 \leq i \leq N-n}{argmin} \, D(T_{in}, Q)$.

We will use DTW as a similarity measure.

2.2   The Intel Xeon Phi Architecture and Programming Model

The Intel Xeon Phi coprocessor is an x86 many-core coprocessor of 61 cores, connected by a high-performance on-die bidirectional interconnect where each core supports 4× hyperthreading and contains 512-bit wide vector processor unit (VPU). Each core has two levels of cache memory: a 32 Kb L1 data cache, a 32 Kb L1 instruction cache, and a core-private 512 Kb unified L2 cache. The Intel Xeon Phi coprocessor is to be connected to a host computer via a PCI Express system interface. PCI Express is used for data transfer between CPU and the coprocessor.

Being based on Intel x86 architecture, the Intel Xeon Phi coprocessor supports the same programming tools and models as a regular Intel Xeon processor.

The Intel Xeon Phi coprocessor supports three programming modes: native, offload and symmetric. In native mode the application runs independently, on the coprocessor only. In offload mode the application is running on the host and offloads computationally intensive part of work to the coprocessor. The symmetric mode allows the coprocessor to communicate with other devices by means of Message Passing Interface (MPI).

2.3   Related Work

Currently DTW is considered as best similarity measure for many applications [3], despite the fact that it is very time-consuming [8,16]. Research devoted to acceleration of DTW computation includes the following.

The SPRING algorithm [13] uses computation-reuse technique. However, this technique squeezes the algorithm's applications because data-reuse supposes non-normalized sequence. In [11] indexing technique to speed up the search was used, which need to specify the query length in advance. Authors of [9] suggested multiple indices for various length queries. Lower bounding [7] allows one to discard unpromising subsequences using the lower bound of DTW distance estimated in a cheap way. The UCR-DTW algorithm [12] integrates all the possible existing speedup techniques and most likely it is the fastest of the existing subsequence matching algorithms.

All the aforementioned algorithms aim to decrease the number of calls of DTW subroutine, not accelerating DTW itself. However, because of its complexity, DTW still takes a large part of the total application runtime [18]. There are approaches exploiting the allocation of DTW computation of different subsequences into different processing elements. In [15] subsequences starting from different positions of the time series are sent to different Intel Xeon processors, and each processor computes DTW. In [16] different queries are distributed onto different cores, and each subsequence is sent to different cores to be compared with different queries. GPU implementation [18] parallelize the generation of the warping matrix but still process the path search serially. GPU implementation proposed in [14] utilizes the same ideas as in [15]. FPGA implementation described in [14] focuses on the naive subsequence similarity search, and do not exploit any pre-processing techniques. It is generated by a C-to-VHDL tool and should be recompiled if length of query is changed. This algorithm supports 8-bit data precision and can not supports queries longer than 2^{10}, so it can not be applied in big-scale tasks. To address these problems in [17] a stream oriented framework was proposed. It implements coarse-grained parallelism by reusing data of different DTW computations and uses a two-phase precision reduction technique to guarantee accuracy while reducing resource cost.

In this work we present a parallel algorithm of the time series subsequence similarity search under DTW on the Intel Xeon Phi many-core coprocessor where the UCR-DTW serial algorithm is used as a basis.

3   Acceleration by the Intel Xeon Phi Coprocessor

3.1   Serial Algorithm

The UCR-DTW serial algorithm [12] is depicted in the Fig. 1. It uses a cascade of three lower bounding of DTW distance, namely LB_{Kim} [10,12], LB_{Keogh} [8] and $LB_{KeoghEC}$ [12]. If the lower bound has exceeded some threshold, the DTW distance also exceeds the same threshold, so the subsequence can be pruned off. Here the bsf (best-so-far) variable stores the distance to the most similar subsequence.

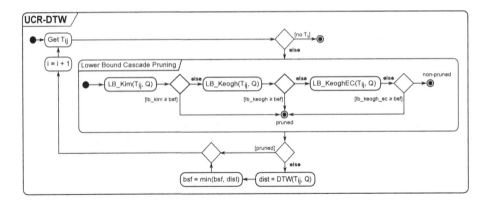

Fig. 1. Serial algorithm

3.2 Parallel Algorithm

Fig. 2 depicts a parallel version of the UCR-DTW algorithm. Parallelization of the original algorithm was performed through the OpenMP technology.

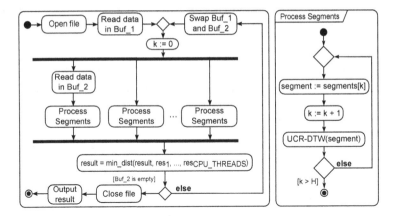

Fig. 2. Parallel algorithm for CPU

The source time series T is partitioned into H equal-length segments. Let P denotes the number of OpenMP-threads, S denotes a maximum length of segment, then H is defined as

$$H = \lceil \frac{N}{P \cdot S} \rceil \cdot P$$

A k-th segment, $0 \le k \le H - 1$, is defined as a subsequence T_{sl}, where

$$s = \begin{cases} 1 & , k = 0 \\ k \cdot \lfloor \frac{N}{H} \rfloor - n + 2 & , else \end{cases}$$

$$l = \begin{cases} \lfloor \frac{N}{H} \rfloor & , k = 0 \\ \lfloor \frac{N}{H} \rfloor + n - 1 + (N \bmod H) & , k = H - 1 \\ \lfloor \frac{N}{H} \rfloor + n - 1 & , else \end{cases}$$

It means that the head part of every segment except first overlaps with the tail part of previous segment in $n-1$ data points, where n is length of the query. This prevents from losing of possible resulting subsequences, which start at tail part of previous segment.

The number of segments H is divisible by the number of threads P for better load balancing.

The algorithm is based on dynamic distribution of segments across threads. We use k variable, which is shared among all threads and identifies first unprocessed segment. The k variable initialized by 0 and while there are unprocessed segments (i.e. $k \leq H$), a thread gets k-th segment, increments k by 1 and processes the segment by means of UCR-DTW subroutine, which implements an original serial algorithm. To provide correct processing of shared data we use critical section to prevent multiple threads from accessing the critical section's code at the same time, i.e. only one active thread can get k-th segment and update the k variable.

We reject static distribution of segments across threads (where each thread is assigned by its own segments before calculations) due to the following reason. Static distribution could result in worse load balancing because of unpredictable amount of pruned and early abandoned subsequences for each thread. So, overhead costs to provide the critical section in case of dynamic distribution is a lesser evil than highly probable load imbalance in case of static distribution.

In contrast with the serial version the bsf variable is shared among the threads. This allows each thread to prune off unpromising subsequence using lower bounding.

This algorithm is ready-to-use on the Intel Xeon Phi coprocessor in native mode. However, experiments have shown (Fig. 5) that the algorithm is slower than on CPU. This implementation does not provide sufficient floating point operations per byte of data to be effectively processed on the coprocessor. To overcome this we combined CPU and coprocessor to process time series as described in the next section.

3.3   Combining CPU and the Intel Xeon Phi

The parallel algorithm for CPU and the Intel Xeon Phi is depicted in Fig. 3.

The idea of the algorithm is that the coprocessor should be exploited only for DTW computations whereas CPU performs lower bounding, prepares subsequences for the coprocessor and computes DTW in case if it really does not have another job. CPU supports a queue of candidate subsequences and the coprocessor computes DTW for each candidate. Queue stores a tuple (i, A) corresponding a candidate subsequence T_{in}, where A is an n-element array containing LB_{Keogh} lower bounds for each position of the subsequence which is used for early abandoning of DTW [12].

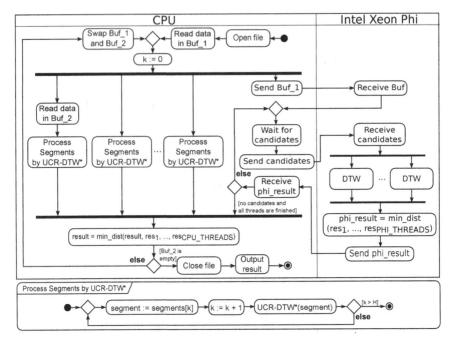

Fig. 3. Parallel algorithm for CPU and the Intel Xeon Phi

To reduce the amount of data transferred to the coprocessor, CPU offloads current buffer of the time series once whereas queue is offloaded each time it is full. The number of elements in the queue is calculated as $C \cdot h \cdot W$, where C is a number of cores of the coprocessor, h is a hyperthreading factor of the coprocessor and W is a number of candidates to be processed by a coprocessor's thread.

The algorithm could be described in the following way. One of the CPU threads is declared as a master and the rest as workers. At start master sends a buffer with the current buffer with the time series to the coprocessor. If queue is full then master offloads it to the coprocessor to perform DTW computation for the corresponding subsequences by the coprocessor's threads.

Worker's activity is similar to activity of threads in parallel algorithm for CPU only. Each worker processes segments by UCR-DTW* (see Fig. 4) subroutine. The UCR-DTW* subroutine calculates cascade of lower bounds for the subsequence. If it is dissimilar to the query then the worker prunes it off otherwise worker pushes this subsequence to the queue. If the queue is full (and data previously transferred to the coprocessor have not been processed yet), the worker calculates DTW by itself.

At the end of offload section the information about most similar subsequence found on the coprocessor is transferred to the CPU. The final result is calculated among the most similar subsequence found on the CPU and same that found on the coprocessor.

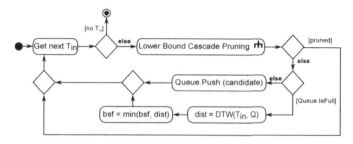

Fig. 4. UCR-DTW* subroutine

4   Experiments

Hardware. To evaluate the developed algorithm we performed experiments on the Tornado SUSU[1] supercomputer's node (see Tab. 1 for its specifications).

Table 1. Specifications of the Tornado SUSU supercomputer's node

Specifications	Processor	Coprocessor
Model	Intel Xeon X5680	Intel Xeon Phi SE10X
Cores	6	61
Frequency, GHz	3.33	1.1
Threads per core	2	4
Peak performance, TFLOPS	0.371	1.076
Memory, Gb	24	8
Cache, Mb	12	30.5

Data Sets. Experiments have been performed on three time series, which are summarized in Tab. 2. The PURE RANDOM data set was generated by a random function. The RANDOM WALK data set is one-dimensional random walk time series. The ECG (electrocardiographic) data set represents approximately 22 hours of one ECG channel sampled at 250 Hz.

Table 2. Data sets used in experiments

Time series	Category	Length
PURE RANDOM	synthetic	10^6
RANDOM WALK	synthetic	10^8
ECG [12]	real	$2 \cdot 10^7$

Goals. In the experiments we investigated a) performance of our algorithm, b) impact of the queue size on the speedup and c) runtime of our algorithm in comparison with analogues for GPU and FPGA.

---

[1] supercomputer.susu.ru/en/computers/tornado/

4.1   Performance

On the PURE RANDOM data set our algorithm shows (Fig. 5a) a two times higher performance than the parallel algorithm for CPU only.

Experimental results on RANDOM WALK data set (Fig. 5b) show that our algorithm is more effective for longer queries. In case of shorter queries the algorithm has the same performance as parallel algorithm for CPU only.

For the experiments on ECG data set we used a subsequence T_{N-nn} of the whole time series T as a query to prevent from finding the most similar subsequence at the early stage of computations and, in turn, to provide sufficient amount of work on DTW computation. Our algorithm shows (Fig. 5c) almost three times higher performance than the parallel algorithm for CPU only.

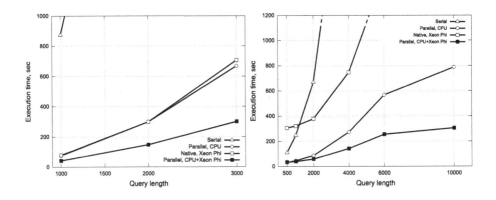

(a) PURE RANDOM data set   (b) RANDOM WALK data set

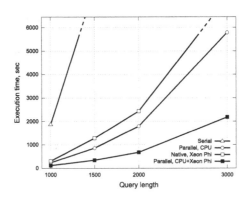

(c) ECG data set

Fig. 5. Performance of the algorithm

4.2   Impact of Queue Size

Results of the experiments are depicted in Fig. 6. In the current experimental environment, i.e. number of cores of the coprocessor C is 60^2, hyperthreading factor of the coprocessor h is 4, optimal number of candidates to be processed by a coprocessor's thread W is 10, so optimal number of the elements in the queue is 2400. Experimental results described above have been achieved with this queue size.

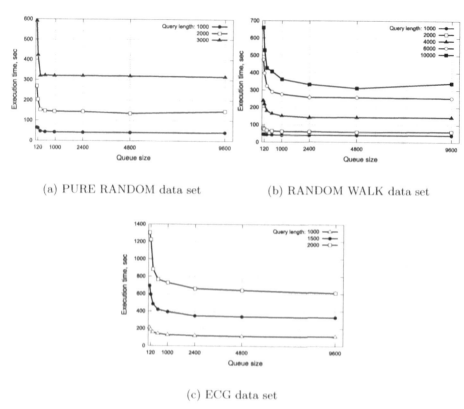

(a) PURE RANDOM data set   (b) RANDOM WALK data set

(c) ECG data set

Fig. 6. Impact of queue size on the speedup.eps

4.3   Comparison with Algorithms for GPU and FPGA

We compared the performance of our algorithm with analogues for GPU and FPGA developed in [14] (there is no comparison with results in [17] because that research was devoted to a little bit different problem of search a set of local-best-match subsequences). We repeated the experiments presented in that paper using the same data set and query length.

---

[2] One core is not involved in computations as it is recommended by the Intel Xeon Phi programmer's manual.

The results of the experiments are depicted in Fig. 7, here percentage on the top of the bar indicates a part of subsequences that have not been pruned and subjected to the DTW computation in our experiments. We also add to the chart results of experiments on random walk and ECG data sets.

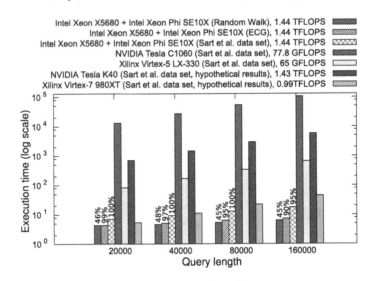

Fig. 7. Comparison of performance

We took into account that the peak performance of the hardware we used is significantly greater than its counterparts of that paper, i.e. overall peak performance of our hardware was 1.44 TFLOPS whereas GPU as NVIDIA Tesla C1060 had 77.8 GFLOPS and FPGA as Xilinx Virtex-5 LX-330 had 65 GFLOPS. To provide more "fair" comparison we added to the chart *hypothetical* results for modern NVIDIA Tesla K40 (1.43 TFLOPS)[3] and Xilinx Virtex-7 980XT (0.99 TFLOPS)[4] multiplying real results of NVIDIA Tesla C1060 and Xilinx Virtex-5 LX-330 by a respective scaling factor. As we can see our algorithm does not concede to analogous on performance.

5 Conclusion

In this paper we have presented an approach to time series subsequence similarity search under DTW distance on the Intel Many Integrated Core architecture. The parallel algorithm combines capabilities of CPU and the Intel Xeon Phi coprocessor. The coprocessor is exploited only for DTW computations whereas CPU performs lower bounding, prepares subsequences for the coprocessor and computes DTW as a last resort. CPU supports a queue of candidate subsequences and the coprocessor computes DTW for every candidate. Experiments

---

[3] www.nvidia.com/content/tesla/pdf/NVIDIA-Tesla-Kepler-Family-Datasheet.pdf

[4] www.xilinx.com/publications/prod_mktg/Virtex7-Product-Brief.pdf

on synthetic and real data sets have shown that our algorithm does not concede to analogous algorithms for GPU and FPGA on performance.

As future work we plan to extend our research for the cases of several coprocessors and cluster system based on nodes equipped with the Intel Xeon Phi coprocessor(s).

Acknowledgments. This work was financially supported by the Ministry of education and science of the Russian Federation ("Research and development on priority directions of scientific-technological complex of Russia for 2014–2020" Federal Program, contract No. 14.574.21.0035).

References

1. Abdullaev, S., Lenskaya, O., Gayazova, A., Sobolev, D., Noskov, A., Ivanova, O., Radchenko, G.: Short-range forecasting algorithms using radar data: Translation estimate and life-cycle composite display. Bull. of South Ural State University. Series: Comput. Math. and Soft. Eng. **3**(1), 17–32 (2014)
2. Berndt, D.J., Clifford, J.: Using dynamic time warping to find patterns in time series. In: Fayyad, U.M., Uthurusamy, R. (eds.) KDD Workshop, pp. 359–370. AAAI Press (1994)
3. Ding, H., Trajcevski, G., Scheuermann, P., Wang, X., Keogh, E.J.: Querying and mining of time series data: experimental comparison of representations and distance measures. PVLDB **1**(2), 1542–1552 (2008)
4. Duran, A., Klemm, M.: The intel many integrated core architecture. In: Smari, W.W., Zeljkovic,V. (eds.) HPCS, pp. 365–366. IEEE (2012)
5. Dyshaev, M., Sokolinskaya, I.: Representation of trading signals based on Kaufman adaptive moving average as a system of linear inequalities. Bull. of South Ural State University. Series: Comput. Math. and Soft. Eng. **2**(4), 103–108 (2013)
6. Epishev, V., Isaev, A., Miniakhmetov, R., Movchan, A., Smirnov, A., Sokolinsky, L., Zymbler, M., Ehrlich, V.: Physiological data mining system for elite sports. Bull. of South Ural State University. Series: Comput. Math. and Soft. Eng. **2**(1), 44–54 (2013)
7. Fu, A.W.-C., Keogh, E.J., Lau, L.Y.H., Ratanamahatana, C.A.: Scaling and time warping in time series querying. In: Böhm, K., Jensen, C.S., Haas, L.M., Kersten, M.L., Larson, P., Ooi, B.C. (eds.) Proceedings of the 31st International Conference on Very Large Data Bases, Trondheim, Norway, August 30–September 2, 2005, pp. 649–660. ACM (2005)
8. Keogh, E.J., Lau, L.Y.H., Ratanamahatana, C.A., Wong, R.C.-W.: Scaling and time warping in time series querying. VLDB J. **17**(4), 899–921 (2008)
9. Keogh, E.J., Wei, L., Xi, X., Vlachos, M., Lee, S.-H., Protopapas, P.: Supporting exact indexing of arbitrarily rotated shapes and periodic time series under Euclidean and warping distance measures. VLDB J. **18**(3), 611–630 (2009)
10. Kim, S.-W., Park, S., Chu, W.W.: An index-based approach for similarity search supporting time warping in large sequence databases. In: Georgakopoulos, D., Buchmann, A. (eds.) Proceedings of the 17th International Conference on Data Engineering, Heidelberg, Germany, April 2–6, 2001, pp. 607–614. IEEE Computer Society (2001)
11. Lim, S.-H., Park, H.-J., Kim, S.-W.: Using multiple indexes for efficient subsequence matching in time-series databases. In: Li Lee, M., Tan, K.-L., Wuwongse, V. (eds.) DASFAA 2006. LNCS, vol. 3882, pp. 65–79. Springer, Heidelberg (2006)

12. Rakthanmanon, T., Campana, B.J.L., Mueen, A., Gustavo, B., Westover, B., Zhu, Q., Zakaria, J., Keogh, E.J.: Searching and mining trillions of time series subsequences under dynamic time warping. In: Yang, Q., Agarwal, D., Pei, J. (eds.) KDD, pp. 262–270. ACM (2012)

13. Sakurai, Y., Faloutsos, C., Yamamuro, M.: Stream monitoring under the time warping distance. In: Chirkova, R., Dogac, A., Özsu, M.T., Sellis, T.K. (eds.) Proceedings of the 23rd International Conference on Data Engineering, ICDE 2007, The Marmara Hotel, Istanbul, Turkey, April 15–20, 2007, pp. 1046–1055. IEEE (2007)

14. Sart, D., Mueen, A., Najjar, W.A., Keogh, E.J., Niennattrakul, V.: Accelerating dynamic time warping subsequence search with GPUs and FPGAs. In: Webb, G.I., Liu, B., Zhang, C., Gunopulos, D., Wu, X. (eds.) ICDM, pp. 1001–1006. IEEE Computer Society (2010)

15. Sharanyan, S., Arvind, K., Rajeev, G.: Implementing the dynamic time warping algorithm in multithreaded environments for real time and unsupervised pattern discovery. In: Department of Computer Science and Motial Nehru National Institute of Technology Engineering, ICCCT, pp. 394–398. IEEE Computer Society (2011)

16. Takahashi, N., Yoshihisa, T., Sakurai, Y., Kanazawa, M.: A parallelized data stream processing system using dynamic time warping distance. In: Barolli, L., Xhafa, F., Hsu, H.-H. (eds.) 2009 International Conference on Complex, Intelligent and Software Intensive Systems, CISIS 2009, Fukuoka, Japan, March 16–19, 2009, pp. 1100–1105. IEEE Computer Society (2009)

17. Wang, Z., Huang, S., Wang, L., Li, H., Wang, Y., Yang, H.: Accelerating subsequence similarity search based on dynamic time warping distance with FPGA. In: Hutchings, B.L., Betz, V. (eds.) The 2013 ACM/SIGDA International Symposium on Field Programmable Gate Arrays, FPGA 2013, Monterey, CA, USA, February 11–13, 2013, pp. 53–62. ACM (2013)

18. Zhang, Y., Adl, K., Glass, J.R.: Fast spoken query detection using lower-bound dynamic time warping on graphical processing units. In: 2012 IEEE International Conference on Acoustics, Speech and Signal Processing, ICASSP 2012, Kyoto, Japan, March 25–30, 2012, pp. 5173–5176. IEEE (2012)

Subspace Nearest Neighbor Search - Problem Statement, Approaches, and Discussion

Position Paper

Michael Hund[1]([⊠]), Michael Behrisch[1], Ines Färber[2], Michael Sedlmair[3],
Tobias Schreck[4], Thomas Seidl[2], and Daniel Keim[1]

[1] University of Konstanz, Konstanz, Germany
{michael.hund,michael.behrisch,daniel.keim}@uni-konstanz.de
[2] RWTH Aachen University, Aachen, Germany
{faerber,seidl}@informatik.rwth-aachen.de
[3] University of Vienna, Wien, Austria
michael.sedlmair@univie.ac.at
[4] Graz University of Technology, Graz, Austria
tobias.schreck@cgv.tugraz.at

Abstract. Computing the similarity between objects is a central task for many applications in the field of information retrieval and data mining. For finding k-nearest neighbors, typically a ranking is computed based on a predetermined set of data dimensions and a distance function, constant over all possible queries. However, many high-dimensional feature spaces contain a large number of dimensions, many of which may contain noise, irrelevant, redundant, or contradicting information. More specifically, the relevance of dimensions may depend on the query object itself, and in general, different dimension sets (subspaces) may be appropriate for a query. Approaches for feature selection or -weighting typically provide a global subspace selection, which may not be suitable for all possibly queries. In this position paper, we frame a new research problem, called *subspace nearest neighbor search*, aiming at multiple query-dependent subspaces for nearest neighbor search. We describe relevant problem characteristics, relate to existing approaches, and outline potential research directions.

Keywords: Nearest neighbor search · Subspace analysis and search · Subspace clustering · Subspace outlier detection

1 Introduction

Searching for similar objects is a crucial task in many applications, such as image or information retrieval, data mining, biomedical applications, and e-commerce. Typically *k-nearest neighbor queries* are used to compute *one result* list of similar objects derived from a given set of data dimensions and a distance function.

© Springer International Publishing Switzerland 2015
G. Amato et al. (Eds.): SISAP 2015, LNCS 9371, pp. 307–313, 2015.
DOI: 10.1007/978-3-319-25087-8_29

However, the consideration of all dimensions and a single distance function may not be appropriate for all queries, as we will discuss in the following.

For datasets with a high number of dimensions, similarity measures may loose their discriminative ability since similarity values concentrate about their respective means. This phenomenon, known as the *curse of dimensionality* [2], leads to an instability of nearest neighbor queries in high-dimensional spaces. The instability increases with the proportion of irrelevant or conflicting dimensions.

Consider the following clinical example: A physician is treating a patient with an unknown disease and wants to retrieve similar patients along with their medical history (treatment, outcome, etc.). In the search process, the physician is confronted with a high number of unrelated diseases and respective symptoms. The most similar patients (nearest neighbors, $\mathcal{NN}$) based on all features are often not suited to guide the diagnostic process as irrelevant dimensions, such as the hair color, may dominate the search process. Meaningful conclusions can only be drawn if the *characteristic* dimensions for the particular disease are considered. The challenging question is therefore, what is the relevant subset of dimensions (=*subspace*) specific for a certain query? Do multiple relevant subspaces exist? Many other application examples can be found, where $\mathcal{NN}$ search in query-dependent subspaces is potentially relevant, e.g., in multimedia retrieval a query may depend on the input object type; in recommender systems a query may depend on user preferences; or a kNN-classifier may depend on the class label.

Consequently, we can derive a novel research challenge, which we call *subspace nearest neighbor search*, for short $\mathcal{SNNS}$. Its central idea is to incorporate a *query-dependency focus* into the relevance definition of subspaces. As one example, $\mathcal{SNNS}$ allows deriving discriminative subspaces in which the $\mathcal{NN}$ of a query can be separated from the rest of the data. Alternatively, in the above example, the physician will focus on a large number of dimensions to maximize the semantic interpretability of the $\mathcal{NN}$ along with the query-dependent subspace.

$\mathcal{SNNS}$ is inspired by works in subspace clustering and -search. However, it differs from these fields, as the goal is to derive query-dependent subspaces. Therefore, we define a novel problem definition. In $\mathcal{SNNS}$, our goal is to (1) detect *query dependent* and *previously unknown subspaces* that are relevant, and (2) derive the corresponding nearest neighbor set to the query within that corresponding subspace. This paper addresses the following questions: "What is a relevant subspace for a given query?", "How can we computationally extract this relevance information?", and "How can we adapt ideas from subspace clustering, outlier detection, or feature selection for $\mathcal{SNNS}$?"

2   Related Problems

Next, we give a concise overview of the fields related to $\mathcal{SNNS}$. An overview about the fields and its relation to $\mathcal{SNNS}$ is also given in Fig. 1 Ⓐ - Ⓓ.

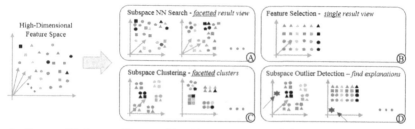

Fig. 1. Focus of Subspace Nearest Neighbor Search ($\mathcal{SNNS}$) and related approaches: While $\mathcal{SNNS}$ aims at multiple, query-dependent subspaces, related fields focus on a single result or on subspaces with different properties.

Feature Selection, Extraction and Weighting. The aim of feature selection [10] is to determine one subspace that improves a global optimization criterion (e.g., classification error). As shown in (B), there are two main differences to $\mathcal{SNNS}$: Feature selection derives a single subspace (result view) for all analysis tasks, and the resulting subspace is query independent. In contrast, $\mathcal{SNNS}$ is aiming at a *faceted result view* of multiple, query-dependent subspaces.

Subspace Clustering. Subspace clustering aims at finding clusters in different axis-parallel or arbitrarily-oriented subspaces [9]. The approaches are based on subspace search methods and heuristics to measure the subspace cluster quality. The computation of clusters and subspaces can be tightly coupled or decoupled, see e.g., [8]. As shown in (C), subspace clustering and $\mathcal{SNNS}$ both aim at a facetted result, but differ in their relevance definition of a subspace: dense clusters vs. query-dependent nearest neighbors in multiple subspaces.

Subspace Outlier Detection. Methods in this area search for subspaces in which an arbitrary, or a user-defined object is considered as outlier [13]. As before, the search process consists of subspace search methods and criteria to measure the subspace quality, e.g., by item separability [11]. Subspace outlier detection is similar to $\mathcal{SNNS}$ as both approaches aim for query-dependent subspaces (D), however, the relevance definition of a subspace differs significantly as $\mathcal{SNNS}$ searches for objects that are similar to the query, while subspace outlier detection seeks for objects dissimilar to all other objects.

Query-Dependent Subspace Search. In [5] it was proposed to determine one query-dependent subspace to improve $\mathcal{NN}$-queries. The authors describe an approach to measure the quality of a subspace by the separability between all data records and the $\mathcal{NN}$ of a query. In their evaluation, they show that a query-dependent subspace reduces the error of a $\mathcal{NN}$-classification substantially. The work can be seen as initial approach on $\mathcal{SNNS}$ and, therefore, most closely relates to our work. However, the general aims of [5] differ, as it does not search for a facetted result view, i.e. different $\mathcal{NN}$ sets in multiple, different subspaces.

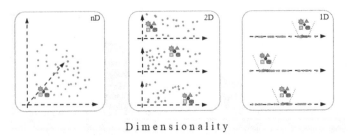

Dimensionality

Fig. 2. Illustration of our subspace model: A subspace is considered *relevant*, iff the nearest neighbors are similar to the query in *all* dimensions of the subspace.

Other Related Problems. Besides these main lines, another related field is that of recommender systems [1], which focuses on similarity aspects to retrieve items of interest. Intrinsic dimensionality estimation [3] shares the intuition of a minimum-dimensional space that preserves the distance relationships. One other recent work focuses on the efficient $\mathcal{NN}$ retrieval in subspaces [7].

3   Definition of Subspace Nearest Neighbor Search

In the following we define characteristics of the $\mathcal{SNNS}$ problem and introduce an initial model to identify relevant candidate subspaces.

The aim of $\mathcal{SNNS}$ can be divided into two coupled tasks: (a) detect <u>all</u> previously unknown subspaces that are *relevant* for a $\mathcal{NN}$ search of a given query, and (b) determine the respective set of $\mathcal{NN}$ within each relevant subspace. Different queries may change the relevance of subspaces and affect the resulting $\mathcal{NN}$-sets. Therefore, the characteristics of the query need to be considered for the subspace search strategy and the evaluation criterion (c.f. Section 4).

We propose an initial subspace model[1] to derive the relevance of a subspace w.r.t. a $\mathcal{NN}$-search. As illustrated in Fig. 2, a subspace is considered *relevant*, iff the following holds: "A set of objects a, b, c are $\mathcal{NN}$ of the query q in a subspace s, iff a, b, and c are a $\mathcal{NN}$ of q in *all* dimensions of s." More formally:

$$\forall_{n \in nn(q,s)} \ and \ \forall_{d \in dim(s)} : \ n \in nn(q,d)$$

whereby $nn(q, s)$ indicates the $\mathcal{NN}$ of q in s, and $dim(s)$ the set of dimensions of the subspace. This principle of a common set of $\mathcal{NN}$ in different dimensions is similar to the concept of the *shared nearest neighbor distance* [6] or consensus methods. The intuition is that the member dimensions of a subspace agree (to a certain minimum threshold) in their $\mathcal{NN}$ rankings, when considered individually.

This *item-based* subspace concept is different to the distance distribution-based model presented in [5], or most subspace clustering approaches. Besides

---

[1] Our model assumes *axis-parallel* subspaces. Further research is necessary to analyze the usefulness of *arbitrarily-oriented* subspaces for $\mathcal{NN}$ search.

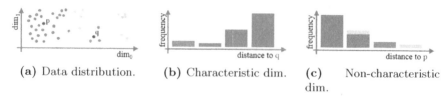

(a) Data distribution.   (b) Characteristic dim.   (c)   Non-characteristic dim.

Fig. 3. Distance distribution based measure to determine the characteristic of a dimension w.r.t. a $\mathcal{NN}$ search of a given queries p and q.

the advantage of a semantic $\mathcal{NN}$ interpretability, the model allows to compute heterogeneous subspaces. The relevance of a subspace is independent of a global distance function, but relies on individual $\mathcal{NN}$ computations in all dimensions.

Not every subspace, considered relevant by our model, is necessarily *interesting* in all application scenarios. In the medical example from the beginning, a physician will focus on the semantic interpretability of the results, while accepting potential redundant information. In other scenarios, the minimal description of a subspace may be preferred (c.f. intrinsic dimensionality [3]). Alternative interestingness definitions, such as focusing on subspaces with a minimum – respectively maximum– number of $\mathcal{NN}$ could be possible, too. Generally, the *quality criterion* for nearest neighbor subspaces, has to be regarded as application dependent.

4   Discussion and Open Research Questions

While initial experiments[2] hint on the usefulness of $\mathcal{SNNS}$, we have identified six central research directions that should be explored in the future.

Determine $\mathcal{NN}$ per Dimension. A central question that arises from the model definition is when a data record is considered as $\mathcal{NN}$ to q. Whenever similarity is modeled by a distance function we need to define, detect, or learn an appropriate $\mathcal{NN}$ membership threshold.

Efficient Search Strategy. The number of axis-parallel subspaces is $2^d - 1$ for a d-dimensional dataset. Consequently, an efficient search strategy is necessary to quickly detect relevant subspaces. *Top-down* approaches, based on a *locality criterion* [9], assume that relevant subspaces can be approximated in full space. Yet, our initial tests lead to the assumption that shared $\mathcal{NN}$ in independent dimensions, as required by our model, can benefit from a *bottom-up* strategy starting from $\mathcal{NN}$ in individual dimensions. Our model fulfills the *downward closure property* [9] which allows to make use of *APRIORI-like* algorithms.

Query-Based Interestingness for Dimensions. The subspace search strategy can further benefit by focusing on interesting dimensions. We propose a

---

[2] C.f. supplementary material on our website: http://files.dbvis.de/sisap2015/.

measure for single dimensions, based on the idea described in [5] that extracts the characteristic of dimension w.r.t. the query. As shown in Fig. 3, dimensions in which most data records are similar to the query are considered as non-characteristic, hence they are less interesting for possible subspaces.

Subspace Quality Criterion. Novel criteria are needed to rank the detected subspaces by their interestingness. The intuition to measure a subspace's quality differs significantly from earlier approaches, as outlined in Section 2. In addition, novel user interfaces and visualizations are necessary to understand and interpret multiple, partially redundant, subspaces and their different rankings [4].

Evaluation. Evaluating subspace analysis methods is challenging, as obtaining real-world dataset with annotated subspace information is expensive [12]. Likewise, synthetic data for the evaluation of subspace clustering (e.g., *OpenSubspace Framework* [12]), differs in the analysis goals (c.f. Section 2). Hence, research will benefit from a established ground-truth dataset for the evaluation of $\mathcal{SNNS}$.

Multi-input $\mathcal{SNNS}$. In many scenarios such as in the medical domain, a small set of query records needs to be investigated by means of $\mathcal{SNNS}$. One challenge for *multi-input $\mathcal{SNNS}$* are dimensions in which the set of queries differ.

5 Conclusion

This position paper outlines a novel research problem, called subspace nearest neighbor search ($\mathcal{SNNS}$), which aims at determining *query-dependent* subspaces for nearest neighbor search. Initial experiments have proven the usefulness and that it is beneficial to drive research in this field.

Acknowledgments. We would like to thank the German Research Foundation (DFG) for financial support within the projects A03 of SFB/Transregio 161 "Quantitative Methods for Visual Computing" and DFG-664/11 "SteerSCiVA: Steerable Subspace Clustering for Visual Analytics".

References

1. Adomavicius, G., Tuzhilin, A.: Toward the next generation of recommender systems: A survey of the state-of-the-art and possible extensions. IEEE TKDE **17**(6), 734–749 (2005)
2. Beyer, K.S., Goldstein, J., Ramakrishnan, R., Shaft, U.: When is "nearest neighbor" meaningful? In: Proc. 7th Int. Conf. Database Theory, pp. 217–235 (1999)
3. Camastra, F.: Data dimensionality estimation methods: a survey. Pattern Recognition **36**(12), 2945–2954 (2003)
4. Gleicher, M., Albers, D., Walker, R., Jusufi, I., Hansen, C.D., Roberts, J.C.: Visual comparison for information visualization. Information Visualization **10**(4), 289–309 (2011)
5. Hinneburg, A., Keim, D.A., Aggarwal, C.C.: What is the nearest neighbor in high dimensional spaces? In: Proc. 26th Int. Conf. on VLDB, Cairo, Egypt (2000)

6. Houle, M.E., Kriegel, H.-P., Kröger, P., Schubert, E., Zimek, A.: Can shared-neighbor distances defeat the curse of dimensionality? In: Gertz, M., Ludäscher, B. (eds.) SSDBM 2010. LNCS, vol. 6187, pp. 482–500. Springer, Heidelberg (2010)
7. Houle, M.E., Ma, X., Oria, V., Sun, J.: Efficient algorithms for similarity search in axis-aligned subspaces. In: Traina, A.J.M., Traina Jr, C., Cordeiro, R.L.F. (eds.) SISAP 2014. LNCS, vol. 8821, pp. 1–12. Springer, Heidelberg (2014)
8. Kailing, K., Kriegel, H.-P., Kröger, P., Wanka, S.: Ranking interesting subspaces for clustering high dimensional data. In: Lavrač, N., Gamberger, D., Todorovski, L., Blockeel, H. (eds.) PKDD 2003. LNCS (LNAI), vol. 2838, pp. 241–252. Springer, Heidelberg (2003)
9. Kriegel, H.P., Kröger, P., Zimek, A.: Clustering high-dimensional data: A survey on subspace clustering, pattern-based clustering, and correlation clustering. ACM TKDD **3**(1), 1 (2009)
10. Liu, H., Motoda, H.: Computational Methods of Feature Selection. Data Mining and Knowledge Discovery Series. Chapman & Hall/CRC Press (2007)
11. Micenkova, B., Dang, X.H., Assent, I., Ng, R.: Explaining outliers by subspace separability. In: 13th. IEEE ICDM, pp. 518–527 (2013)
12. Müller, E., Günnemann, S., Assent, I., Seidl, T.: Evaluating clustering in subspace projections of high dimensional data. In: VLDB, vol. 2, pp. 1270–1281 (2009)
13. Zimek, A., Schubert, E., Kriegel, H.P.: A survey on unsupervised outlier detection in high-dimensional numerical data. Statistical Analysis and Data Mining **5**(5), 363–387 (2012)

Query-Based Improvement Procedure and Self-Adaptive Graph Construction Algorithm for Approximate Nearest Neighbor Search

Alexander Ponomarenko[✉]

National Research University Higher School of Economics, Nizhny Novgorod, Russia
aponomarenko@hse.ru

Abstract. The nearest neighbor search problem is well known since 60s. Many approaches have been proposed. One is to build a graph over the set of objects from a given database and use a greedy walk as a basis for a search algorithm. If the greedy walk has an ability to find the nearest neighbor in the graph starting from any vertex with a small number of steps, such a graph is called a navigable small world. In this paper we propose a new algorithm for building graphs with navigable small world properties. The main advantage of the proposed algorithm is that it is free from input parameters and has an ability to adapt on the fly to any changes in the distribution of data. The algorithm is based on the idea of removing local minimums by adding new edges. We realize this idea to improve search properties of the structure by using the set of queries in the execution stage. An empirical study of the proposed algorithm and comparison with previous works are reported in the paper.

Keywords: Nearest neighbor search · Non-metric search · Approximate search

1 Introduction

The nearest neighbor search problem has naturally appeared in many fields of science. The problem is formulated as follows. Let D be a domain, $X \subset D$ be a finite set of objects (database), $d : D \times D \to R^{[0;+\infty)}$ be a distance function. We need to preprocess X in such a way that for a given query $q \in D$ the k-closest objects from X can be found as fast as possible. Many methods have been proposed for searching an exact nearest neighbor as well as an approximate. Some recent methods are [6], [7]. Also a good overview of methods for the exact nearest neighbor search can be found in [1] and an empirical comparison of several approximate state of the art methods can be found in [3]. One of the recent promising approaches for the approximate nearest neighbor search problem is to build a graph $G(X, E)$ over the set of objects X and use the greedy walk algorithm as a base for the search algorithm. Thus, the search of k-closest objects takes the form of the search of vertices in the graph G, where each object from the set X uniquely corresponds to a different vertex of the graph. Several works based on this approach have been proposed [2,4,5]. Methods [2, 5]

© Springer International Publishing Switzerland 2015
G. Amato et al. (Eds.): SISAP 2015, LNCS 9371, pp. 314–319, 2015.
DOI: 10.1007/978-3-319-25087-8_30

are based on the idea of connecting a set of approximate Voronoi regions into a network where one region corresponds to one data point. Voronoi regions are approximated by k-closest points together with a set of points for which the current point is one of the k-closest. The main drawback of this method is that it requires initial setting of the parameter k which in turn requires a priori knowledge about the properties of the input data set (a small value of the parameter leads to a small accuracy of search; too large value causes excessive search complexity and needs more memory to store the graph).

In this paper we explore an idea how to form a graph without any a priori knowledge of input data. Instead of finding k-closest neighbors we explicitly try to build a graph in such a way that the greedy walk algorithm is able to find a new data point starting from any other random vertex of the current graph G. Moreover we propose a way to improve search properties of the structure with a similar idea by using the set of incoming queries.

2   Greedy Walk Algorithm

As mentioned above we use an approach where each object from the set X is uniquely mapped to a different vertices of the graph G. Thus, the search of k-closest objects takes the form of the search of vertices in the graph. As a search algorithm we suggest to use a simple greedy walk algorithm. The pseudo code of the greedy walk algorithm is presented below. The greedy walk algorithm is quite simple. It starts from some vertex v_{start}; calculates the distance between the query and each neighbor of v_{start}; goes to the vertex v_{curr} for which the distance is minimal; calculates the distance d between the query q and each object from neighbors $N(v_{curr})$ of vertex v_{curr} and so on, until the algorithm cannot improve the distance to the query. The vertices in which the greedy walk stops we call "local minimums". Note that the Greedy_Walk algorithm with local minimum also returns the set P. This set P contains all vertices that were contacted during a search (lines 2 and 5).

```
Greedy_Walk( q∈ D , G(V,E) , v_start ∈ V )
1    v_curr ← v_start
2    P ← P∪N(v_curr)
3    while   min   (d(q,x)) < d(q, v_curr) do
              x∈N(v_curr)
4         v_curr ← arg min(d(q,x))
                    x∈N(v_curr)
5            P ← P∪N(v_curr)
6    end while
7    return   v_curr , P
```

3   Insertion Algorithm

The insertion algorithm is based on the idea, that each time when we insert a new vertex to the graph G, we can explicitly check the possibility of finding the new vertex starting from other vertices by the greedy algorithm. We also exploit a trick of keeping alive old links. Since time this links starting to be used as long links by the greedy search algorithm. So, this trick allows us to produce the navigable small world property for our graph.

At first we will describe the `Get_Local_Minimums` function. It returns the set of local minimums in which randomly started greedy walks stop. The `Get_Local_Minimums` algorithm has parameter τ which means how many times we allow a randomly started greedy walk not to bring a new local minimum. Also the `Get_Local_Minimums` function returns the set P which we consider as global view of the all greedy walks.

```
Get_Local_Minimums( q∈ D , G(V,E) , τ ∈ N )
01  L ← {} ; τ'← 0
02   while  τ'< τ  do
03     V_start ← Random( V ) //put to  V_start  random vertex from  V
04     v,P ← Greedy_Walk( q , G , V_start )
05      if  v∉ L  then  τ'← 0 ;  L ← L∪v
08      else  τ'← τ'+1
09   end while
11   return  L , P
```

```
Insert_By_Repairing( x∈ X , G(V,E) , τ ∈ N )
01   V ← V ∪ x ; P ← {}
02   L,P ← Get_Local_Minimums( X , G , τ )
03   P ← P ∪ x
04   for each  z∈ L  do
05      P'← {y∈ P : d(y,x) < d(z,x)}
06      x'← argmin(d(z, y))
             y∈P'
07      E ← E ∪ (x',z) ∪ (z,x')
08   end for
```

Now we ready to describe `Insert_By_Repairing` procedure. At first, we collect in the set L all local minimums (line 02) which we found by a number of randomly started greedy walks. After that, we remove these local minimums in the following way. For every local minimum z in the set of all viewed vertices (points) P we select the vertex x' such that the distance from x' to x is less than the distance from z to x, and the distance from z to x' is minimal. So, we make the local

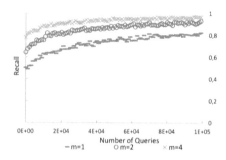

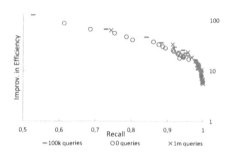

Fig. 1. Recall after Improving by queries applying `Repair_By_Query` procedure

Fig. 2. The overall efficiency of the structure after applying `Repair_By_Query` procedure

minimum z to be able to route the greedy walk to the vertex x' which is closer to the destination point x by adding edge between z and x'.

Finally, we present procedure `Add_All` which builds a data structure over the set X. It sequentially selects a random object from the set X and inserts it to the graph by running the `Insert_By_Repairing` procedure.

```
Add_All ( X ⊂ D , τ ∈ N )
01    G ← ({ },{ })
02    while X ≠ { } do
03        x ← Random ( X ) ; X ← X \ x
04        G ← Insert_By_Repairing ( x , G , τ )
05    end while
```

4   Improvement Based on Queries

The idea of repairing by removing local minimums can be extended to improving the search properties of the structure by queries at the query execution stage. In the procedure `Repair_By_Query` we do the same as in `Insert_By_Repairing` procedure with only one exception. We search the local minimums relative to the query object q instead of the inserted one.

```
Repair_By_Query ( q ∈ D , G(V,E) , τ ∈ N )
01    L,P ← Get_Local_Minimums ( q , G , τ )
02    x ← arg min(d(y,q))
            y∈L
03    for each z ∈ L \ x  do
04        P' ← { y ∈ P : d(y,q) < d(z,q)}
05        x' ← arg min(d(z, y))
                y∈P'
06        E ← E ∪ (x', z) ∪ (z, x')
07    end for
```

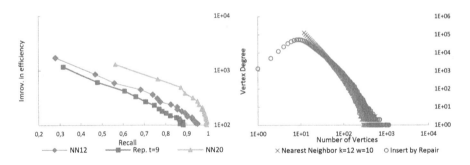

Fig. 3. Comparison with NN [5]   **Fig. 4.** Vertex Degree Distribution

5   Simulations

5.1   Improvement Based on Queries

We have performed simulations in order to verify the idea of improving the data structure using the stream of incoming queries. We have built a graph by Add_All algorithm with parameter $\tau = 9$ on the set of 10,000 points uniformly distributed in the 30-dimensional unit hyper cube. After that, we generate 100 times 1000 queries with the same distribution. Each time after applying 1000 times the Repair_By_Query procedure we have measured the value of recall for searching 5 nearest neighbors. As a search algorithm we have used Multi_Search algorithm proposed in [5]. This algorithm uses the sequences of greedy based searches started from a random vertex and selects the best results. The parameter "m" is the number of searches. We have used m = 1, 2, and 4 correspondingly. As can been seen from Fig. 1 the accuracy of each search has increased after applying the Repair_By_Query procedure.

Also we have measured how the overall efficiency of the structure has changed when applying Repair_By_Query procedure. We have measured the values of improvement in efficiency (how many times less the algorithm needs to calculate distances than the exhaustive search) and values of recall after applying Repair_By_Query procedure 100,000 and 1,000,000 times to the structure built over the set of 100,000 30-dimensional points by Insert_By_Repairing algorithm (Fig. 2). Unfortunately, it has not made a significant contribution to the overall efficiency of the structure.

5.2   Insertion by Repairing

We have made the comparison of the search properties of the graphs produced by Insert_By_Repairing (parameter $\tau = 9$) algorithm with the algorithm of connecting with k-nearest neighbors [5]. The parameters were w=10; k=12 and 20 accordingly. The parameter "w" is the number of searches used by Multi_Search procedure at the construction stage As a data set we have used one million random points uniformly distributed in the unit 30-dimensional hyper cube. As a distance function we use L_2. The result of comparison is presented in Fig. 3 and a valuation of the distribution of the vertex degrees is presented in Fig. 4.

6   Conclusion

In this paper we have explored the idea of using information about local minimums during the insertion and during the search. We have proposed an algorithm which uses this information to improve the accuracy of the search procedure. Also based on this idea we have proposed a new graph construction algorithm. Despite that the proposed algorithm cannot outperform the algorithm of connecting with k-nearest neighbors, it demonstrates the idea that the information about local minimums can be used for data insertion. Moreover we suppose that this idea can be used for tuning parameters of other data insertion algorithms, for example for tuning the parameter k in the algorithm of connection with k-nearest neighbors [5].

Acknowledgements. The work was conducted at National Research University Higher School of Economics and supported by RSF grant 14-41-00039.

References

1. Chávez, E., Navarro, G., Baeza-Yates, R., Marroquín, J.L.: Searching in metric spaces. In: ACM computing surveys (CSUR) 33, vol. 3, pp. 273–321 (2001)
2. Malkov, Y., Ponomarenko, A., Logvinov, A., Krylov, V.: Scalable distributed algorithm for approximate nearest neighbor search problem in high dimensional general metric spaces. In: Navarro, G., Pestov, V. (eds.) SISAP 2012. LNCS, vol. 7404, pp. 132–147. Springer, Heidelberg (2012)
3. Ponomarenko, A., Avrelin, N., Naidan, B., Boytsov, L.: Comparative analysis of data structures for approximate nearest neighbor search. In: DATA ANALYTICS 2014, The Third International Conference on Data Analytics, pp. 125–130 (2014)
4. Lifshits, Y., Shengyu, Z.: Combinatorial algorithms for nearest neighbors, near-duplicates and small-world design. In: Proceedings of the Twentieth Annual ACM-SIAM Symposium on Discrete Algorithms, pp. 318–326. Society for Industrial and Applied Mathematics (2009)
5. Malkov, Y., Ponomarenko, A., Logvinov, A., Krylov, V.: Approximate nearest neighbor algorithm based on navigable small world graphs. Information Systems **45**, 61–68 (2014)
6. Chávez, E., Graff, M., Navarro, G., Téllez, E.S.: Near neighbor searching with K nearest references. Information Systems **51**, 43–61 (2015)
7. Skopal, T.: Unified framework for fast exact and approximate search in dissimilarity spaces. ACM Transactions on Database Systems (TODS) **32**(4), 29 (2007)

Posters

Is There a Free Lunch for Image Feature Extraction in Web Applications

Martin Kruliš[(✉)]

Parallel Architectures/Algorithms/Applications Research Group,
Faculty of Mathematics and Physics, Charles University in Prague,
Malostranské nám. 25, Prague, Czech Republic
krulis@ksi.mff.cuni.cz

Abstract. Feature extraction is one of the essential parts of multimedia indexing in similarity search and content-based retrieval methods. Most applications that employ these methods also implement their client side interface using web technologies. The world wide web has become a well-established platform for distributed software and virtually all personal computers, tablets, and smartphones are equipped with a web browser. In the past, most applications employed a strict client-server approach, where the client part (running in the browser) handles only the user interface and the server side handles data storage and business logic. However, the client-side technologies leaped forward with the new HTML5 standard and the web browser has become capable of handling much more complex tasks. In this paper, we propose a model where the multimedia indexing is handled at the client side, which reduces necessary computational power of the server to run a web application that manages large multimedia database. We have implemented an in-browser image feature extractor and compared its performance with a server implementation.

Keywords: HTML5 · Image · Feature extraction · Multimedia indexing

1 Introduction

The content-based multimedia retrieval [2] in connection with the query by example paradigm has become an integral part of various multimedia retrieval systems (e.g., web portals). The content-based retrieval complements traditional keyword-based retrieval techniques that require expensive manual, automatic, or assistive tagging [15]. In order to search the multimedia data in the content-based way, the systems employ a suitable similarity model comprising data descriptors and similarity measures reflecting particular retrieval needs.

In case of image data, the similarity model attempts to capture visual features, that are easily differentiated by the human eye such as dominant colors or shapes. The process of extracting these features is usually computationally demanding. Even though the extraction is performed only once per each new image, the process itself presents nontrivial costs for the whole system.

© Springer International Publishing Switzerland 2015
G. Amato et al. (Eds.): SISAP 2015, LNCS 9371, pp. 323–331, 2015.
DOI: 10.1007/978-3-319-25087-8_31

On the other hand, the data are often inserted by users via specialized applications or web interface. Therefore, it might be beneficial to offload the indexing process to the client side to reduce demands for computational power at the server side or in the cloud.

In order to harness computational power of the connected users, the application must shift significant amount of code to the client side. In the past, the only way to do that was to provide a native application (i.e., a thick client) which was installed and executed by the user. This approach is quite inconvenient and possibly even dangerous for the user.

At preset, we recognize a massive shift in application platforms as many desktop applications have moved to the web. With the standardization of new HTML5 technologies, the web browser has become a convenient platform that does not only display web pages, but also provide means for complex user interaction and a sandbox that can run scripts downloaded from the Internet in a relatively safe manner. In this work, we have utilized HTML5 technologies for a client-side image feature extraction. Hence, a multimedia application can perform image indexing along with the image upload, which basically shifts its computational cost from the application operator to the application users. We also provide a comparison of the client-side extractor with a highly parallel extractor [9], which could be used at server side.

Section 2 revises related work. The image feature extraction process which was selected as a representative multimedia indexing operation is described in Section 3. In Section 4, we introduce some innovations presented in HTML5 and how they can be used to implement feature extraction at client side. Section 5 summarize our experimental results and compare the speed of the client-side extractor with the GPU extractor. Section 6 concludes our paper.

2   Related Work

Utilizing unused resources of the desktop computers is not a new idea. Many systems, such as Entropia [1] or SETI@home [13], have successfully used idle CPU in the past. With new HTML5 technologies, this task become even easier and could be employed for many various applications.

The idea of utilizing web client as a computational platform emerged as soon as API for asynchronous HTTP transfers (generally called AJAX [4]) has become supported in majority of web browsers. One of the first attempts in this area was made by Merelo-Guervós et al. [11]. The have created a distributed computing system in which the workers were implemented in JavaScript and run in browsers of the users. The system was used to solve a genetic algorithm task, which requires only small amount of data exchange to relatively large portions of computational work. The main problem of the system was that the distributed tasks had to be computed in the main thread of the browser which also needs to handle all user interface events.

A more recent project that utilized web clients as distributed computational nodes is the WeevilScout framework [12]. It also communicates with the server that coordinates the work by the means of AJAX, but it utilizes *Web Workers* [14] to perform the computations. The system was originally used to solve a computationally demanding task from the domain of bio-informatics. Another similar work was presented by Duda et al. [3]. Their framework was designed to solve evolutionary algorithms and it was tested on the basis of permutation flowshop scheduling problem.

New HTML technologies also allows additional resources such as operating memory or persistent storage to be used for the purposes of a distributed system. For instance, we may consider implementing a distributed data storage that utilizes *WebStorage* or *IndexDB* API in the browser to hold the data at the client side [8]. However, these resources are much harder to utilize and exploit than the computational power of the client hardware.

3   Image Feature Extraction

We have selected image feature extraction as a representative of multimedia indexing algorithm. Even though it is not one of the most computationally demanding algorithms, it will suffice as a proof of concept and provide us with general estimates of the client-based implementation overheads. We provide only a brief review of this algorithm while, since it is detailed in our previous work [9].

The feature extraction process takes an image and produces its feature signature descriptor. The signature S^o of an object (image) o is a set of features, which are points from feature space F_s. Our feature space is in fact a discrete subset of $\mathbb{R}^7$, where each dimension (domain) has a particular meaning. A point $f \in F_s$ is interpreted as $f = (x, y, L, a, b, c, e)$. The (x, y) coordinates represent a position within the image, the (L, a, b) properties hold the color information, and (c, e) are the contrast and entropy values of the image texture. Each feature f from the signature S^o is accompanied by a weight value $w_f \in \mathbb{R}^+$, that determines the importance of the feature within the set. A larger area with the same color and texture can be represented by a single feature with greater weight while more heterogeneous areas are represented by multiple features with smaller weights.

 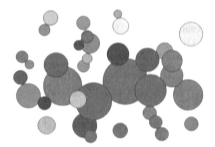

Fig. 1. Example of an image and its feature signature visualization

The feature extraction process consist of the following steps:

1. Image preprocessing
2. Feature sampling
3. Clustering

The image preprocessing can be used to change the size of the image or apply some graphical filters such as blur distortion. These procedures can be used to normalize the images in the database (e.g., to the same size and proportions) and to reduce information noise. In our case, we only resample the image to predefined size.

The feature sampling process uses sampling set of random points from the image and extract the features (color and texture) at those points. The color is simply taken from the pixel at the sampling point coordinates and converted to a CIE Lab [10] color space. The contrast and entropy (c, e values) are computed by scanning a small rectangular window around the sampling point coordinates and statistically analysing differences in pixel illumination.

The final clustering algorithm aggregates the information produced by feature sampling. The produced cluster centroids are used as final feature points and the feature weights are computed from cluster sizes. We employ a slightly modified version of k-means algorithm [5]. It uses a fixed number of iterations instead of producing fixed amount of clusters and the clusters may be dismissed if they are too small or joined when their centroids are closer than specified threshold.

4   Extraction in Web Browser

4.1   JavaScript and HTML5 Technologies

Before we describe our solution, we revise new web technologies and APIs, which were introduced with the new HTML5 standard [6]. These innovations are quite extensive with respect to previous capabilities of client-side scripts, so we narrow our focus on features, which are closely related to our problem.

Accessing Files. Original HTML forms and AJAX API allowed uploading files from the user to server, but the internal contents of the files were not accessible to the client-side scripts. At present, the browsers implement the `FileReader` API which allows reading files from the host system. To ensure security, the reader can access only files that were explicitly selected by the user, either in the `<input>` form element with `file` type or by a drag-and-drop operation.

The file itself is represented as the `File` object, which is an extension of the `Blob` object. The reader can return the loaded contents of the file as a text string, as a data URL (with base64 encoded data), or as an `ArrayBuffer` object that handles binary contents. In case of the `ArrayBuffer` object, the binary data may be subsequently accessed via typed array objects (e.g., `Uint8Array`) or via `DataView` object.

Image Processing. Images are embedded into web pages by `<img>` elements. This concept expects that images are external files located by their URL; however, the image element may receive a data URI with contents of the image (in base64 encoding) instead of traditional HTTP URL.

One of the most important innovations of HTML5 is the introduction of `<canvas>` element. It provides a drawing area which may be used by client-side scripts to render 2D or even 3D graphics. A canvas is tied with a drawing context which provide a predefined set of drawing functions. All canvas implementations support `2d` context – i.e., context for basic drawing. Most current browsers also support `webGL` context which is basically a binding for selected subset of OpenGL functions, so the canvas acts as a viewport for GPU-accelerated 3D graphics rendering engine.

The `2d` context can be used for interaction with images. An `<img>` element may be drawn on the canvas and the canvas contents may be exported to a data URI, which can be subsequently used as source for image element. Furthermore, a rectangular section of canvas may be accessed through `ImageData` object, which holds the color values of its pixels in a binary format.

Intensive Computations. Performing intensive computations in a web browser is quite complicated. Besides the limitations imposed by the JavaScript language (such as the fact that the code is interpreted or that all numeric values are stored as 64-bit float numbers), the code is executed in a single thread in event-driven manner. Therefore, all JavaScript functions invoked in the browser must be sufficiently short (in the terms of computing time), so they will not block the application loop that process events from the user interface.

To provide a more suitable environment for computations, HTML5 introduced *Web Workers* [14] API. A worker code is executed in a separated thread, so it does not interfere with the user interface operations. Furthermore, the browser may utilize multiple threads for multiple workers, so the web application may benefit from running on a multicore CPU or on a multiprocessor system. The workers and the main thread communicate via messages. The message contents must be cloneable to prevent workers from manipulating with the DOM objects of the displayed document. However, read-only binary data (such as image contents) can be cloned quite efficiently.

In addition to Web Workers, the Khronos consortium proposed a WebCL [7] standard. Similarly to WebGL, the WebCL is a modified JavaScript binding for OpenCL technology which is designed for parallel computing on various devices such as multicore CPUs or GPUs. It executes pieces of code called kernels, which are compiled dynamically at runtime, in a parallel manner on available devices; Therefore, it could utilize computational resources of the host system more efficiently or even utilize hardware which is inaccessible by Web Workers. At present, a WebCL plugin is availabile for some browsers, but its wide-scale support has yet to come.

4.2 Proposed Solution

In traditional applications, the feature extraction (or any form of multimedia indexing) is performed at server side after the user uploads its data (i.e., the image). Our objective is to perform the extraction at client side whilst the image is being uploaded and then upload the signature itself. Feature signatures are orders of magnitude smaller than the original image; hence, the subsequent signature upload should take significantly less time than the image upload.

Our implementation works as follows. First, the image is loaded by the means of `FileReader` into a data URI (i.e., into a special URI that holds the entire contents of the image in base64 encoding). This URI is supplied to an `<img>` element and displayed to the user. The `<img>` may remain hidden or it may not be included in the displayed DOM tree, but many users would prefer to see the pictures before they confirm their upload (and the subsequent processing).

The content of the `<img>` element is drawn onto a `<canvas>` in desired size (i.e., the drawing procedure ensures resampling). The entire canvas is then exported to an `ImageData` object, which holds the pixel values in 32-bit RGBA binary format (8-bits per channel). The image data (along with the extraction parameters) are passed on in a message to one of available Web Workers. The worker performs the extraction and sends a feature signature in a message back to the main thread. Finally, the main threads initiates an asynchronous HTTP transfer to upload the signature to the server.

The extraction process inside Web Workers is a quite straightforward implementation of the algorithms described in Section 3. All numerical values are represented as JavaScript numbers (i.e., 64-bit floats). The sampled features (and subsequently the centroids) are represented as an array of object, where each object has attributes named by the attributes of the feature space.

We have considered applying more parallel approach to feature extraction, but the empirical results indicate that it is not necessary in our case. The single-threaded extraction (i.e., extraction performed by only one worker) is quite efficient and in all our tests, the extraction was performed well within the time required to upload the image. Furthermore, if the user uploads multiple image, each image may be processed by a different worker. Finally, in some cases it might be beneficial to perform the extraction multiple times (and thus by multiple workers) with different configurations and then select the best signature produced.

4.3 Security and Reliability Issues

Even though the multimedia indexing can be hardly considered a security issue, the offload model which let the client compute the feature signature creates a vulnerability. A client that does not perform the extraction correctly (no matter whether intentionally or unintentionally) may disrupt the functionality of the similarity model.

A detailed solution to this problem is beyond the scope of this paper. However, we believe that the right approach is to employ redundant extraction or signature verification process. If the images in the application are shared among other users (which is one of the key points of most multimedia web applications), other users may perform feature extraction not only on images they have uploaded, but also on selected images that were sent to them – e.g., as a result of a search process. If the same (or very similar) signature is yielded by multiple different users, it is more likely to be correct.

5   Experiments

We have tested our extractor on a commodity laptop with Intel Core i7-4700HQ CPU (clocked at 2.4GHz) and 16GB of RAM. The laptop used Windows 8.1 as an operating system and the implementation was tested on Firefox and Chrome, which are two leading web browsers. Let us emphasize that the environment of the browser is not ideal for exact time measurements and thus the presented times should be perceived in such context.

A small set of 5Mpix photographs with average size of approximately 1MB (in JPEG format) was selected as the testing data. The extractor resized the images to 150×150 pixel thumbnails, which were sampled by $3,000$ points with normal distribution. The subsequent clustering selected 300 random samples as initial centroids and then performed 10 refining iterations. The parameters match the parameters used in the evaluation of the state-of-the-art extractor [9], so we can compare their result directly.

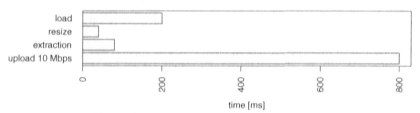

Fig. 2. Average times of individual operations

The measured results indicate, that the whole extraction process is much faster than loading the image from the persistent storage and decoding it from JPEG format. Furthermore, both operations are much faster than upload of the image, even if it was performed by 10 Mbit uplink which is quite fast for most users with DSL or wireless connection.

Finally, we have compared the performance of the in-browser implementation with the state-of-the-art extractor. The JavaScript code is approximately $2.5\times$ slower than serial CPU version written in C++ and several hundred times slower than the GPU extractor. On the other hand, we believe that the situation may be improved when the WebCL technology will become widely available.

6   Conclusions

In this paper, we have proposed a multimedia indexing model which could be adopted by web applications. It reduces performance requirements on the server side as it offloads the indexing process to the client side. As a proof of concept, we have implemented an in-browser version of the image feature extraction algorithm that utilized HTML5 technologies and run completely at the client side. Even though this solution is less efficient than a compiled application, it is applicable in the studied case since the client-side code usually requires to perform only a few extractions.

In our future work, we would like to utilize the hardware of the client more efficiently by the means of WebCL technology, which will hopefully be soon available in all browsers. Furthermore, we are planning to open our prototype to the public, so we get a better overview of the performance of current smartphones and tablets.

Acknowledgments. This paper was supported by Czech Science Foundation (GAČR), project number P103-14-14292P.

References

1. Chien, A., Calder, B., Elbert, S., Bhatia, K.: Entropia: architecture and performance of an enterprise desktop grid system. Journal of Parallel Distributed Computing **65**, 597–610 (2003)
2. Datta, R., Joshi, D., Li, J., Wang, J.Z.: Image retrieval: Ideas, influences, and trends of the new age. ACM Computing Surveys (CSUR) **40**(2), 5 (2008)
3. Duda, J., Dłubacz, W.: Distributed evolutionary computing system based on web browsers with JavaScript. In: Manninen, P., Öster, P. (eds.) PARA 2012. LNCS, vol. 7782, pp. 183–191. Springer, Heidelberg (2013)
4. Garrett, J.J., et al.: Ajax: A new approach to web applications (2005)
5. Hartigan, J.A., Wong, M.A.: Algorithm AS 136: A k-means clustering algorithm. Applied Statistics, 100–108 (1979)
6. Hickson, I., Hyatt, D.: Html5. W3C Working Draft, May 2011
7. Jeon, W., Brutch, T., Gibbs, S.: Webcl for hardware-accelerated web applications. In: TIZEN Developer Conference May, pp. 7–9 (2012)
8. Krulis, M., Falt, Z., Zavoral, F.: Exploiting HTML5 technologies for distributed parasitic web storge. In: Proceedings of the Dateso 2014 Annual International Workshop on DAtabases, TExts, Specifications and Objects, Roudnice nad Labem, Czech Republic, 16 April 2014, pp. 71–80 (2014). http://ceur-ws.org/Vol-1139/poster10.pdf
9. Kruliš, M., Lokoč, J., Skopal, T.: Efficient extraction of feature signatures using multi-GPU architecture. In: Li, S., El Saddik, A., Wang, M., Mei, T., Sebe, N., Yan, S., Hong, R., Gurrin, C. (eds.) MMM 2013, Part II. LNCS, vol. 7733, pp. 446–456. Springer, Heidelberg (2013)
10. McLaren, K.: XIII–The Development of the CIE 1976 (L* a* b*) Uniform Colour Space and Colour-difference Formula. Journal of the Society of Dyers and Colourists **92**(9), 338–341 (1976)
11. Merelo-Guervós, J.J., Castillo, P.A., Laredo, J.L.J., Mora Garcia, A., Prieto, A.: Asynchronous distributed genetic algorithms with javascript and json. In: IEEE Congress on Evolutionary Computation, CEC 2008, (IEEE World Congress on Computational Intelligence), pp. 1372–1379. IEEE (2008)

12. Reginald, C., Putra, G., Belloum, A., Koulouzis, S., Bubak, M., de Laat, C.: Distributed Computing on an Ensemble of Browsers (2013)
13. Univ. of Berkeley: SETI@Home (2006). http://setiathome.ssl.berkeley.edu/
14. W3C: Web Workers. http://www.w3.org/TR/workers/
15. Wang, M., Ni, B., Hua, X.S., Chua, T.S.: Assistive tagging: A survey of multimedia tagging with human-computer joint exploration. ACM Comput. Surv. **44**(4), 25:1–25:24 (2012). http://doi.acm.org/10.1145/2333112.2333120

On the Use of Similarity Search to Detect Fake Scientific Papers

Kyle Williams[1]([✉]) and C. Lee Giles[1,2]

[1] Information Sciences and Technology, The Pennsylvania State University,
University Park, State College, PA 16802, USA
[2] Computer Science and Engineering, The Pennsylvania State University,
University Park, State College, PA 16802, USA
kwilliams@psu.edu

Abstract. Fake scientific papers have recently become of interest within the academic community as a result of the identification of fake papers in the digital libraries of major academic publishers [8]. Detecting and removing these papers is important for many reasons. We describe an investigation into the use of similarity search for detecting fake scientific papers by comparing several methods for signature construction and similarity scoring and describe a pseudo-relevance feedback technique that can be used to improve the effectiveness of these methods. Experiments on a dataset of 40,000 computer science papers show that precision, recall and MAP scores of 0.96, 0.99 and 0.99, respectively, can be achieved, thereby demonstrating the usefulness of similarity search in detecting fake scientific papers and ranking them highly.

Keywords: Similarity search · Fake papers · SciGen

1 Introduction

In recent years their has been increasing pressure on academics to publish large numbers of articles in order to sustain their careers, obtain funding and ensure prestige. As a result, it has been argued that there has been a decrease in the quality of articles submitted for publication [3] as well as a surge in the number of for-profit, predatory, and low quality journals and conferences to meet the demand for venues for publication [2]. As a result, it is reasonable to expect that fraudulent and and fake scientific papers may exist in document collections [8] and their identification and removal is important for many reasons.

In this paper we address the problem of using similarity search to detect fake scientific papers as generated by SciGen[1], which is a computer science paper generator. The benefit of using similarity search as opposed to other methods, such as supervised text classification, is that the latter requires training. When one has the code to automatically generate fake papers, as is the case for Sci-Gen, then it is relatively simple to generate training data. However, in the case

---

[1] http://pdos.csail.mit.edu/scigen/

© Springer International Publishing Switzerland 2015
G. Amato et al. (Eds.): SISAP 2015, LNCS 9371, pp. 332–338, 2015.
DOI: 10.1007/978-3-319-25087-8_32

where the code for generating fake papers is not available, it becomes a tedious task to create training data since training cases first need to be identified. In contrast, similarity search only requires one sample of a fake paper if we make the assumption that there exists some regularity among fake papers generated by the same method.

We investigate the use of several methods for similarity search for detecting SCIGen papers, including state of the art near duplicate detection methods and simpler keyword and keyphrase-based methods and demonstrate their effectiveness in retrieving fake SCIGen papers. One of the challenges of this approach, however, is that it requires documents to have features in common in order for them to be retrieved. We exploit the fact that we expect some regularity to exist among automatically generated documents and devise a pseudo-relevance feedback mechanism to improve the performance of similarity search.

2 Related Work

There has already some work on identifying fake scientific papers. Early work was based on the intuition that in a SciGen generated paper, the references are to fake or non-existent papers [10]. Thus, by analyzing the references, one is able to determine if a paper is fake or not. Thus, the authors extract the references from fake papers and submit them to a public Web search engine and a paper is classified as being fake or not based on the extent to which its references match actual search results. While this method is useful, it can easily be fooled by making the references in fake papers actually refer to real papers.

Labbé and Labbé do an analysis of the extent to which fake and duplicate papers exist in the scientific literature [4]. Their method is based on calculating the inter-textual distances between documents based on the similarity and frequency of the words appearing in the documents. Once the inter-textual differences have been calculated, texts are grouped using agglomerative hierarchical clustering.

A problem related to fake paper detection is plagiarism detection since in both cases the goal is to detect suspicious text. There has already been several efforts to use similarity search to detect plagiarism. For instance, one of the tasks in the annual PAN workshop and competition on uncovering plagiarism, authorship, and social software misuse focuses on the source retrieval problem [7]. In this task, the input is a suspicious document and the goal is to retrieve potential sources of plagiarism from the Web. Most of the approaches in this task view the problem as a similarity search problem where the goal is to retrieve search results that are similar to the query document. Competitive approaches have considered both supervised and unsupervised solutions to the problem [7].

This section has discussed various studies that have dealt with fake academic papers or using similarity search to retrieve content of interest. An important thing to note is that relatively small datasets were used in all of the studies involving SciGen. For instance, in [4], most of the corpora only contained 10s or 100s of documents, though the corpora based on the Arxiv contained a few thousand documents. By contrast, we perform our experiments on a dataset containing over 40,000 real papers to which we add 100 fake papers.

3   Approach

Given a sample SCIGen paper q as a query, we seek to retrieve all SCIGen documents in a collection C. To do this, we perform automatic feature extraction on every document in the collection and index the documents. At query time, we select a retrieval feature and use it to automatically extract features from q, which we then use to retrieve all documents that have at least one feature in common with q and rank the results.

3.1   Feature Extractors

Shingle Features. Shingles are sequences of words that occur in documents and were originally used for calculating the similarity of documents [1] and are considered as state of the art for near duplicate detection. Due to space constrains, we do not describe the method for generating the shingle features but use the same approach as in [9]. We experimented with different shingle lengths and found a length of 5 to work well for this study.

Simhash Features. Simhash is a state of the art algorithm duplicate detection algorithm [5]. For each document, the simhash is calculated as described in [9] and the output is a 64-bit hash. Each hash is partitioned into $k + 1$ sub-hashes and these sub-hashes are indexed [5]. At query time, the simhash of the query document is also partitioned into $k + 1$ sub-hashes that are used to query the index and retrieve documents that have at least one sub-hash in common with the query document. We experimented with different values of k and found $k = 4$ to work well. Thus, we use this value of k in this study.

Keyphrase Features. We extract keyphrases from each document using the Maui tool [6] and use these keyphrases as features. For each document, the top 10 keyphrases are identified for querying. Thus a document will be retrieved if its text contains at least one of the top 10 keyphrases in the query document.

TF-IDF Features. We also investigate the use of features based on TF-IDF. Each term in a query document is scored using TF-IDF. We then form a Boolean OR query with the top 10 TF-IDF scored terms and a document will be retrieved if it contains a term that matches one of the top 10 terms in the query document.

For each document retrieved using the features extracted by one of the feature extractors, we perform full-text based ranking based on cosine similarity.

3.2   Dataset

43,390 ACM papers from the CiteSeerX collection constitute our collection of *real* scientific papers. We then used SciGen to generate 100 *fake* papers and added these to the existing collection of real papers. We then generated an additional 10 fake papers for testing. In our experiments, the goal is to use the testing papers to retrieve the 100 known fake papers in the dataset.

4   Experiments

4.1   Retrieving SCIGen Papers

We consider the use of the four feature extractors for retrieving SCIGen papers using similarity search. For each of the 10 query documents, we extract features which we use to formulate a query and we report the averages over the 10 documents. Figure 1 shows the different metrics for the different feature extractors (the Shingles+Feedback approach is described in Section 4.2).

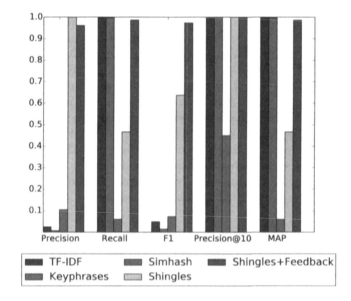

Fig. 1. Performance metrics for different feature extractors.

As can be seen from Figure 1, the different features perform quite differently in their ability to retrieve SCIGen papers. The first thing to notice is that almost perfect recall can be achieved by the TF-IDF and keyphrase-based methods, with average recall values of 0.999 and 0.997, respectively. This clearly indicates that these simple features are very good at identifying SCIGen papers; however, this comes at the cost of precision which, as can be seen from the figure, is very low for these two methods at 0.0251 and 0.0074, respectively. The reason for the very low precision for these methods is that many of the TF-IDF ranked terms and keyphrases are common among computer science papers and thus many documents are retrieved. The F-scores show that these methods perform worst overall in terms of overall retrieval with F-scores of 0.0489 and 0.0147. The TF-IDF scored keyword and keyphrase methods, however, achieve good rankings with Precision@10 of 1.0 and MAP of 0.999 and 0.997, respectively.

For the shingles method, the overall precision is perfect thereby implying that only SCIGen papers were retrieved. The downside of this approach, however, is that the recall is relatively low at 0.467. Overall though, the shingles method achieves the highest F1 score. Shingles also lead to perfect Precision@10; however, MAP is 0.467 since not all 100 SCIGen documents were retrieved.

The simhash method performs worst overall and achieves precision of 0.1052, recall of 0.06, Precision@10 of 0.45 and MAP of 0.06. This is somewhat expected since the simhash method is based on a single hash that represent a full document whereas the other methods are based on sub-documents. Since simhash is state of the art for near duplicate detection there is sufficient evidence to conclude that SCIGen documents are not similar enough to be called near duplicates.

The metrics that take into consideration the ranking of results are all relatively good and one can deduce from this that, in general, the cosine similarity-based ranking function is suitable since it places almost all retrieved SCIGen documents in the top 100 documents. While this is highly desirable, the one shortcoming is that, in this case, we know that there are 100 fake SCIGen documents and thus calculating MAP among the top 100 makes sense. However, in the general case, we do not know how many documents need to be detected. We are faced with the situation where we can achieve high recall at the cost of precision as is the case for the keyword and keyphrase-based methods, or we can achieve high precision at the cost of recall as is the case with the shingles-based method. In the next section, we describe a method whereby we can address this shortcoming. Due to space constrains, we focus on the case of shingles but the method is applicable to similarity search in general.

4.2 Improving Performance Through Pseudo-Relevance Feedback

In information retrieval, feature mismatch occurs when the terms that a user uses to describe a document do not match the terms used by the document authors. The standard way to address this problem is through query reformulation. We extend this approach to the detection of SCIGen papers where we expect some feature regularity among a sufficiently large number of SCIGen documents. We devise a pseudo-relevance feedback mechanism whereby after the initial query is submitted, we select the top k returned documents and submit each of them as a query using the same method as for the original document. We then combine and rank all the search results returned from the different query documents. The motivation behind this approach is that, while the initial query document may not have features in common with all relevant documents in the collection, documents that are retrieved might. The Shingles+Feedback bar in Figure 1 shows the effect of performing this pseudo-relevance feedback on the top 10 documents returned by the initial query with shingle features.

As can be seen in Figure, the effect of the pseudo relevance feedback has a large effect on recall, which was the initial shortcoming of the original shingle-based method. When the pseudo-relevance feedback is included, their is a slight decrease in overall precision from 1.0 to 0.96, however this comes at the benefit of an almost 2-fold increase in recall from 0.467 to 0.987. As a result, the recall becomes competitive with that achieved by the keyword and keyphrase-based methods. This increase in recall is reflected in the change in the F-score which increases from 0.64 to 0.97. The pseudo-relevance feedback has no effect on Precision@10, which remains at 1.0, but leads to a large increase in MAP which, like recall, goes from 0.467 to 0.987. Thus, there is clear evidence from this experiment that exploiting the expected regularity among automatically generated documents is a reasonable approach in order to improve retrieval performance.

5   Conclusions

We have described a method whereby similarity search can be used to detect fake scientific papers, which have increasingly become a problem as a result of the increasing pressure on academics to publish or perish. We described several methods for extracting features for similarity search and evaluated their use in detecting SCIGen papers. Inspired by the fact that we expect some form of regularity to exist among automatically generated documents, we devised a pseudo-relevance feedback mechanism to improve the performance of similarity search and showed how precision, recall and MAP scores of 0.96, 0.99 and 0.99, respectively, can be achieved. We only presented an evaluation of the pseudo-relevance feedback mechanism with shingle features; however, the approach is general enough that it can be applied to any set of features for similarity search.

Acknowledgments. We gratefully acknowledge partial support by the National Science Foundation.

References

1. Broder, A., Glassman, S., Manasse, M., Zweig, G.: Syntactic clustering of the Web. Computer Networks and ISDN Systems **29**(8–13), 1157–1166 (1997)
2. Butler, D.: Investigating journals: The dark side of publishing. Nature **495**(7442), 433–435 (2013)
3. Gad-el Hak, M.: Publish or perish - an ailing enterprise? Physics Today **57**(3), 61–62 (2004)
4. Labbé, C., Labbé, D.: Duplicate and fake publications in the scientific literature: how many SCIgen papers in computer science? Scientometrics **94**(1), 379–396 (2012)
5. Manku, G., Jain, A., Sarma, A.D.: Detecting near-duplicates for web crawling. In: WWW, pp. 141–149 (2007)

6. Medelyan, O., Frank, E., Witten, I.H.: Human-competitive tagging using automatic keyphrase extraction. In: EMNLP, vol. 3, pp. 1318–1327 (2009)
7. Potthast, M., Hagen, M., Beyer, A., Busse, M., Tippmann, M., Rosso, P., Stein, B.: Overview of the 6th international competition on plagiarism detection. In: CLEF (2014)
8. Van Noorden, R.: Publishers withdraw more than 120 gibberish papers. Nature, February 2014
9. Williams, K., Giles, C.L.: Near duplicate detection in an academic digital library. In: DocEng, pp. 91–94 (2013)
10. Xiong, J., Huang, T.: An effective method to identify machine automatically generated paper. In: KESE, pp. 101–102. IEEE (2009)

Reducing Hubness for Kernel Regression

Kazuo Hara[1]([✉]), Ikumi Suzuki[2], Kei Kobayashi[2], Kenji Fukumizu[2],
and Miloš Radovanović[3]

[1] National Institute of Genetics, Mishima, Shizuoka, Japan
kazuo.hara@gmail.com
[2] The Institute of Statistical Mathematics, Tachikawa, Tokyo, Japan
[3] University of Novi Sad, Novi Sad, Serbia

Abstract. In this paper, we point out that *hubness*—some samples in
a high-dimensional dataset emerge as *hubs* that are similar to many
other samples—influences the performance of kernel regression. Because
the dimension of feature spaces induced by kernels is usually very high,
hubness occurs, giving rise to the problem of *multicollinearity*, which is
known as a cause of instability of regression results. We propose hubness-
reduced kernels for kernel regression as an extension of a previous approach for kNN classification that reduces *spatial centrality* to eliminate
hubness.

1 Introduction

Recently, *hubness*, a phenomenon occurring in high-dimensional datasets as a
result of *curse of dimensionality* [5], has attracted the attention of researchers
in the artificial intelligence community, especially for data mining and machine
learning. For instance, a new clustering algorithm was presented by taking advantage of hubness [9]. The performance of k-nearest-neighbor (kNN) classification
was improved by eliminating hubness [7,6,8,3].

In this paper, we point out that the hubness influences the performance of
kernel regression as well. Because the dimension of feature spaces induced by
kernels is usually very high, hubness occurs. Therefore, in the learning phase
of kernel regression, hubs in training samples that are similar to many other
training samples provide highly correlated information to the learning model.
The problem caused by such correlation of input variables is known as *multicollinearity*, which degrades the generalization error on test samples [4,1].

We then propose *hubness-reduced kernels* for kernel regression as an extension
of *localized centering* [3], a technique for kNN classification, which transforms
similarity measures to reduce *spatial centrality* to get rid of hubness.

2 Multicollinearity in Kernel Regression

Let us assume that n pairs of input object and a scalar output value $\{x_i, y_i\}_{i=1}^{n}$
are given as training samples, and that the goal is to predict the output y_{test} for

© Springer International Publishing Switzerland 2015
G. Amato et al. (Eds.): SISAP 2015, LNCS 9371, pp. 339–344, 2015.
DOI: 10.1007/978-3-319-25087-8_33

a new input x_{test}. In ordinary linear regression, an input object x is represented as a d-dimensional vector. More formally,

$$x \mapsto f(x) = (f_1(x), \ldots, f_d(x))^{\mathrm{T}}, \tag{1}$$

where a set of d functions $\{f_i(x)\}_{i=1}^d$ are ad-hoc feature extractors designed by domain experts. Then, weight w_f of the functions, which is a d-dimensional vector, is determined using training samples to minimize the least-squares loss calculated as $\sum_{i=1}^n (y_i - w_f^{\mathrm{T}} f(x_i))^2$. Also, the output y_{test} for a new input x_{test} is predicted as $\hat{y}_{test} = w_f^{\mathrm{T}} f(x_{test})$.

In the situation described above, *multicollinearity* is related to cases in which the functions $\{f_i(x)\}_{i=1}^d$ are highly correlated, or to cases in which a function extracts feature values from input objects in a very similar way to that of the other functions. In such a case, a problem occurs by which the prediction becomes less reliable because the weight w_f tends to be overfitted to noisy training samples [4,1].

In contrast, with kernel regression, an input object x is mapped to an n-dimensional vector, such that

$$x \mapsto k(x) = (k_1(x), \ldots, k_n(x))^{\mathrm{T}}, \tag{2}$$

where $\{k_i(x)\}_{i=1}^n$ are kernel functions that give similarity between input x and a training object x_i. Some examples include $k_i(x) = \langle f(x), f(x_i) \rangle$ for linear kernels, and $k_i(x) = exp(-\frac{1}{2\gamma^2}||f(x) - f(x_i)||^2)$ for Gaussian kernels. These kernel functions $k_i(x)$ take larger values for input objects lying in the neighbor of the training object x_i, and smaller values for other input objects. If neighbor objects are different for each of the training objects $\{x_i\}_{i=1}^n$, then corresponding kernel functions $\{k_i(x)\}_{i=1}^n$ take large values for different objects. Therefore, no correlation emerges between the kernel functions.

However, because a vector space (i.e., a reproducing kernel Hilbert space) induced by kernels is usually high-dimensional, *hub* objects that are similar to many other objects tend to occur, and the kernel function $k_h(x)$ corresponding to a hub object x_h gives large values for many objects. This fact implies that $k_h(x)$ correlates with other kernel functions, and thereby produces multicollinearity.

3   Hubness in Kernel-Induced Spaces

The phenomenon of *hubness* is known to emerge when the nearest neighbors (NNs) in a high-dimensional dataset are considered [5]. Let $D \subset \mathbb{R}^d$ be a dataset in d-dimensional space and let $N_k(x)$ denote the number of times a sample $x \in D$ occurs in the kNNs of other samples in D, under some similarity measure. When the dimension is high, the shape of the N_k distribution skews to the right. A small number of samples take large N_k values. Such samples, similar to many other samples, are called *hubs*. This phenomenon is called *hubness*.

We next demonstrate the emergence of hubness using artificial data. we generate a dataset from a mixture of two Gaussian distributions with sample size

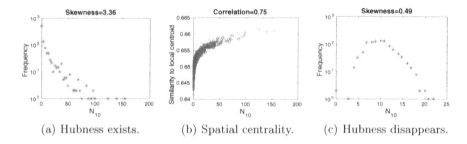

(a) Hubness exists.   (b) Spatial centrality.   (c) Hubness disappears.

Fig. 1. (a) Hubness occurs: samples with a large N_{10} value occur, and the N_{10} distribution is skewed to the right. (b) Correlation between the N_{10} value and the similarity to the local centroid is strong. (c) Hubness is reduced (lower N_{10} value and smaller skewness) by the transformation according to Equation (7).

$n = 1000$ and dimension $d = 1000$. Specifically, we generate 500 samples each from i.i.d. *Gaussian*$(\mathbf{0}, \mathbf{I})$ and i.i.d. *Gaussian*$(\mathbf{1}, \mathbf{I})$, where $\mathbf{0}, \mathbf{1}$ respectively denote d-dimensional vectors of zeros and ones and $\mathbf{I}$ is the $d \times d$ identity matrix.

To determine kNN samples, we use a positive definite kernel that is equal to the inner-product on a Hilbert space induced by the kernel. Note that each sample $x \in D$ is mapped to the Hilbert space according to a function denoted as $\phi(x)$. Here, we use a Gaussian kernel,[1] where similarity between two samples $x_i, x_j \in D$ is given as

$$K(x_i, x_j) = \langle \phi(x_i), \phi(x_j) \rangle = exp(-\frac{1}{2\gamma^2} ||x_i - x_j||^2) \qquad (3)$$

with a deviation parameter γ set as the median of pairwise distances among samples in D. The distribution of N_{10} is shown in Figure 1(a). We can observe the presence of hubs, i.e., samples with a particularly large N_{10} values.

Following Radovanović *et al.* [5], we evaluate the degree of hubness by the *skewness* of the N_k distribution. A large skewness indicates strong hubness in a dataset. Indeed, skewness is large (i.e., 3.36) in Figure 1(a).

3.1 Origin of Hubness: Spatial Centrality

For the artificial dataset described above, we form a scatter plot of samples with respect to the N_{10} value and the similarity to the local centroid (Figure 1(b)). A local centroid is defined for each sample $\phi(x)$ as

$$c_\kappa(\phi(x)) \equiv \frac{1}{\kappa} \sum_{\phi(x') \in \kappa NN(\phi(x))} \phi(x'), \qquad (4)$$

which is the mean of the κ-nearest neighbor samples of $\phi(x)$ under some local neighborhood size $\kappa \in [1, n-1]$ [3]. It is noteworthy that the local centroid is

---

[1] Gaussian kernel is a shift-invariant kernel, where the NN based on the inner-product $\langle \phi(x_i), \phi(x_j) \rangle$ is equivalent to that based on the norm $||x_i - x_j||$.

not always obtained explicitly using Equation (4), because it is computed not in the original d-dimensional space but in a space induced by the kernel that defines $K(\cdot, \cdot)$. Therefore, the similarity to the local centroid is calculated as

$$\langle \phi(x), c_\kappa(\phi(x)) \rangle = \frac{1}{\kappa} \sum_{\phi(x') \in \kappa \mathrm{NN}(\phi(x))} K(x, x'). \tag{5}$$

In the dataset generated from two Gaussians, strong correlation exists between the N_{10} value and the similarity to the local centroid ($\kappa = 20$) as shown in Figure 1(b). This is called *spatial centrality* of the dataset.

4   Reducing Hubness for Kernel Regression

Because the existence of spatial centrality is considered to be an ingredient of hubness [5], hubness is expected to be suppressed by removing the spatial centrality. Following this idea, Hara et al. [3] proposed a hubness-reduction method called *localized centering* for kNN classification. The method transforms the similarity measure such that the transformed similarity does not generate spatial centrality. The transformation is given by subtracting similarity to the local centroid in Equation (5) from the original similarity in Equation (3), such that

$$Sim^{\mathsf{LCENT}}(x_i, x_j) \equiv \langle \phi(x_i), \phi(x_j) \rangle - \langle \phi(x_j), c_\kappa(\phi(x_j)) \rangle \tag{6}$$

$$= K(x_i, x_j) - \frac{1}{\kappa} \sum_{\phi(x') \in \kappa \mathrm{NN}(\phi(x_j))} K(x_j, x'). \tag{7}$$

After this transformation, the similarity to the local centroid for any database sample x_j becomes the same because substituting $\phi(x_i) = c_\kappa(\phi(x_j))$ in Equation (6) yields a constant value (i.e., zero). This fact indicates that no spatial centrality with respect to the local centroid exists after the transformation. Therefore, the hubness is expected to be reduced. Indeed, by applying the transformation to the dataset used to draw Figure 1, the skewness decreases from 3.36 (Figure 1(a)) to 0.49 (Figure 1(c)).

It should be noted that the resulting similarity measure $Sim^{\mathsf{LCENT}}(\cdot, \cdot)$ is not symmetric with respect to x_i and x_j. This does not matter for kNN classification because it is assumed that similarity is computed between two samples that have different roles, i.e., a query and a database sample. Indeed, x_i and x_j in Equation (6) respectively correspond to a query and a database sample.

We now propose hubness-reduced kernels by a symmetrization of Sim^{LCENT}, as follows.

$$K^{\mathsf{HR}}(x_i, x_j) \equiv \langle \phi(x_i) - c_\kappa(\phi(x_j)), \phi(x_j) - c_\kappa(\phi(x_i)) \rangle$$
$$= \langle \phi(x_i), \phi(x_j) \rangle - \langle \phi(x_i), c_\kappa(\phi(x_i)) \rangle - \langle \phi(x_j), c_\kappa(\phi(x_j)) \rangle + \langle c_\kappa(\phi(x_i)), c_\kappa(\phi(x_j)) \rangle$$
$$= K(x_i, x_j) - \frac{1}{\kappa} \sum_{\phi(x')} K(x_i, x') - \frac{1}{\kappa} \sum_{\phi(x'')} K(x_j, x'') + \frac{1}{\kappa^2} \sum_{\phi(x')\phi(x'')} K(x', x'')$$

$$\tag{8}$$

The transformed kernels are not always positive definite. For such a case, we replace all negative eigenvalues of the transformed kernels with zeros.[2]

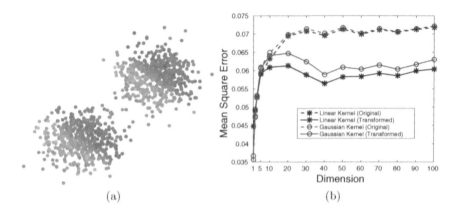

Fig. 2. (a) An illustration of the multi-centered dataset (generated with sample size $n = 1000$, dimension $d = 2$, and $\delta = 4$). Red and blue points respectively correspond to samples with positive and negative outputs. (b) Mean squared error of kernel regression, using multi-centered datasets with $n = 100$, $d \in [1, 100]$, and $\delta = 1$.

5   Experiment

We demonstrate that the proposed kernels K^{HR} reduce the correlation between input variables, i.e., multicollinearity. Thereby, they improve prediction accuracy. We used artificial data generated from a mixture of two Gaussian distributions[3] with a fixed sample size $n = 100$ and variation in dimension d from 1 to 100. More precisely, the dataset consisted of $\{x_i\}_{i=1}^{\frac{n}{2}}$ generated from i.i.d. $Gaussian(\mathbf{0}, \mathbf{I})$ with output $y_i = sin(z_i)exp(-|z_i|) + \epsilon$, where $z_i = \langle x_i, \frac{1}{||\mathbf{1}||} \rangle$, and $\{x_i\}_{i=\frac{n}{2}+1}^{n}$ generated from i.i.d. $Gaussian(\delta\mathbf{1}, \mathbf{I})$ with output $y_i = sin(z_i)exp(-|z_i|) + \epsilon$, where $\delta = 1$, $z_i = \langle x_i - \delta\mathbf{1}, \frac{1}{||\mathbf{1}||} \rangle$, and ϵ is noise generated from $Gaussian(\mathbf{0}, \sigma^2\mathbf{I})$ where $\sigma^2 = 0.01$. The dataset is illustrated in Figure 2(a).

In the manner described above, we generated samples of size n each for training, validation, and testing. However, they were centered or shifted such that the mean of the training samples was zero.

For kernel regression, we used a linear kernel $K(x_i, x_j) = \langle x_i, x_j \rangle$ and a Gaussian kernel $K(x_i, x_j) = exp(-\frac{1}{2\gamma^2}||x_i - x_j||^2)$ with a deviation parameter γ set as the median of pairwise distances among training samples, following Gretton *et al.* [2]. For each kernel, we tested the proposed transformation according to Equation (8).

---

[2] It is important to pursue if this is a good way of fixing the non-positive definiteness as a future work.

[3] We used multi-centered datasets, because the correlation between training samples is not reduced by *centering*, a common pre-processing to avoid multicollinearity.

To avoid overfitting to training samples, we used kernel ridge regression. Given a kernel matrix K and a vector y of output values with respect to training samples, the model parameter α was learned as a solution $\alpha = (K + \lambda I)^{-1}y$, where λ is a hyper-parameter of ridge regression. We selected λ as well as the hyper-parameter κ of the proposed method using validation samples.

The output y_t of a test sample x_t was predicted as $\hat{y}_t = \sum_{i=1}^{n} \alpha_i k(x_i, x_t)$ using training samples $\{x_i\}_{i=1}^{n}$. The methods were evaluated according to the mean squared error (MSE) $(y_t - \hat{y}_t)^2$ over the test samples.

For each setting of dimension d ranging from 1 to 100, we repeated the process described above 100 times. The average of the MSE obtained is shown in Figure 2(b).

Figure 2(b) shows that when the number of dimensions is large (i.e., more than 10), the proposed kernel transformation improves MSE for both linear and Gaussian kernels. The result suggests that hubness emerges in high-dimensional data, and affects kernel regression through the resulting multicollinearity. However, hubness, and hence the MSE can be reduced using our proposed kernels.

6   Conclusion

After pointing out that hubness gives rise to the multicollinearity problem, and that it therefore influences the performance of kernel regression, we proposed hubness-reduced kernels for kernel regression as an extension of a previous approach for kNN classification. We demonstrated that reduction of hubness produces an effect on kernel regression for multi-centered datasets.

References

1. Chatterjee, S., Hadi, A.S., Price, B.: Regression Analysis By Example. Wiley Series In Probability And Statistics. Wiley, New York (2000)
2. Gretton, A., Fukumizu, K., Teo, C., Song, L., Schölkopf, B., Smola, A.: A kernel statistical test of independence. Advances in Neural Information Processing Systems **20**, 585–592 (2008)
3. Hara, K., Suzuki, I., Shimbo, M., Kobayashi, K., Fukumizu, K., Radovanović, M.: Localized centering: reducing hubness in large-sample data. In: AAAI (2015)
4. Montgomery, D.C., Peck, E.: Introduction to linear regression analysis. Wiley-Interscience Publication, John Wiley & sons, New York (1992)
5. Radovanović, M., Nanopoulos, A., Ivanović, M.: Hubs in space: Popular nearest neighbors in high-dimensional data. Journal of Machine Learning Research **11**, 2487–2531 (2010)
6. Schnitzer, D., Flexer, A., Schedl, M., Widmer, G.: Local and global scaling reduce hubs in space. Journal of Machine Learning Research **13**(1), 2871–2902 (2012)
7. Suzuki, I., Hara, K., Shimbo, M., Matsumoto, Y., Saerens, M.: Investigating the effectiveness of laplacian-based kernels in hub reduction. In: AAAI (2012)
8. Suzuki, I., Hara, K., Shimbo, M., Saerens, M., Fukumizu, K.: Centering similarity measures to reduce hubs. In: EMNLP, pp. 613–623 (2013)
9. Tomasev, N., Radovanovic, M., Mladenic, D., Ivanovic, M.: The role of hubness in clustering high-dimensional data. IEEE Transactions on Knowledge and Data Engineering **26**(3), 739–751 (2014)

Demo Papers

FELICITY: A Flexible Video Similarity Search Framework Using the Earth Mover's Distance

Merih Seran Uysal[✉], Christian Beecks, Daniel Sabinasz, and Thomas Seidl

Data Management and Exploration Group, RWTH Aachen University,
Aachen, Germany
{uysal,beecks,sabinasz,seidl}@cs.rwth-aachen.de

Abstract. In this paper, we demonstrate our novel system, called FELICITY, which is capable of processing user-adaptive content-based k-nearest-neighbor queries efficiently by utilizing video signatures and the Earth Mover's Distance (EMD). To this end, we implement an optimal multi-step query processing algorithm by approximating the EMD between any two video signatures by utilizing lower-bounding filter approximation techniques. The system enables the user to adapt query parameters and interact with the system, helping him to find out the best parameter combination for the desired similarity tasks. Moreover, our system incorporates a user-friendly visualization interface for the EMD flow between video signatures, providing an intuitive understanding of the EMD and video similarity.

Keywords: Earth mover's distance · Video similarity search · Query processing · Feature extraction

1 Introduction

Recently, the rapid increase in the dissemination of multimedia capture devices and social networking websites have attracted the attention of researchers with respect to query processing and similarity search in video databases. Not only the utilization of efficient and effective query processing techniques, but also video representation models are of crucial importance in order to carry out required automatic similarity search tasks. In this demo, we present our novel *Flexible Video Similarity Search Framework Using the Earth Mover's Distance (FELICITY)* which utilizes the well-known similarity measure Earth Mover's Distance (EMD) [3] on flexible, compact object representation models, i.e. signatures, in order to process k-nearest-neighbor (k-nn) queries in video databases.

A signature X which models the inherent content-based properties of the corresponding video basically comprises a set R_X of representatives each of which is coupled with a real number, the so-called *weight*, denoting the number of features assigned to that representative. Unlike frame-based and sequence-based video models [1,2], the signature model utilized in this paper is not contingent upon frames or keyframes, attaining great flexibility via exploiting all requested feature types, such as color, position, contrast, and coarseness [6,7].

© Springer International Publishing Switzerland 2015
G. Amato et al. (Eds.): SISAP 2015, LNCS 9371, pp. 347–350, 2015.
DOI: 10.1007/978-3-319-25087-8_34

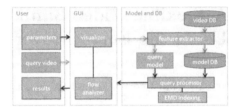

Fig. 1. Illustration of the system architecture of the FELICITY-framework

Given two signatures X and Y, the EMD computes the minimum amount of work required to transform one signature into another one, where representatives of X and Y with their weights are represented as earth hills and earth holes, respectively. For the EMD computation, a ground distance d is applied to any pair of representatives $x \in R_X$ and $y \in R_Y$. The dimensions of the underlying feature space comprising all representatives play an important role in the EMD computation. In particular, if the user is not familiar with feature space dimensions, he needs an interactive system enabling flexible parameter selection to attain the k-nn query processing with the desired parameter combination. To this end, our new framework offers the user the flexibility to specify the individual weighting of dimensions, such as position, color, texture, and temporal dimension, for the signature generation phase. In addition, our framework exploits the state-of-the-art lower-bounding approximation techniques including the recently introduced Independent Minimization for Signatures (IM-Sig) [5] in order to overcome the limitation of super-cubic time complexity of the EMD. Moreover, the user is able to interactively analyze the flows of the EMD via individual illustrations, providing insights into its working principle with the utilization of various parameters, such dimension weightings, number of representatives, and k-nn query parameter k. Thus, the user has the opportunity to analyze the EMD flows between the query signature and any signature in the returned result list, which enables him to comprehend why returned videos match the query video best. Last but not least, our framework provides a user-friendly web-based interface.

2   System Overview

To the best of our knowledge, our framework is the first demonstration which illustrates the EMD and its lower-bounding distance functions on signatures in the context of video similarity search, offering various user parameters. As depicted in Figure 1, the user can determine required parameters and load a query video to the system which is displayed to the user by the *visualizer* module. The *feature extractor* first extracts a sampling of video features in the given feature space and then clusters them via k-means clustering algorithm to generate a video signature. The previously generated signatures of the videos in the underlying database and the query signature are gathered by the *query processor* which utilizes a filter-and-refine architecture. The lower-bounding distance

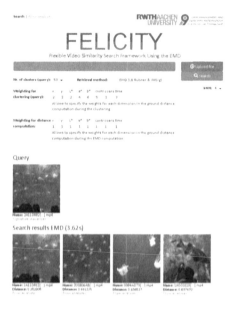

Fig. 2. Illustration of the system architecture of the FELICITY-framework

functions Rubner [3] and IM-Sig [5] are applied to generate a candidate set for which the exact EMD distance computation is performed in the refinement step later on. In addition, the *query processor* performs k-nn search by applying the optimal multi-step algorithm [4]. The results are transferred to the *visualizer* which displays them in ascending order with respect to the EMD value. The *flow analyzer* offers the opportunity to analyze the EMD flows among representatives of the query signature and another signature in the result set. The *visualizer* not only visualizes the signatures for the flow analysis, but also the videos via a video player embedded in the GUI.

We used HTML 5, CSS, and Javascript, and JAVA for the development of the user interface, video signature extraction, and database-related functionalities. As data sets, we downloaded video clips from YouTube (www.youtube.com) and Vine (https://vine.co) and extracted signatures with different dimensionalities.

3   A Demonstration Scenario and GUI

The user can either use the *upload file* button or directly drag and drop the video file in the interface to load the query video. The parameters of dimension weightings, number of representatives for signature extraction, and k-nn query parameter k are specified by the user. Figure 2 depicts a screenshot with respect to 4-nn query results. Furthermore, after displaying query results, the system enables the user to interactively play the query video and result videos. When the user intends to understand how the EMD is computed between the query signature and any returned video in the result list, he can deal with the flow

(a) Videos and their signatures   (b) Flow analysis

Fig. 3. Illustration of the flow analyzer.

analyzer, as depicted in Figure 3. In this way, he can gain more insights into the feature space dimensions and find out the best dimension weightings, i.e. parameter combination, for the desired k-nn query processing.

4   Conclusion

We showcase a novel system processing user-adaptive k-nearest-neighbor queries efficiently by utilizing signatures and the Earth Mover's Distance (EMD). Our system allows for flexible query parameter selection and EMD flow analysis, helping the user to find out the best parameter combination for the desired similarity tasks. The demonstrator will be attractive for database users and researchers interested in query parameter selection for k-nearest-neighbor processing in video databases. We anticipate to discuss with the SISAP audience the parameter selection, flow analysis, and lower-bounding techniques for the EMD, as well as the integration of further modules, such as efficiency improvement techniques.

Acknowledgments. This work is funded by DFG grant SE 1039/7-1.

References

1. Huang, Z., Shen, H.T., Shao, J., Zhou, X., Cui, B.: Bounded coordinate system indexing for real-time video clip search. ACM Trans. Inf. Syst. **27**(3), 17:1–17:33 (2009)
2. Huang, Z., Wang, L., Shen, H.T., Shao, J., Zhou, X.: Online near-duplicate video clip detection and retrieval: an accurate and fast system. In: ICDE, pp. 1511–1514 (2009)
3. Rubner, Y., Tomasi, C., Guibas, L.: A metric for distributions with applications to image databases. In: ICCV 1998, pp. 59–66 (1998)
4. Seidl, T., Kriegel, H.-P.: Optimal multi-step k-nearest neighbor search. In: SIGMOD, pp. 154–165 (1998)
5. Uysal, M.S., Beecks, C., Schmücking, J., Seidl, T.: Efficient filter approximation using the earth mover's distance in very large multimedia databases with feature signatures. In: CIKM, pp. 979–988 (2014)
6. Uysal, M.S., Beecks, C., Schmücking, J., Seidl, T.: Efficient similarity search in scientific databases with feature signatures. In: SSDBM, pp. 30:1–30:12 (2015)
7. Uysal, M.S., Beecks, C., Seidl, T.: On efficient query processing with the earth mover's distance. In: PIKM@CIKM, pp. 25–32 (2014)

Searching the EAGLE Epigraphic Material Through Image Recognition via a Mobile Device

Paolo Bolettieri[1], Vittore Casarosa[1(✉)], Fabrizio Falchi[1], Lucia Vadicamo[1],
Philippe Martineau[2], Silvia Orlandi[3], and Raffaella Santucci[3]

[1] CNR-ISTI, Pisa, Italy
casarosa@isti.cnr.it
[2] Eureva, Paris, France
[3] Università di Roma La Sapienza, Rome, Italy

Abstract. This demonstration paper describes the mobile application developed by the EAGLE project to increase the use and visibility of its epigraphic material. The EAGLE project (European network of Ancient Greek and Latin Epigraphy) is gathering a comprehensive collection of inscriptions (about 80 % of the surviving material) and making it accessible through a user-friendly portal, which supports searching and browsing of the epigraphic material. In order to increase the usefulness and visibility of its content, EAGLE has developed also a mobile application to enable tourists and scholars to obtain detailed information about the inscriptions they are looking at by taking pictures with their smartphones and sending them to the EAGLE portal for recognition. In this demonstration paper we describe the EAGLE mobile application and give an outline of its features and its architecture.

Keywords: Mobile application · Image recognition · Similarity search · Epigraphy · Latin and Greek inscriptions

1 The EAGLE Project

One of the main motivations of the project EAGLE (Europeana network of Ancient Greek and Latin Epigraphy [1], a Best Practice Network partially funded by the European Commission) was to collect in a single repository information about the thousands of Greek and Latin inscriptions presently scattered in a number of different institutions (museums and universities) across all Europe. The collected information, about 1,5 million digital objects (texts and images), representing approximately 80% of the total amount of classified inscriptions in the Mediterranean area, is being ingested into Europeana and is also made available to the scholarly community and to the general public, for research and cultural dissemination, through a user-friendly portal supporting advanced query and search capabilities.

In addition to the query capabilities (full text search a la Google, fielded search, faceted search and filtering), the EAGLE portal supports two applications intended to make the fruition of the epigraphic material easier and more useful. A Story Telling application provides tools to assemble epigraphy-based narratives to be made available

© Springer International Publishing Switzerland 2015
G. Amato et al. (Eds.): SISAP 2015, LNCS 9371, pp. 351–354, 2015.
DOI: 10.1007/978-3-319-25087-8_35

at the EAGLE portal, intended for the fruition of the epigraphic material by less knowledgeable users or young students. A Flagship Mobile Application (FMA) enables a user to get information about one visible inscription by taking a picture with a mobile device, and sending it to the EAGLE portal for recognition. This demo will show the EAGLE Flagship Mobile Application (presently implemented on Android) and the next sections will briefly describe the functionality and the architecture of the FMA.

2   The Flagship Mobile Application

The FMA enables a user to get information about one visible inscription by taking a picture with a mobile device, and sending it to the EAGLE portal, specifying the recognition mode. In "Similarity Search Mode" the result is a list of inscriptions (just thumbnails and some summary information) ranked in order of similarity to the image sent to the EAGLE server; by clicking on one of the thumbnails the user will receive all the information associated with that inscription. In "Exact Match Mode" the result is all the information associated with the image, if recognized, or a message saying that the image was not recognized.

The Graphical User Interface (GUI) of the FMA, available on the touch screen of the mobile device gives access to the functions listed below. The user can navigate through the different functions with tabs, and at any moment has access to the initial page.

- Search EAGLE content using image recognition in Similarity Search mode
- Search EAGLE content using image recognition in Exact Match mode
- Search EAGLE content using text search
- Login to the mobile application using an account already existing at the EAGLE portal
- For logged-in users, annotate and save queries and their results
- For logged-in users, annotate and save pictures taken with the mobile device
- For logged-in user, access and review the navigation history.

The mobile application communicates (through the Internet) with the Flagship Mobile Application (FMA) server, which in turn communicates with the EAGLE server using the specific APIs supporting the mobile application. Figure 1 shows the main functionality blocks of the EAGLE portal and the communication APIs between the FMA server and the EAGLE server. Complete details of the architecture and the mobile application can be found in [2].

The Image Recognizer (middle block on the right in the EAGLE server) has three main functions: (i) Image Feature Extractor, (ii) Image Indexer and support of Similarity Search Mode, (iii) Support of Exact Match Mode.

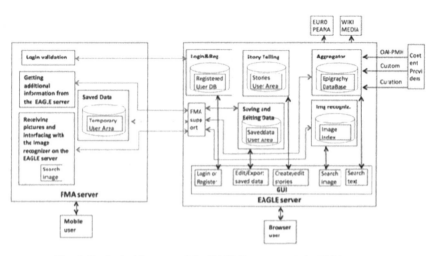

Fig. 1. Basic Architecture of the EAGLE server and the FMA server

2.1   Image Feature Extractor

The Image Feature Extractor analyses the visual content of the EAGLE images and captures certain local visual properties of an image (features). Local features are low level descriptions of Keypoints (or salient points), which are interest points in an image, whose description is invariant to scale and orientation. The result of extraction of visual features is a mathematical description of the image visual content that can be used to compare different images, judge their similarity, and identify common content. The Image Recognizer in EAGLE has a multi-threaded architecture for fast extraction of features and for taking advantage of multicore processors. It has a plug-in architecture, so that it is easy to add or delete the mathematical libraries supporting the many different algorithms for the extraction of local visual features, such as SIFT, SURF, ORB, etc. and their aggregations, such as BoF and VLAD [4].

2.2   Indexer and Support of Similarity Search and Exact Match Modes

The Image Indexer leverages the functionality of the Melampo CBIR System. Melampo stands for Multimedia Enhancement for Lucene to Advanced Metric PivOting [3]. It is an open source Content Based Image Retrieval (CBIR) library developed at CNR-ISTI that allows efficient comparison of images by visual similarity through the use of local features.

After the visual feature extraction, the local features are encoded using an approach called "Bag of Features", where a vocabulary of visual words is created starting from all the local descriptors of the whole dataset. The set of all the local descriptors of all the images is divided into a number of clusters (depending on the algorithms used, this number can go from a few hundreds to tens of thousands) and a textual tag is assigned to each cluster (usually in a random fashion). The set of all the textual tags becomes the "vocabulary" of visual words related to the whole set of images. At this point each image can be described by a set of "words" in this vocabulary, corresponding to the clusters containing the visual features of the image.

The support of Similarity Search Mode is based on the use of the Lucene search engine. Each image is represented by a set of words (the textual tags of the visual vocabulary), and Lucene builds the index of those words. At query time, the query image is transformed into a set of words, and then Lucene performs a similarity search, returning a list of images ranked according to the similarity with the query image.

The support of Exact Match Mode is based on a set of classifiers, each one recognizing a specific epigraph. The construction of the classifiers is done off-line, selecting from the complete database those epigraphies for which several images are available. The set of images representing the same epigraph is the training set used for building the classifier of that epigraph. At query time, the recognizer performs a similarity search for the image to be recognized and then takes from the result list the first k results for which there is also a classifier. The recognizer uses the RANSAC algorithm to perform geometry consistency checks [5] and assign a score to each class. We decided to assign to each class the highest matching score (i.e., percentage of inliers after the RANSAC) between the query image and all the image in the classifier. If the score is above a given threshold, the image is recognized.

3   Results

The Flagship Mobile Application has been tested on a preliminary database of about 17 thousand images for Similarity Search and 70 training sets for Exact Match, using different vocabulary size and visual features representation. Presently, the best results have been obtained using VLAD for visual features aggregations, with a codebook size of 256.

References

1. The EAGLE Project. http://www.eagle-network.eu/
2. The EAGLE Project, Deliverable D4.1 – Aggregation and Image Retrieval system (AIM) Infrastructure Specification
3. Gennaro, C., Amato, G., Bolettieri, P., Savino, P.: An approach to content-based image retrieval based on the Lucene search engine library. In: Lalmas, M., Jose, J., Rauber, A., Sebastiani, F., Frommholz, I. (eds.) ECDL 2010. LNCS, vol. 6273, pp. 55–66. Springer, Heidelberg (2010)
4. Jégou, H., Perronnin, F., Douze, M., Sanchez, J., Perez, P., Schmid, C.: Aggregating local image descriptors into compact codes. IEEE Transactions on Pattern Analysis and Machine Intelligence 34(9), 1704–1716 (2012)
5. Giuseppe, A., Falchi, F., Gennaro, C.: Geometric consistency checks for kNN based image classification relying on local features. In: Proceedings of the Fourth International Conference on SImilarity Search and APplications, pp. 81–88. ACM (2011)

Author Index

Printed in the United States
By Bookmasters